Text Book of

COMPUTER FUNDAMENTALS

For
First Year Diploma – Semester I
All Branches of Diploma in Engineering

As Per New Syllabus of SBTE, Jharkhand

Jyoti P. Mirani

TTTI (Bhopal), DCT
Lecturer, Electronics Deptt.
Institute of Technology
Ulhasnagar

Vikas B. Agarwal

B.E. (Computer)
Ex. Lecturer, Comp. Tech. Deptt.
Thakur Polytechnic
Kandivali (E), Mumbai

N0670

<table>
<tr><td>Computer Fundamentals</td><td align="right">ISBN: 978-93-86943-44-6</td></tr>
</table>

Second Edition : **August 2019**
© : **Authors**

Published By :
NIRALI PRAKASHAN

Abhyudaya Pragati, 1312, Shivaji Nagar

Off J.M. Road, PUNE – 411005

Tel - (020) 25512336/37/39, Fax - (020) 25511379

Email : niralipune@pragationline.com

➢ **DISTRIBUTION CENTRES**

PUNE

Nirali Prakashan : 119, Budhwar Peth, Jogeshwari Mandir Lane, Pune 411002, Maharashtra
(For orders within Pune) Tel : (020) 2445 2044; Mobile : 9657703145
Email : niralilocal@pragationline.com

Nirali Prakashan : S. No. 28/27, Dhayari, Near Asian College Pune 411041
(For orders outside Pune) Tel : (020) 24690204; Mobile : 9657703143
Email : bookorder@pragationline.com

MUMBAI

Nirali Prakashan : 385, S.V.P. Road, Rasdhara Co-op. Hsg. Society Ltd.,
Girgaum, Mumbai 400004, Maharashtra; Mobile : 9320129587
Tel : (022) 2385 6339 / 2386 9976, Fax : (022) 2386 9976
Email : niralimumbai@pragationline.com

➢ **DISTRIBUTION BRANCHES**

JALGAON

Nirali Prakashan : 34, V. V. Golani Market, Navi Peth, Jalgaon 425001, Maharashtra,
Tel : (0257) 222 0395, Mob : 94234 91860; Email : niralijalgaon@pragationline.com

KOLHAPUR

Nirali Prakashan : New Mahadvar Road, Kedar Plaza, 1st Floor Opp. IDBI Bank, Kolhapur 416 012
Maharashtra. Mob : 9850046155; Email : niralikolhapur@pragationline.com

NAGPUR

Nirali Prakashan : Above Maratha Mandir, Shop No. 3, First Floor,
Rani Jhanshi Square, Sitabuldi, Nagpur 440012, Maharashtra
Tel : (0712) 254 7129; Email : niralinagpur@pragationline.com

DELHI

Nirali Prakashan : 4593/15, Basement, Agarwal Lane, Ansari Road, Daryaganj
Near Times of India Building, New Delhi 110002 Mob : 08505972553
Email : niralidelhi@pragationline.com

BENGALURU

Nirali Prakashan : Maitri Ground Floor, Jaya Apartments, No. 99, 6th Cross, 6th Main,
Malleswaram, Bengaluru 560003, Karnataka; Mob : 9449043034
Email : niralibangalore@pragationline.com

Other Branches : Hyderabad, Chennai

niralipune@pragationline.com | www.pragationline.com
Also find us on www.facebook.com/niralibooks

Preface ...

We take an opportunity to present this book entitled as **"Computer Fundamentals"** to the students of First Semester Diploma in Engineering. The book is written according to the New Syllabus of State Board of Technical Education (SBTE), Jharkhand.

The book covers theory of Fundamentals of Computer, Introduction to MS Office, Introduction to Internet, Introduction to HTML and Software, Information Technology.

A special words of thank to Shri. Dineshbhai Furia, Mr. Jignesh Furia for showing full faith in us to write this text book. We also thank to Mr. Amar Salunkhe and Mr. Akbar Shaikh of Nirali Prakashan for their excellent co-operation.

We also thank Ms. Chaitali Takale, Mr. Ravindra Walodare, Mr. Sachin Shinde, Mr. Ashok Bodke, Mr. Moshin Sayyed and Mr. Nitin Thorat.

Although every care has been taken to check mistakes and misprints, any errors, omission and suggestions from teachers and students for the improvement of this text book shall be most welcome.

... Authors

𝒮𝓎𝓁𝓁𝒶𝒷𝓊𝓈 ...

Chapter 1 : Fundamentals of Computer [4 Hours, 6 Marks]

1.1 Introduction

1.2 Type of Computer

1.3 Components of PC

1.4 Inputs and Output Devices

1.5 Computer Languages

1.6 Memory of Computer

Chapter 2 : Introduction to MS Office [8 Hours, 12 Marks]

2.1 MS-Word: Introduction, Starting MS-Word Screen and its Components, Elementary Working with MS-Word

2.2 MS-Excel: Introduction, Starting MS-Excel, Basics of Spreadsheet, MS-Excel Screen and its Components, Elementary Working with MS-Excel

2.3 MS-PowerPoint: Introduction, Starting MS-PowerPoint, Basics of PowerPoint, MS-PowerPoint Screen and its Components, Elementary Working with MS-PowerPoint

Chapter 3 : Introduction to Internet [4 Hours, 6 Marks]

3.1 What is Internet?

3.2 Computer Communication and Internet

3.3 WWW and Web Browsers

3.4 Creating Own Email Account

3.5 Networking and Types

Chapter 4 : Introduction to HTML and Software [8 Hours, 10 Marks]

4.1 Introduction to HTML, Working of HTML

4.2 Creating and Loading HTML Pages, Tags

4.3 Structure of HTML, Document, Stand Alone Tags

4.4 Formatting Text, Adding Images, Creating Hyper Links, Tables

4.6 Cyber Security

4.7 Computer Virus

Chapter 5 : Information Technology [6 Hours, 6 Marks]

6.1 Current IT Tools

6.2 Social Networking, Mobile Computing, Cloud Computing

6.3 Introduction of IOT and IOE

6.4 Computer Application in various fields like Data Analysis, Database Management, Artificial Intelligence

Contents ...

Fundamentals of Computer

Contents

1.1 INTRODUCTION

- Today's world is an information rich world. In today's world, computers have become an integral part of our lives; computers are being used in every sphere of human activity whether it is at home, at office or at bank.
- Fields such as education, entertainment, medicine, banking, military, weather forecasting and telecommunications have been greatly influenced by the use of computers. This pervading presence of computers has made it necessary for everyone to have a fundamental knowledge of computers.
- A computer is basically a programmable computing machine. Computing is the process of utilizing computer technology to complete a task. Computing machine is a machine for performing particular tasks automatically.

1.1.1 What is Computer/Meaning of Computer

- The term computer is derived from the Latin word "compute" means "to calculate".
- A computer is basically a programmable computing machine that is used to store, retrieve and manipulate data.
- A computer is an electronic device that performs a given task (operation) on the basis of given instructions.

- The word COMPUTER can be analyzed as follows:

C	**C**alculate
O	**O**perate
M	**M**emorize
P	**P**rint
U	**U**pdate
T	**T**abulate
E	**E**dit
R	**R**esponse

1.1.2 Definition and Working of Computer

Definition of Computer:

- A computer can be defined as, "an electronic device which processes the information supplied (inputs) and produces the desired result (output) according to the given instructions (programs)". **OR**

- A computer can be defined as, "an information processing machine that can perform mathematical and logical manipulations (data processing) in accordance with a pre-defined set of instructions (programs) with capacity to store data temporarily and/or permanently".

Working of the Computer:

- Fig. 1.1 shows work environment of a computer.

- A computer mainly works on the following principle (See Fig. 1.1):

  1. **Input:** Taking the input in the form of instruction and data.

  2. **Process:** Processing the instruction and data and store the data.

  3. **Output:** Display the stored data or output into the print format.

Fig. 1.1: Work Environment of the Computer

1.1.3 Characteristics/Features of Computers

- The main characteristics of the computers, which makes them powerful and useful are listed below:

  1. **Automation:** An automatic machine works by itself without human intervention. Computers have automation power that means computer can perform the task automatically by using programs.

  2. **Speed:** Computers are of high speed in its operations. The speed is measured in terms of Instructions Per Second (IPS). All modern computers can process information at a speed of a couple of Million Instructions Per Second (MIPS).

  3. **Accuracy:** Computers are highly accurate in its operations. They either give correct answer or do not answer at all. Errors can occur in computers but these are mainly due to human rather than technological weakness.

  4. **Reliability:** It is the ability of the computers to perform the same job exactly in the same way in any numbers of times.

  5. **Versatility:** A computer is capable of performing almost any task provided that the task can be reduced to a series of logical steps.

  6. **Integrity:** It is the ability of the computers to carry out a sequence of instructions.

  7. **No Feelings:** Computers are devoid of emotions. They have no feeling because they are machines.

8. **Diligence Continuity:** A computer is free from monotony, tiredness, lack of concentration etc. It can work for hours without creating any error.

9. **Power of Remembering:** Computers can store and use any amount of information because of its storage capability.

1.1.4 Block Diagram of Computer

- Fig. 1.2 shows block diagram of a computer.

- The basic computer structure (block diagram) explains/describes the way in which different units of computer are interconnected with each other. Every computer system has essential four important units i.e., input unit, output unit, CPU and storage unit.

- The basic functional units of a computer in Fig. 1.2 are described below:

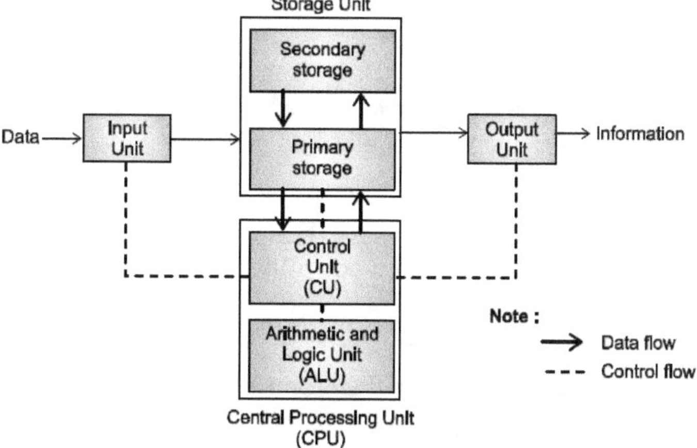

Fig. 1.2: Block Diagram of a Computer

1. Input Unit:

- Input is the process of entering data and programs (instructions) in to the computer system with the input device.

- The device that accepts data from the user and communicates the same to the CPU is called as an input device.

- Some common input devices are keyboard, mouse, joystick, light pen, track ball, scanner, graphic tablet, microphone, Magnetic Ink Card Reader (MICR), Optical Character Reader (OCR), Barcode reader, Optical Mark Reader (OMR) etc.

- Functions of input unit are listed below:

    (i) It accepts or reads data/instructions from outside world.

    (ii) Input unit converts these data/instructions in computer acceptable form.

    (iii) Input unit supplies the converted data/instructions to the storage unit for storage and further processing.

2. CPU:

- The Central Processing Unit (CPU) is referred to as "brain" of a computer system. CPU converts data (input) into meaningful information (output).

- The CPU controls all the internal and external devices, performs arithmetic and logic operations, and operates only on binary data (1's and 0's).

- In addition, CPU also controls the usage of main memory to store data and instructions and controls the sequence of operations.

- The CPU consists of three main subsystems, the Arithmetic Logic Unit (ALU), the Control Unit (CU), and the Registers.

- ALU performs the arithmetic and logic operations on the data that is made available to it.

- CU is responsible for organizing the processing of data and instructions. CU also controls and co-ordinates the activity of the other units of computer.

- CPU uses the registers to store the data, instructions during processing.

3. Memory or Storage Unit:

- The process of saving data and instructions permanently or temporary is known as storage.

- Memory unit can store instruction, data and intermediate results. This unit supplies information to the other units of the computer when needed.
- There are two types of memories i.e., **Volatile Memory** (whose contents are erased when the system's power is turned OFF) and **Non-volatile Memory** (whose contents will be saved regardless if the power to the computer is ON or OFF).
- The memory unit consists of primary memory and secondary memory.
    - **(i) Primary Memory** (main memory) of the computer is used to store the data and instructions during execution of the instructions. Random Access Memory (RAM) and Read Only Memory (ROM) are the primary memories.
    - **(ii) Secondary Memory** (Auxiliary memory) is non-volatile and is used for permanent storage of data and programs. Magnetic tape, disks are the examples of secondary storage.

4. Output Unit:

- The result of computer processing is called as output.
- This result is communicated to user through a devices called output devices such as monitor, plotter, printer etc.
- Functions of output unit are listed below:
    - (i) Output unit accepts the produced results, which are in the coded form.
    - (ii) It converts these coded results to human acceptable form.
    - (iii) Output unit supplies the converted results to outside world.

1.1.5 Generations of Computers

- Evolution of modern computer is commonly considered in terms of generations of computers.
- Each new generation has made the changes in computer characteristics such as increase in speed, increase in storage capacity, increase in reliability, reduction in system cost, decreasing in size, etc.
- According to the technology used, there are five generations of computers, which are discussed below:

    ### 1. First Generation Computers (1942-1955):

    - o   The first generation computers were using vacuum tubes and machine languages were used for giving instructions.
    - o   The computers of this generation were very large in size and their programming was a difficult task.
    - o   The first commercial electronic digital computer capable of using stored programs was called "Universal Automatic Calculator" (UNIVAC) built by Macuchy and Eckert in 1951.
    - o   The major first generation computers are UNIVAC-1, IBM-701, IBM-650, ENIAC (Electronic Numerical Integrator And Calculator), EDVAC (Electronic Discrete Variable Automatic Computer), EDSAC (Electronic Delay Storage Automatic Calculator), etc.

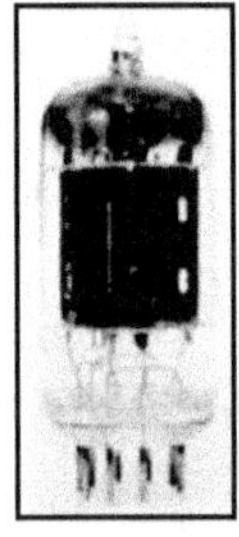

(a) Vacuum Tube

(b) 1st Generation Computer

Fig. 1.3

Advantages	Disadvantages
(i) First generation computers were fastest calculating devices of their time.	(i) Bulky in size (required large rooms) and time consuming for assembly and installation.
(ii) They support parallel processing.	(ii) Vacuum tube required very high power consumption.

2. Second Generation Computers (1955-1964):

- Computers are entered into second generation by the introduction of transistors. Vacuum tubes were replaced by tiny solid-state components called as transistors.
- Transistors were highly reliable, requires less power and faster than vacuum tubes.
- High Level Languages such as FORTRAN, COBOL, BASIC etc. were introduced. The practice of writing programs in machine languages were replaced by high level languages.
- Punched cards were used for input-output operations.
- Major second generation computers were IBM-1400 series, IBM-7000 series, Honeywell 200, CDC 3600, UNIVAC 1108 etc.

(a) Transistors

(b) 2nd Generation Computer

Fig. 1.4

Advantages	Disadvantages
(i) Transistors are faster and more reliable than vacuum tubes.	(i) Time consuming for assembly and installation.
(ii) Cheaper in cost and less power consumption also smaller in size.	(ii) Maintenance is high.
	(iii) Difficult and costly for commercial production.

3. Third Generation Computers (1964-1975):

- The third generation computers used the new technology i.e., Integrated Circuits (ICs) in place of transistors.
- All electronic components like transistors, resistors and capacitors were fabricated on silicon chips i.e., ICs.
- IC has higher speed, larger storage capacity and smaller size.
- Operating systems were introduced for use in computers. Significant advances in hardware technology made the introduction of keyboards and monitors for data input and output. More high level languages like Pascal, RPG were also introduced in the generation.
- Major third generation computers were PDP-8, PDP-11, IBM-360 series, ICL -2900 series, CDC's CYBER -175, TDC-316, IBM 370/168 etc.

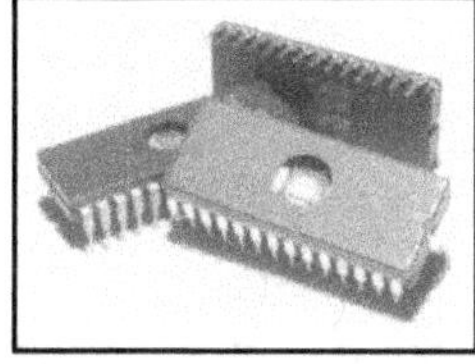

(a) IC's

(b) 3rd Generation Computer

Fig. 1.5

Advantages	Disadvantages
(i) Required small space (portable). (ii) More reliable and faster in speed. (iii) Support high level languages. (iv) Installation is required in less time.	(i) Cost is more than fourth generation computers. (ii) Highly sophisticated technology required for the manufacturing chips.

4. Fourth Generation Computers (1975-1989):

- o The ICs used in third generation computers had about 10 to 100 transistors per unit. This technology was called Small-Scale Integration (SSI).
- o Later, with the advancement of technology for manufacturing ICs, it is possible to integrate 10,000 transistors in an IC. This technology is called Large-Scale Integration (LSI).
- o Very Large Scale Integration (VLSI) can pack a million or more transistors on a single chip. LSI and VLSI technologies led to the introduction of Microprocessors.
- o Computers which are designed using Microprocessors become the fourth generation computers.
- o Intel introduced the first microprocessor 4004 using LSI. The languages C, LISP, Prolog become popular in this generation.
- o Major fourth generation computers are APPLE II, IBM 4341, DEC 10, IBM System 370, CRAY–MPC, WIPRO 860, IBM AS/400/B60, IBM ps/2 MODEL 80, HCL Magnum, etc.

(a) Microprocessor   (b) 4th Generation Computers

Fig. 1.6

Advantages	Disadvantages
(i) Portable in size and easy for installation. (ii) Cheaper in cost and more reliable. (iii) Supports high level language and networking.	(i) Expensive. (ii) Single user oriented.

5. Fifth Generation Computers (1989 onwards):

- o In the fifth generation, the VLSI technology became ULSI (Ultra Large Scale Integration) technology, resulting in the production of microprocessor chips having ten million electronic components.
- o Fifth generation is based on parallel processing hardware and AI (Artificial Intelligence) software.
- o AI is an emerging branch in computer science, which interprets means and method of making computers think like human beings. All the high-level languages like C and C++, Java, .Net etc., are used in this generation.

Fig. 1.7: 5th Generation Computer

- o Quantum computation and molecular and nanotechnology will be used. So we can say that the fifth generation computers will have the power of human intelligence.
- o Some computer types of fifth generation are UltraBook and ChromeBook.

Advantages	Disadvantages
(i) Very less power is required. (ii) More smaller and handy than computers of fourth generation computers. (iii) Faster in speed and more reliable.	(i) AI and the overall advanced technology. So, it requires a powerful learning curve. (ii) They tend to be sophisticated and complex tools.

1.1.6 Advantages and Disadvantages of Computers

Advantages of the Computers:

1. **Speed:** When data, instructions, and information flow along electronic circuits in a computer, they travel at incredibly fast speeds. Many computers process billions or trillions of operations in a single second.

2. **Reliability:** The electronic components in modern computers are dependable and reliable because they rarely break or fail.

3. **Consistency:** Given the same input and processes, a computer will produce the same results consistently.

4. **Storage:** A computer can transfer data quickly from storage to memory, process it, and then store it again for future use. Many computers store enormous amounts of data and make this data available for processing anytime it is needed.

5. **Communications:** Most computers today can communicate with other computers, often wirelessly.

Disadvantages of the Computers:

1. **Health Risks:** Prolonged or improper computer use can lead to injuries or disorders of the hands, wrists, elbows, eyes, neck, and back.

2. **Impact on Environment:** Computer manufacturing processes and computer waste are depleting natural resources and polluting the environment.

3. **Public Safety:** Adults, teens, and children around the world are using computers to share publicly their photos, videos, journals, music, and other personal information. Some of these unsuspecting, innocent computer users have fallen victim to crimes committed by dangerous strangers.

4. **Impact on Labor Force:** Although computers have improved productivity in many ways and created an entire industry with hundreds of thousands of new jobs, the skills of millions of employees have been replaced by computers.

5. **Violation of Privacy:** Nearly every life event is stored in a computer somewhere like in medical records, credit reports, tax records, etc. In many instances, where personal and confidential records were not protected properly, individuals have found their privacy violated and identities stolen.

1.2 TYPES OF COMPUTER

- Computers can be classified according to purpose, data handling and functionality as shown in Fig. 1.8.

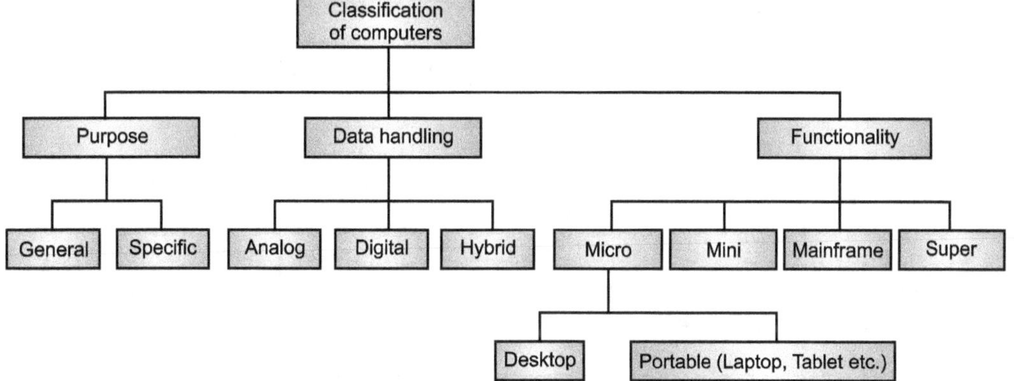

Fig. 1.8: Types or Classification of Computers

(A) Types of Computers According to Type of Data Handling Techniques: Different types of computers process the data in a different manner. According to the basic data handling principle, computers can be classified into three categories i.e., analog, digital and hybrid as discussed below:

1. **Analog Computers:**

   o A computing machine that works on the principle of measuring, in which the measurements obtained are translated into desired data (information) is known as analog computer.

- o Modern analog computers usually employ electrical parameters, like pressures, voltages, resistances or currents, temperatures, to represent the quantities being manipulated.
- o The earliest computers were analog computers.
- o Analog computers are used for scientific and engineering purposes. Slide rule, Antikythera mechanism, astrolabe, differential analyzer, Deltar, Kerrison Predictor are the examples of analog computers.

Advantages	Disadvantages
(i) They provides continuous representation of all data. (ii) They are faster in speed and inexpensive.	(i) Analog computers are not accurate. (ii) Analog computers has lack memory.

2. Digital Computers:

- o A computer that operates with information, numerical or otherwise, represented in a digital form (0's an 1's) is known as digital computer.
- o Digital computers process data including text, sound, graphics and video into a digital value in 0's and 1's.
- o The desktop PC at home, banks, offices etc., are the examples of digital computers.

Advantages	Disadvantages
(i) Digital computers are accurate. (ii) They can store large amount of data.	(i) Digital computers are slower than analog computers. (ii) They have higher cost and complexity.

3. Hybrid Computers:

- o Those computers which employ both the features of analog and digital computers are known as hybrid computer.
- o EAI 180, HPLC and EAI 185 are the examples of hybrid computers.

Advantages	Disadvantages
(i) Less expensive than the digital computer. (ii) Hybrid computers have tremendous computing speed.	(i) Hybrid computers required detailed knowledge of operation for both the analog and digital computers. (ii) Simulations using hybrid computers were extremely time consuming.

- • Following table compares analog, digital and hybrid computers:

Sr. No.	Analog Computer	Digital Computer	Hybrid Computer
1.	Analog computers are used to process analog data.	Digital computers are used to process digital (letters, numerals, special symbols) data.	A hybrid computer can process both digital and analog data.
2.	Speed is a faster than digital computer.	Speed is slower than hybrid computer and to analog computer.	It has high speed than digital and analog computers.
3.	Analog computer do not requires any storage capability because they measure quantities in a single operation.	Digital computers requires storage capability.	Hybrid computer requires storage capability.
4.	Data in analog computer is of continuous in nature.	Data in digital computer is discrete in nature.	Hybrid computer process both continuous and discrete data.
5.	Analog computer can process only numeric data.	Digital computer process numeric as well as non-numeric data.	It process both numeric and non-numeric data.
6.	Examples: (i) Slide rule (ii) Astrolabe	Examples: (i) Desktop PC (ii) UltraBook	Examples: (i) HPLC (ii) EAI 180

(B) Types of Computers According to Functionality: Based on physical size, performance and application areas, the computers can be divided into four major categories i.e., micro, mini, mainframe and super computers.

1. **Super Computers:**

Fig. 1.9: Super Computer

   - The super computer were first presented in the year 1960 by Seymour Cray at CDC (Control Data Corporation).
   - A super computer is the fastest type of computer that can perform complex operations at a very high speed. These computers can process billions of instructions per second.
   - Examples of super computers are CRAY 3, Cyber 205, NEC SX-3 and PARAM.

Advantages	Disadvantages
(i) They are very fast in speed.	(i) They require very high power and cost is high.
(ii) They have very high storage capacity.	(ii) Super computers are larger in size.

   Applications of Super Computers:

   (i) Weather forecasting, (ii) Animated graphics, (iii) Fluid mechanics, (iv)Nuclear energy research etc.

2. **Mainframe Computers:**

Fig. 1.10: Mainframe Computer

   - The mainframes are 32-bit or 64-bit computers and have large storage capacity and can support a large number of terminals (ranges from 64 to 100).
   - Mainframe computer consists of a high-end computer processor, with related peripheral devices, capable of supporting large volumes of data processing, high performance on-line transaction processing systems, and extensive data storage and retrieval.
   - A mainframe computer is a multi-user computer, capable for hundreds of users simultaneously.
   - Examples of mainframe computers are DEC-1090, IBM 308-580 series, IBM 4300, ICIM 2904, etc.

Advantages	Disadvantages
(i) They have huge memory and data processing.	(i) Cost of hardware is high and more resource consumption.
(ii) They have high speed compared to volume of data.	(ii) Special operating systems/software is required.

   Applications:

   (i) Normally, they are used in banking, airlines, military and railways etc.

   (ii) Both e-business and e-commerce use mainframe computers to perform business functions and exchange money over the Internet.

3. **Mini Computer:**

Fig. 1.11: Mini Computer

   - A mini computer is a small digital computer. It is a multi-user computer.
   - Mini computer sometimes called a mid-range computer. Mini computer is designed to meet the computing needs for several peoples (users) simultaneously in a small to medium size business environment. It is capable of supporting from 4 to about 200 simultaneous users.
   - Some of the widely used mini computers are PDP 11, IBM (8000 series), and VAX 7500.

Advantages	Disadvantages
(i) Smaller in size, more reliable and faster in speed.	(i) Cost is more than micro computer.
(ii) Larger storage capacity and support standardized high level languages.	(ii) Air conditioning is required.

Applications:
  (i) Mini computers can be used as a communications tool in a larger system.
  (ii) Mini computers used for data management can be employed to acquire data, as in process control, generate data, or simply as a storage system for information.

4. Micro Computers:
  o Micro computers are small, low-cost and single-user digital computer.
  o A micro computer usually consists of a microprocessor, a storage unit, an input device, an output device and an Operating System (OS).
  o The micro computer is generally the smallest of the computer family. Originally, they were designed for individual users only, but now-a-days they have become powerful tools for many businesses that, when networked together, can serve more than one user.
  o IBM-PC Pentium 100, IBM-PC Pentium 200, and Apple Macintosh are the examples of micro computers.
  o Micro computers include desktop, (See Fig. 1.12) laptop (See Fig. 1.13) and hand-held (See Fig. 1.14) computers.

Advantages	Disadvantages
(i) They are smaller in size and cheaper in cost.	(i) They are single user oriented.
(ii) They are faster in speed and easy for installation.	(ii) They are non-portable.
(iii) They can share resources in networking.	

Applications of Micro Computers:
  (i) Micro computers are used for education, business, organisations, universities, banks etc.
  (ii) Micro computers are also used in book-keeping, inventory, medical records and communication.

Types of Micro Computers:

(I) Desktop Computers:
  o Desktop computer or Personal Computer (PC) is the most common type of micro computer.
  o Desktop computer is a stand-alone machine that can be placed on the desk.

Fig. 1.12: Desktop Computer

  o Externally, it consists of three units i.e., keyboard, monitor and a system unit containing the CPU, memory, hard disk drive, etc., (See Fig. 1.12).
  o The PC are suited to the needs of a single user at home, schools, small business units and organisations.
  o Microsoft, Apple, HP, Lenovo and Dell are some of the PC manufacturers.

Advantages	Disadvantages
(i) Easy to upgrade and faster in speed.	(i) Not portable.
(ii) Cheaper than laptops and tablets.	(ii) Big and heavy and require physical space.

(II) Laptop Computers:
  o Laptop computers are portable and light weight personal computers.
  o Laptop computer is also known as notebook computer or notepad.
  o The laptop computer is a small-size computer that incorporates all the features of a typical desktop computer. Laptop computers are provided with a rechargeable battery that removes the need of continuous external power supply.

Fig. 1.13: Laptop Computer

  o The different manufacturers of laptop computers are Acer, Apple, Panasonic, Sony and HP.

Advantages	Disadvantages
(i) Light weight and portable.	(i) Easy to steal or loose or theft.
(ii) Easy to carry because of smaller in size.	(ii) Very expensive.

(III) Hand-held Computers:

- o Hand-held PCs are small computing devices which we can hold them in the hands for this reason they are called as hand-held devices or hand-held PCs.
- o Hand-held computers also called as Personal Digital Assistant (PDA). It is a computer that can conveniently be stored in a pocket (of sufficient size) and used while the user is holding it.
- o A PDA user generally uses a pen or electronic stylus, instead of a keyboard for input.

Fig. 1.14: Hand-held Computer

- o Hand-held computer are also known as palmtop computers.
- o Some examples of PDAs are Apple Newton, Casio Cassiopeia, and Franklin eBookMan.
- o Over the last few years, PDAs have merged into mobile phones to create smart phones. Smart phone a mobile phone that performs many of the functions of a computer.
- o Smart phones are a hand-held device that integrates mobile phone capabi-lities with the more common features of a handheld computer or PDA.

Advantages	Disadvantages
(i) PDAs have increased in power and decreased in size.	(i) PDAs display data on small screens which is difficult to navigate data on them.
(ii) PDAs have GPS, Wi-Fi, E-mail and Bluetooth connectivity.	(ii) PDAs are expensive.
	(iii) PDAs are limited in terms of memory.

(IV) Tablet Computers:

- o A tablet is a wireless, portable personal computer with a touch screen interface. In this screen interface we can write or type the data and instructions.
- o It uses digitizing tablet technology to accept data and information for processing.
- o The tablet form factor is typically smaller than a notebook computer but larger than a smart phone.

Fig. 1.15: Tablet Computer

- o Tablet computer has features of the notebook computer but it can accept input from a stylus or a pen instead of the keyboard or mouse.

Advantages	Disadvantages
(i) Small in size and light weight.	(i) Easy to damage.
(ii) Recognizes handwriting.	(ii) High cost.

(C) Types of Computers according to Purpose:

- There are two types of computers according to their purpose i.e., general-purpose computers and special-purpose computers.

  1. General-purpose Computers:
  - o A general-purpose computer, as the name suggests, is designed to perform a range of tasks.
  - o The computer that use in schools, banks and homes are general-purpose computers.

  2. Special-purpose Computers:
  - o Special-purpose computers are designed to perform a single specific task.
  - o A set of instructions for the specific task is built into the machine. Hence, they cannot be used for other applications unless their circuits are redesigned i.e., they lacked versatility.
  - o Most analog computers are special purpose computers. Special-purpose computers are used for airline reservations, satellite tracking, air traffic control and so on.

1.3 COMPONENTS OF PC

- Computers are information processing machines viewed as a system which consist of a number of components like hardware, software etc., that work together with the common objective to processing data into information.

1.3.1 Computer System

- A system is a group of element or components that work together to carry out some common tasks, to achieve a common goal or objective.

- A computer can be viewed as a system, which consists of a number of interrelated components or elements that work together with the objective/goal of converting data into information.

- A complete computer system consists of four components/elements i.e., Hardware, Software, Data and User as shown in Fig. 1.16.

Fig. 1.16: Parts/Elements/Components of a Computer System

- The mechanical devices/parts/components that make up the computer are called as hardware. Software is a set of instructions (programs). Software tells the computer what to do.

- Data is the raw instructions or raw facts that computer can process. A processed data is called as information. Data can be text, number, graphics, audio or video that the computer manipulates. People who use computers are called computer users.

1.3.2 Computer Hardware and Software

- Computers work through an interaction of hardware (physical parts like keyboard, mousse etc.) and software (a set of instructions that instructs hardware what to do).

1. Hardware:

- The mechanical/electronic devices/parts/components (keyboard, mouse, hard disk, cabinet, cables etc.) that make up the computer are called as hardware.

- A computer's hardware consists of interconnected electronic and/or mechanical devices that a computer user can use to control the computer's operation, input and output.

2. Software:

- A sequence of instructions needs to be given to the computer in a programming language that the computer understands to make hardware work, such a set of instructions is known as software.

- Software is a set of instructions that makes the computer perform tasks.

- In other words, software tells the computer what to do.

- Computer software can be defined as, "a set of instructions that directs a computer to perform specific tasks or operations".

- Software can be categorized as, system software and application software.

  (i) System Software refers to that collection of programs which co-ordinate the activities at hardware and all programs running on the computer system. Examples, Operating System (OS), Device Drivers, Utility Programs (Norton's Anti-virus and MCafee Anti-virus), Language Translators Programs (Compiler, Assembler and Interpreter) etc.

  (ii) Application Software is a general purpose program or a collection of programs written by the users to solve a specific problem. Examples, Word Processors (WordStar and MS-Word), Spreadsheet (Lotus 1-2-3 and Excel), Desktop Publishing (DTP) Software (Coral Draw and PageMaker) etc.

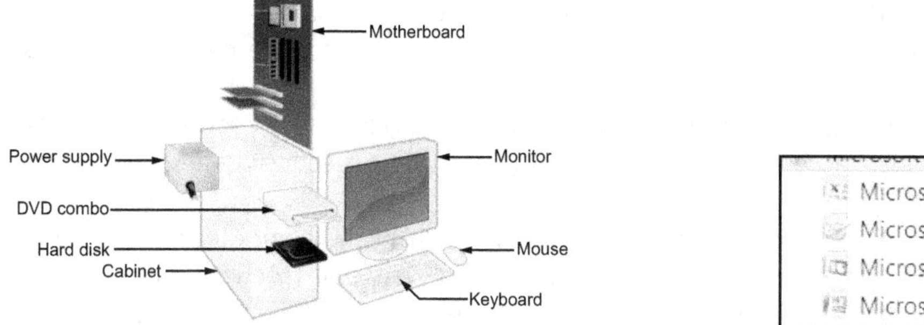

(a) Hardware of Computer   **(b) Softwares**

Fig. 1.17

- The basic hardware parts of computers are shown in Fig. 1.18. There are numerous other devices on the market but this is the general (common) list of parts.

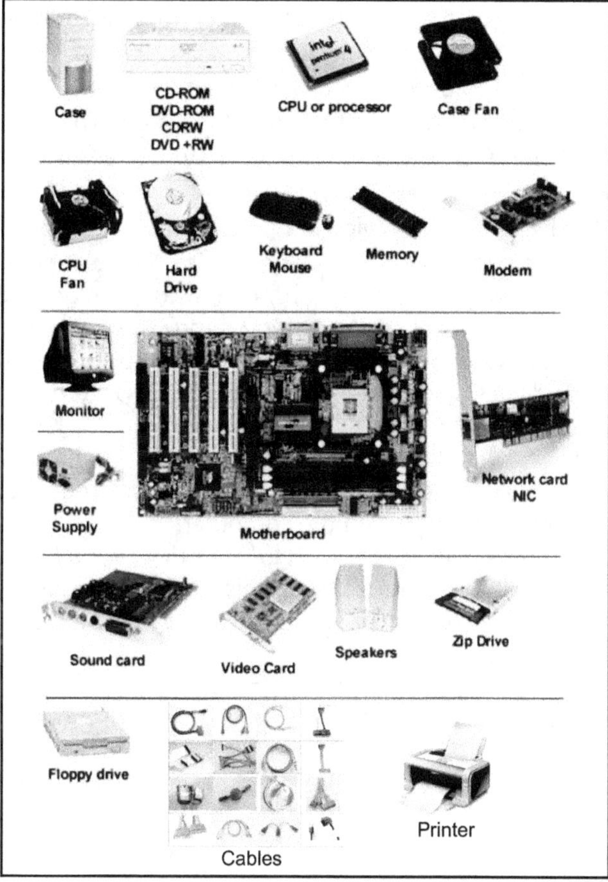

Fig. 1.18

- Various computer hardware parts in Fig. 1.18 are explained below:
- A **computer case**, also known as a computer chassis, tower, system unit, cabinet, base unit, or simply case, is the enclosure that contains most of the components of a computer like RAM, Hard disk, processor etc.
- A **combo drive** is a type of optical drive that combines CD-R/CD-RW recording capability with an ability to read (but not write) DVD media; some manufacturers refer this as CD-RW/DVD-ROM drive.
- A **processor** is an integrated electronic circuit that performs the calculations that run a computer. A processor performs arithmetical, logical, Input/Output (I/O) and other basic instructions that are passed from an Operating System (OS).
- The **motherboard** is the main circuit board that all other computer components either placed in or are connected to. It is also called as main board or system board.

- **Random Access Memory (RAM)** is a type of data storage used in computers that is generally located on the motherboard. RAM is the temporary storage for data and programs that are being accessed by CPU.
- A **hard disk** is part of a memory unit, often called a "disk drive," "hard drive," or "hard disk drive," that store and provides relatively quick access to large amounts of data on an electromagnetically charged surface or set of surfaces.
- The **power supply** is generally located in one corner of the back side of the computer case. A power supply is a hardware component that supplies power to an electrical device. It receives power from an electrical outlet and converts the current from AC (Alternating Current) to DC (Direct Current), which is what the computer requires. It also regulates the voltage to an adequate amount, which allows the computer to run smoothly without overheating.
- **Keyboard** is most common input device. The data and instructions are input by typing on the keyboard.
- A **graphics card** in computer is used to convert data in machine to useful images on the monitor. The video card is an expansion card that allows the computer to send graphical information to a video display device such as a monitor, TV, or projector.
- **Sound cards** enable the computer to output, record and manipulate sound.
- **Monitor** is the most important output device of a computer system. The monitor is the display screen of a computer. Cathode Rays Tube (CRT), Liquid Crystal Display (LCD), Light Emitting Diode (LED), Plasma Panel are the types of monitors.
- A **printer** is an external hardware output device that accepts text and graphic output from a computer and transfers the information to paper, usually to standard size sheets of paper also called as Hard copy.
- **Computer cables** alternatively referred to as a cord, connector or plug, a cable is one or more wires covered in a plastic covering that connects a computer to a power source or other device. A data cable is a cable that provides communication between devices. For example, the data cable that connects your monitor to your computer and allows your computer to display a picture on the monitor. A power cable is any cable that powers the device.
- **Network cards** also known as Network Interface Cards (NICs) are hardware devices that connect a computer with the network. They are installed on the mother board. They are responsible for developing a physical connection between the network and the computer.
- A **modem** (modulator-demodulator) is a network hardware device that modulates one or more carrier wave signals to encode digital information for transmission and demodulates signals to decode the transmitted information. The goal is to produce a signal that can be transmitted easily and decoded to reproduce the original digital data.
- A **floppy disk** is a removable disk and is read and written by a Floppy Disk Drive (FDD).
- **Zip disk** is similar to 3 ½ inch floppy disk. But it can store 100 MB or more data. Zip disk drive is used to read and write data on a zip disk.
- **Computer speakers** are speakers external to a computer. Speakers contain amplifiers which vibrate to produce the sound.
- **Case fan** is a fan located on the side of a computer case, inside the case. Helps to circulate air within the computer case, as well as blow hotter air out of the case.
- **CPU fan** is a fan located on top of a computer processor. Helps to pull and blow hot air off the processor, helping keep it cooler.

Comparison between Computer Hardware and Computer Software:

Sr. No.	Hardware	Software
1.	Hardware works based on the instructions of the software.	Software tells the hardware what to do.
2.	User can see, touch and feel the hardware.	User cannot touch the software.
3.	These are the physical components of a computer system.	These are the logical components of a computer system.
4.	Hardware components are less expensive.	Software is generally costlier and expensive.

1.4 INPUT/OUTPUT (I/O) DEVICES

- Input/Output (I/O), in computing, is a communication process between a computer and the outside world i.e., user.
- Input refers to the data or instructions sent to the computer for processing while output refers to the data or instructions sent out from the computer after processing.
- An input device sends information to a computer system for processing, and an output device displays the results of that processing.
- In short, Input/Output devices abbreviated as I/O devices and they provides way of computer and user interaction.
- Fig. 1.19 shows the role of I/O devices in a computer system.

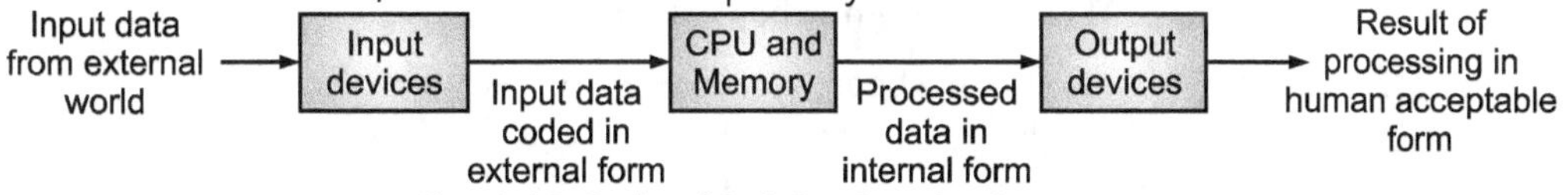

Fig. 1.19: Role of I/O Devices in Computer

1. **Input Devices:**
   - o The input unit of computer is used for entering the data and instruction into the computer. The devices used for input purpose are called as input devices.
   - o Keyboard, Mouse, Light pen, Joystick, Track ball, Scanner etc., are some of the common input devices.

2. **Output Devices:**
   - o Output is data that has been processed into useful information.
   - o Output devices provide output to the user, which is generated after processing the input data. The processed data, presented to the user via the output devices could be text, graphics, audio or video.
   - o Generally, there are two basic categories of output:
     - (i) The output, which can be readily understood and used by the humans, and
     - (ii) The output which is stored on secondary storage devices so that the data can be used as input for further processing.
   - o The output which can be easily understood and used by human beings are of the following two forms.
     - **(i) Hard Copy:** The physical form of output is known as hard copy. In general, it refers to the recorded information copied from a computer onto paper (printouts).
     - **(ii) Soft Copy:** The electronic version of an output, which usually resides in computer memory and/or on disk is known as soft copy. Unlike hard copy, soft copy is not a permanent form of output. It is transient and is usually displayed on the screen.
   - o Monitors, printers, plotters etc., are some common examples of the output devices.

1.4.1 Input Devices

- In this section we study various input devices of computers.

1. **Keyboard:**
   - o A keyboard is standard input device used to send instructional data into computer.
   - o A layout of the computer keyboard is identical to the conventional typewriter keyboard. However, it has more keys than the typewriter keyboard.
   - o The keys of keyboard are shown in Fig. 1.20.
   - o The data and instructions are input by typing on the keyboard. The message typed on the keyboard reaches the memory unit of a computer. It is connected to a computer via a cable.
   - o Keyboard is of two sizes 84 keys or 101/102 keys, but now 104 keys or 108 keys keyboard is also available for Windows and Internet.

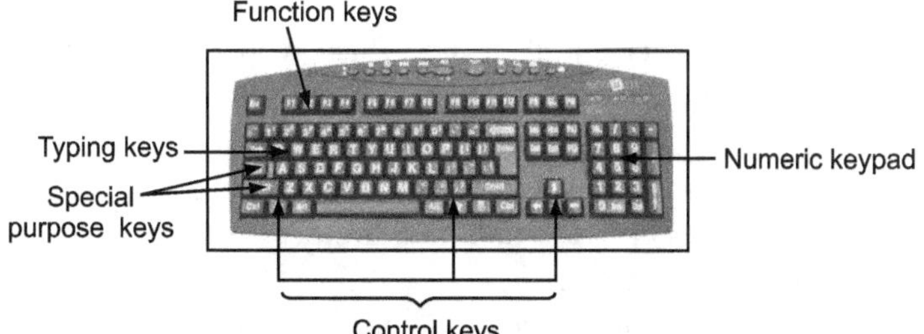

Fig. 1.20: Computer Keyboard with different Types of Keys

2. Mouse:

- o Mouse is an input device.
- o It is a small hand-held pointing device with a rubber ball embedded at its lower side and buttons on the top.
- o Usually, a mouse contains two or three buttons, which can be used to input commands or information.
- o Mouse may be classified as:

  (i) Mechanical Mouse: It uses a rubber ball at the bottom surface, which rotates as the mouse is moved along a flat surface, to move the cursor. It is the most common and least expensive pointing device. Fig. 1.21 (a) shows a mechanical mouse.

  (ii) Optical Mouse: It uses a light beam instead of a rotating ball (scroll) to detect movement across a specially patterned mouse pad. As the user rolls the mouse on a flat surface, the cursor on the screen also moves in the direction of the mouse's movement. Fig. 1.21 (b) shows a optical mouse.

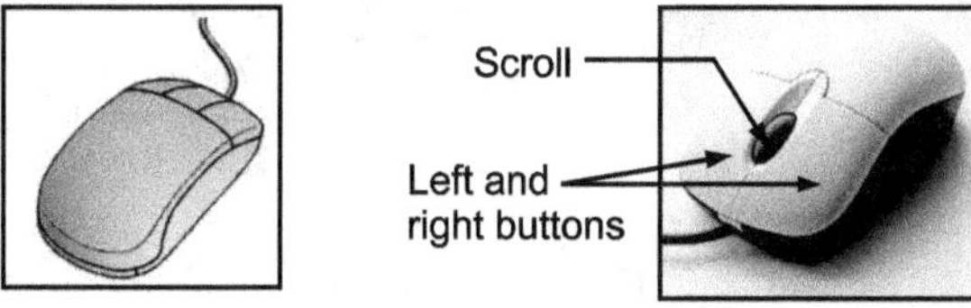

(a) Mechanical Mouse　　　　**(b) Optical Mouse**

Fig. 1.21

3. Joystick:

- o Joystick is a pointing device which is used to move cursor position on a monitor screen.
- o Joystick is a stick having a spherical ball at its both lower and upper ends. The lower spherical ball moves in a socket. The Joystick can be moved in all four directions.
- o The function of joystick is similar to that of a mouse.
- o A joystick is generally used to control the velocity of the screen cursor movement rather than its absolute position. It is used for computer games.
- o The other applications in which it is used are training simulators, CAD/CAM systems, and for controlling industrial robots.

Fig. 1.22: Joystick

4. Light Pen:

- o Light pen is an input device which is used to draw lines or figures on a computer screen.
- o Light pen is a pointing device which is similar to a pen. It is used to select a displayed menu item or draw pictures on the monitor screen.
- o Light pen consists of a photocell and an optical system placed in a small tube. When light pen's tip is moved over the monitor screen and pen button is pressed, its photocell sensing element detects the screen location and sends the corresponding signal to the CPU.

Fig. 1.23: Light Pen

5. Scanners:

- o Scanner is an input device which works more like a photocopy (Xerox) machine.
- o Scanners are peripheral devices used to digitize (convert to electronic format) artwork, photographs, text, or other items from hard copy.
- o Scanner is used for direct data entry form the source document into the computer system.

Fig. 1.24: Scanner

- o It converts the document image into digital form so that it can be fed into the computer. Capturing information like this reduces the possibility of errors typically experienced during large data entry.

- o There are several types of scanners, all of which serve a variety of functions and purposes:
    - **(i) Flat-bed Scanner:** This scanner allows the user to place a full piece of paper, book, magazine, photo or any other object onto the bed of the scanner and have the capability to scan that object.
    - **(ii) Sheet-fed Scanner:** This scanner allows user to scan pieces of paper. The sheet-fed scanner is a less expensive solution when compared to the flat-bed scanner.
    - **(iii) Hand-held Scanner:** The hand-held scanner allows the user to drag over select sections of pages, books, magazines, and other objects scanning only sections.

6. OCR:

- o Optical Character Recognition (OCR) is the mechanical or electric conversion of images of typed, hand-written or printed text into machine encoded text.
- o OCR is a software tool that allows to convert scanned documents into text searchable files.
- o Fig. 1.25 shows an OCR system.
- o If the documents to be scanned contains text, then we need OCR software. This is because when the scanner scans a document the scanned document is stored as a bitmap in the computer's memory.

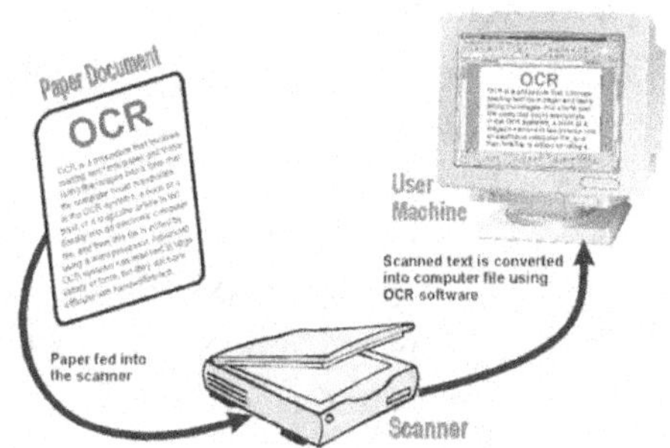

Fig. 1.25: An OCR System

- o The OCR software translates the bitmap image of text to the ASCII codes that the computer can interpret as letters, numbers and special characters.

7. MICR:

- o Magnetic Ink Character Recognition (MICR) is a technique that enables special characters printed in magnetic ink to be read and input rapidly to a computer.
- o When a document that contains this ink needs to be read, it passes through a machine, which magnetizes the ink and then translates the magnetic information into characters.
- o MICR is used extensively in banking because magnetic-ink characters are difficult to forge and are therefore ideal for marking and identifying cheques.

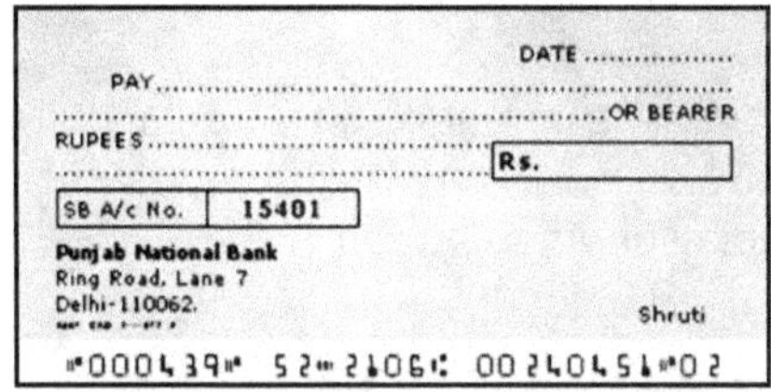

(a) Cheque Number written in MICR Font

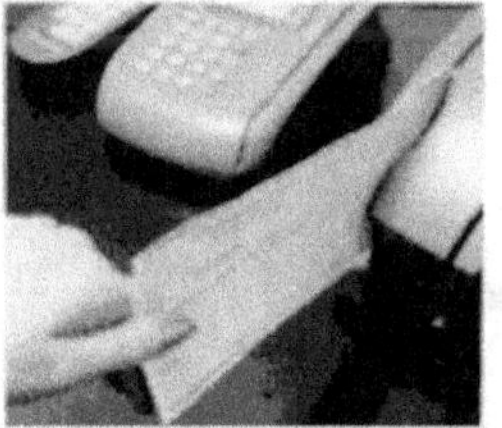

(b) MICR Processing

Fig. 1.26

Comparison between OCR and MICR:

Sr. No.	OCR	MICR
1.	It is used for reading any type of printed text material.	It is used in banks for reading special information from cheques.
2.	The fonts scanned usually can be of a variety of shapes.	The fonts scanned should be of a pre-defined shape like the E-13B font.
3.	The ink in the text document scanned can be of any type.	The ink used in MICR is a magnetic ink made using particles of iron oxide.
4.	The scanned document is usually stored as a text file for editing or printing.	Information from the scanned data is used for processing cheques in a bank.

8. Barcode Reader:

- o Barcode is a machine-readable code in the form of a pattern vertical line of varying widths.
- o It is commonly used for labelling goods that are available in super markets and numbering books in libraries.
- o Bar code reader scans a bar code image, converts it into an alphanumeric value which is then fed to the computer to which bar code reader is connected.
- o A barcode reader (or barcode scanner) is an electronic device for reading printed barcodes.

Fig. 1.27: Barcode Reader

9. Touch Screen:

- o A touch screen is a display screen that is an input device.
- o A touch screen allows the direct selection of a menu item or the desired icon with the touch of finger. Essentially, it registers the input when a finger or other object is touched to the screen.

Fig. 1.28: Touch Screen

10. Digitizers:

- o Digitizer is an input device which converts analog information into a digital form.
- o Digitizer is also known as Tablet or Graphics Tablet (See Fig. 1.15) because it converts graphics and pictorial data into binary inputs.
- o A graphic tablet as digitizer is used for doing fine works of drawing and image manipulation applications.
- o Digitizing is the process of identifying, locating or selecting a menu item, entity or point through an input device.

1.4.2 Output Devices

- In this section we study various types of computer output devices.

1. Monitor:

- o Monitor commonly called as Visual Display Unit (VDU) is the main output device of a computer.
- o VDUs are television like screens that provide the User Interface (UI) in the form of display of text, numbers and images.
- o Monitor forms images from tiny dots, called pixels that are arranged in a rectangular form. The sharpness of the image depends upon the no. of the pixels.
- o The monitor is also known as screen, display, video display or video screen.
- o Monitors comes in LCD, CRT and plasma panel as explained below.

(i) CRT Monitor:

- o CRT (Cathode Ray Tube) monitors look much like old-fashioned televisions and contains a cathode ray tube, hardware to control an electronics beam and a power supply.
- o A CRT is used to display numbers, letters and graphics. In the CRT display is made up of small picture elements called pixels.
- o The most screens are capable of displaying 80 characters of data horizontally and 25 lines vertically.

Fig. 1.29: CRT Monitor

Advantages	Disadvantages
(i) They operate at any resolution without the need for rescaling the image.	(i) They are large, heavy weight, and bulky in size.
(ii) CRTs run at the highest pixel resolutions.	(ii) They consume a lot of electricity and produce a lot of heat.
(iii) CRTs are less expensive.	

(ii) LCD:

- o LCD (Liquid Crystal Display) monitors are much thinner, use less energy, and provide a greater graphics quality. It uses liquid crystals technology to display the images.

- o The liquid crystals are actually the molecules of the liquid filled in the LCD. These LCD molecules easily flow in different directions and have the capability to bend a beam of light.

- o Each pixel on the LCD screen contains a number of liquid molecules layered between two transparent electrodes and crossed polarising filters.

- o The liquid molecules uphold their directions and remain in the same position with respect to each other. As the light falls on the molecules, they bend the light and direct them towards a polarising filter.

- o The polarising filter absorbs the light making the polariser appear dark for the display of images.

Fig. 1.30: LCD

Advantages	Disadvantages
(i) LCD is small in size and low cost.	(i) LCD has narrow viewing angle.
(ii) LCD has less power consumption.	(ii) LCD does not refresh the pixels very quickly.
(iii) LCD supports multiple video resolutions.	

Differentiation between CRT and LCD Monitors:

Sr. No.	CRT Monitor	LCD Monitor
1.	Designed with Cathode Ray Tube (CRT) technologies.	Designed with Liquid Crystal Display (LCD) technologies.
2.	Less expensive and lower cost.	More expensive and higher cost.
3.	Requires more power to run.	Requires less power to run.
4.	CRT is weighted, bulky and large in size.	LCD is light weighted, smaller in size.
5.	Operate at any resolution.	Each panel of LCD has a fixed pixel.

(iii) Plasma Display:

- o Plasma Display Panel (PDP) is a type of flat panel display common to large TV displays 30 inches (76 cm) or larger.

- o PDPs are also called as gas-discharge displays.

- o They are essentially a matrix of very small fluorescent tubes with RGB phosphors.

Fig. 1.31

Advantages	Disadvantages
(i) Larger screen size availability in PDPs.	(i) They require more power than LCDs.
(ii) They have better color accuracy and saturation.	(ii) They have shorter display life span than LCD.

2. Printer:

- o A printer is an output device. A printer prints information and data from the computer onto a paper.

- o The speed of the printer is a measure of the number of characters/lines printed per unit time.

- o The speed of the printer is measured in terms of CPS (Characters Per Second) and LPM (Lines Per Minute).

o   Fig. 1.32 shows types of printers.

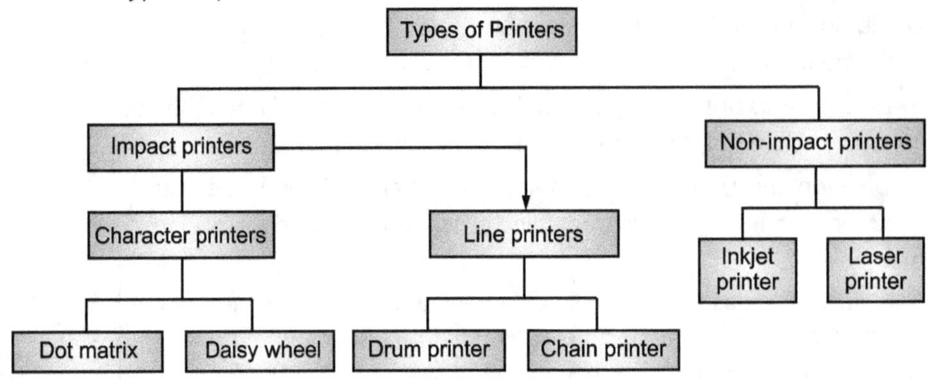

Fig. 1.32: Types of Computer Printers

(i) Impact Printers:

o   The printers that print the characters by striking against the ribbon and onto the paper are called impact printers.

o   An impact printer is like a typewriter and the characters are formed by physically striking the type devices against an inked ribbon.

o   Impact printers can produce a page, a line, or a character at a time. Print quality is low, but these printers are mainly used for printing backup copies of large amounts of data.

o   Impact printers are of following two types:

   (a) Character Printers: Character printers are printers which print one character at a time. These are of further two types Dot-matrix Printer (DMP) and Daisy wheel printer.

   (b) Line Printers: A line printer is a high-speed printing device that is able to store and print a complete line of information at a time. Line printers are printers which print one line at a time. Large computer system typically use line printer. Line printers are of further two types Drum printer and Chain printer.

Advantages of Impact Printers	Disadvantages of Impact Printers
(i) Design and functioning of this kind of printer is easier than that of non-impact printer. (ii) In impact printers, multiple copies can be produced by the use of carbon paper.	(i) They are noisy in operation. (ii) The wear and tear of printer head causes the periodical replacement of the printer head.

(ii) Non-impact Printers:

o   Non-impact printers generally, use specially coated or sensitized papers that respond to thermal or electrostatic stimuli to form an image.

o   The printers that print the characters without striking against the ribbon and onto the paper, are called non-impact printers.

o   Non-impact printers print a complete page at a time, also called as page printers.

o   Non-impact printers, used almost everywhere now, are faster and more quiet than impact printers because they have fewer moving parts.

o   Non-impact printers are of two types Laser printers and Inkjet printers.

Advantages of Non-impact Printers	Disadvantages of Non-impact Printers
(i) They have soundless operation.	(i) Multiple copies cannot be produced.
(ii) They have high quality output.	(ii) They are costly.

Differentiation between Impact and Non-impact Printers:

Sr. No.	Impact Printers	Non-impact Printers
1.	They print by hammering a set of metal pins or character sets on an ink ribbon.	These print by depositing ink drops or dry powder ink from an ink cartridge.
2.	These printers can be used to produce multiple copies by using carbon papers.	These printers cannot be used to produce multiple copies.
3.	Usually produce output of a single colour.	These can produce outputs of different colours.
4.	Noisy in printing.	Much less noisy in printing.
5.	These printers are slower than non-impact printers.	These are faster than impact printers.
6.	Examples: Dot-matrix, Daisy-wheel printers, etc.	Examples: Ink-jet, laser printers etc.

- Let us learn various impact and non-impact printers in detail.

(A) Dot-matrix Printer:

- The dot-matrix printer is so named because it creates the appearance of fully formed characters from dots placed on the edge.
- A dot-matrix printer is also known as an impact matrix printer, works similar to a ribbon of typewriter.
- Fig. 1.33 (a) shows a dot-matrix printer while Fig. 1.33 (b) shows working of dot-matrix printer.
- Dot-matrix printers also called impact printers print the characters by striking the combinations of pins on the print ribbon; the ribbon in turn strikes the paper which is placed against the drum and at the point of the impact characters gets printed on the paper.
- The characters are printed when the pins strike the ribbon, the number of pins can be said to be directly proportional to the print quality.
- Earlier dot-matrix printers used to have 9 pins but the print quality was not that great, then the companies started manufacturing printers with 18 or 24 print heads. 24 head printer used to print quickly and better quality and graphics.
- The speed of dot-matrix printers can vary from 50 to 500 cps.

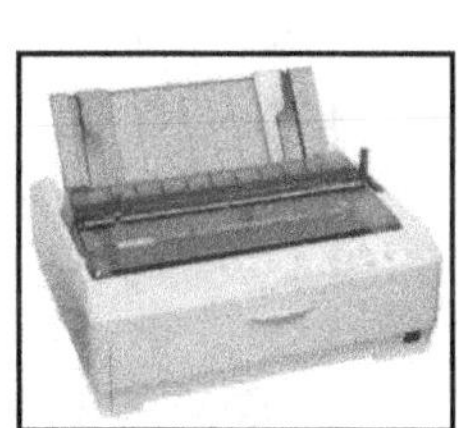

(a) Dot-matrix Printer

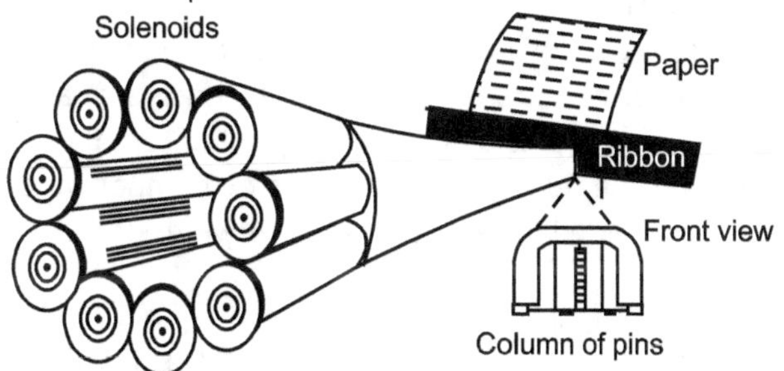

(b) Working of Dot-matrix Printer

Fig. 1.33

Advantages	Disadvantages
(i) Low printing cost.	(i) These printers are noisy and have low resolution.
(ii) Long life.	(iii) They have slow printing speed.

(B) Daisy-wheel printer:

- A daisy-wheel printer uses a printing mechanism known as a "daisy-wheel," which consists of numerous raised letters and numbers arrayed in a circle, (See Fig. 1.34 (a)).
- Head is lying on a wheel and pins corresponding to characters are like petals of Daisy (flower name) that is why it is called daisy-wheel printer.
- These printers are generally used for word-processing in offices which require a few letters to be send here and there with very nice quality representation.
- Fig. 1.34 (b) (ii) shows working of daisy wheel printer. Daisy wheel printers operate in much the same fashion as a typewriter.
- A hammer strikes a wheel with petals (the daisy wheel), each petal containing a letter form at its tip. The letter form strikes a ribbon of ink, depositing the ink on the page and thus printing a character.
- By rotating the daisy-wheel, different characters are selected for printing. The fastest letter-quality printers printed 30 cps.

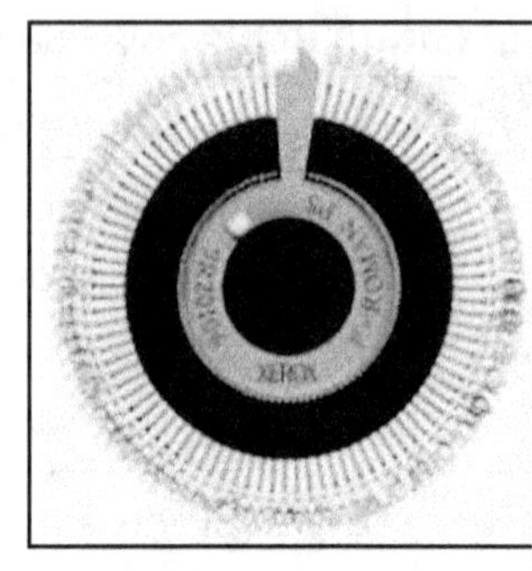

Fig. 1.34: (a) Daisy-wheel

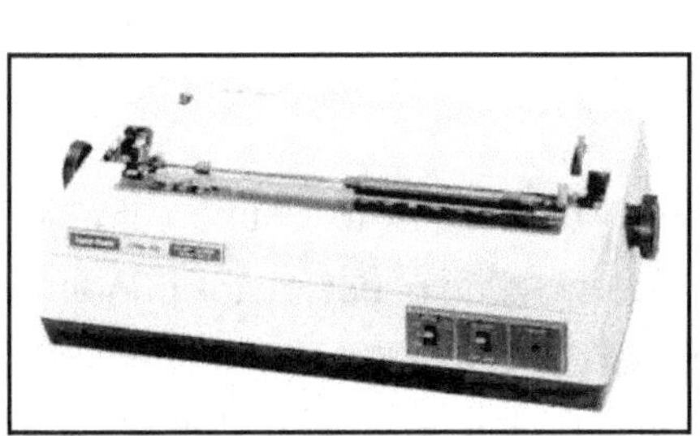

(i) Daisy-wheel Printer

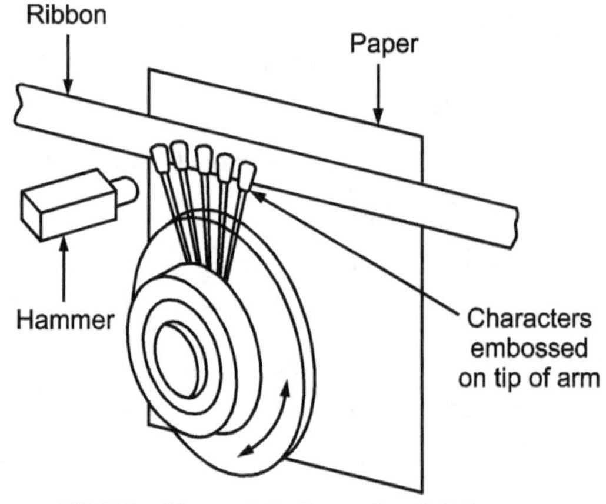

(ii) Working of Daisy-wheel Printer

Fig. 1.34 (b)

Advantages	Disadvantages
(i) More reliable than dot-matrix printers.	(i) Slower than dot-matrix printers.
(ii) It has better printing quality.	(ii) Noisy in operation.
(iii) The fonts of character can be easily changed.	(iii) More expensive than dot-matrix printers.

(C) Drum Printer:

- Drum printer is like a drum in shape so it called drum printer.
- Drum printer consists of a drum which consists of a number of characters; those are printed on the drum. And the number of characters or number of tracks are divided, after examining the width of the paper.
- The surface of drum is divided into number of tracks. Total tracks are equal to size of paper i.e. for a paper width of 132 characters, drum will have 132 tracks.
- A character set is embossed on track. The different characters sets are available in market 48 character set, 64 and 96 characters set. One rotation of drum prints one line.

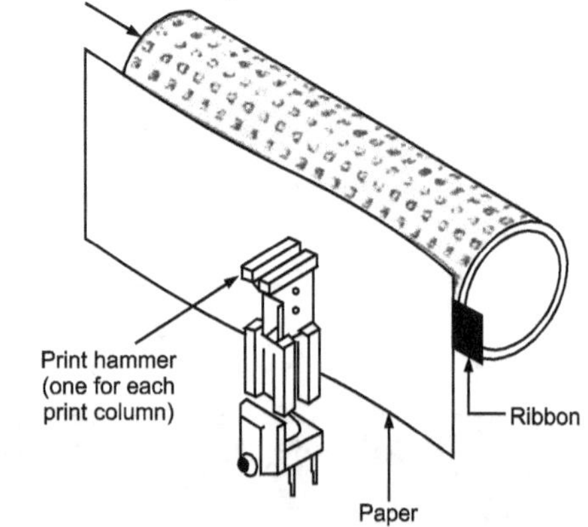

Fig. 1.35: Working of Drum Printer

- Drum printers are fast in speed and speed is in between 300 to 2000 lines per minute.

Advantages	Disadvantages
(i) Very high speed.	(i) Very expensive.
(ii) Low cost.	(ii) Poor quality of printing.

(D) Chain Printer:

- In chain printer chain of character sets are used so it is called chain printers.

- These are also line printers, which print one line at a time.

- All the characters are printed on the chain and the set of characters are placed on the chain. There are 48 and 64 and 96 characters set printers are available.

- There are also some hammers, those are placed in front of the chain, and paper is placed between the hammer and the inked ribbon.

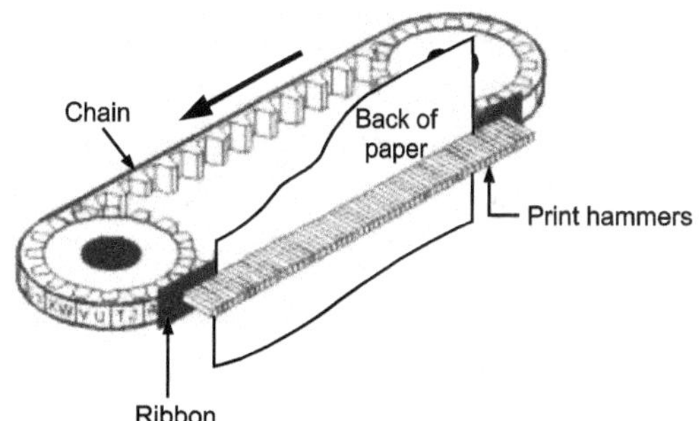

Fig. 1.36: Working of Chain Printer

- Chain printers (also known as train printers) placed the type on moving bars (a horizontally-moving chain).

- As with the drum printer, as the correct character passed by each column, a hammer was fired from behind the paper.

Advantages	Disadvantages
(i) Character fonts can easily be changed.	(i) Noisy in operation.
(ii) Different languages can be used with the same printer.	(ii) Do not have the ability to print any shape of characters.

(E) Inkjet Printer:

- Inkjet printers are non-impact character printers based on a relatively new technology.

- Inkjet printers produce high quality output with presentable features. Using inkjet printer, colour printing is also possible.

Working of Inkjet Printers:

- Inkjet printers are those that place extremely small droplets of ink onto paper to create an image.

- They use a reservoir of aqueous ink, a pump and an ink nozzle to accomplish this. These dots are extremely small and can have different colors combined together to create photo-quality images. They essentially work by shooting ink onto paper.

- Various parts of a typical inkjet printer are shown in the Fig. 1.38.

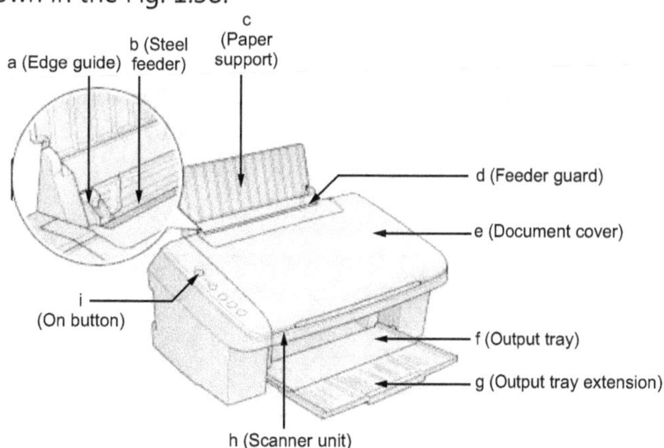

Fig. 1.37: Inkjet Printer

Fig. 1.38: Parts of Inkjet Printer

- Fig. 1.38 shows following parts of inkjet printer.

   (a) Edge guide: Helps load the paper straight. Adjust the left edge guide so that it fits snugly to the width of the paper.

   (b) Sheet feeder: Feeds a stack of paper automatically.

   (c) Paper support: Supports the paper loaded in the sheet feeder.

   (d) Feeder guard: Prevents objects placed on the document cover from falling inside the printer when opening the document cover.

   (e) Document cover: Open and close when we place a photo or document.

   (f) Output tray: Receives ejected paper.

   (g) Output tray extension: Supports the ejected paper.

   (h) Scanner unit: Open and close when we replace an ink cartridge.

   (i) On button: Turns the printer ON and OFF.

Advantages	Disadvantages
(i) High quality printing.	(i) Expensive as cost per page is high.
(ii) More reliable.	(ii) Slow as compare to laser printer.
(iii) Compact size.	(iii) Lifetime of inkjet prints produced by inkjet printer is limited.
(iv) Low noise.	(iv) Easily get blur if get water drop.

(F) Laser Printer:

- Laser printing is the most advance technology. Laser printers are non impact printers.

- Laser printers use laser lights to produces the dots needed to form the characters to be printed on a page.

Working of Laser Printer:

- Static electricity is the principle behind laser printers. A revolving drum or cylinder builds up an electrical charge.

Fig. 1.39: Laser Printer

- A tiny laser beam pointed at the drum discharges the surface in the pattern of the letters and images to be printed creating a surface with positive and negative areas.

- The surface is then coated with toner, a fine powder that is positively-charged so it clings only to the negatively-charged areas, and is then passed onto the paper to form the positive image.

- The paper then passes through heated rollers fusing the toner to the paper. Color lasers make multiple passes, in order to mix the different color toners.

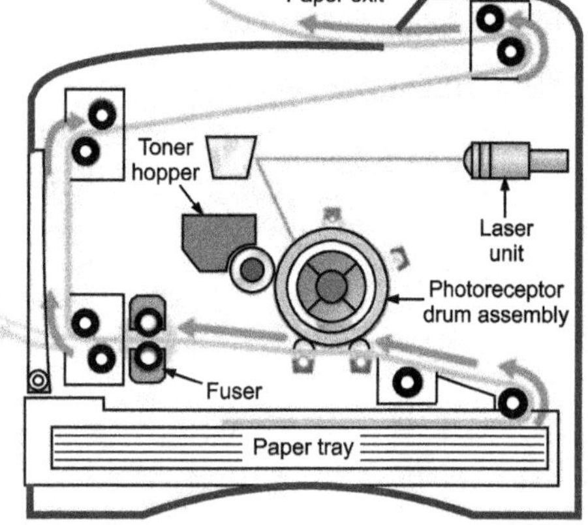

Fig. 1.40: Working of Laser Printer

Advantages	Disadvantages
(i) Very high speed.	(i) Laser printers are more expensive.
(ii) Very high quality printing output.	(ii) Cannot be used to produce multiple copies of a document in a single printing.
(iii) Low noise in printing operation.	(iii) Their size is generally larger.
(iv) Give good and high graphics quality.	

(G) Thermal Printer:

- A thermal impact printer or electro-thermal printer is a printer that uses heated pins to "burn" images onto heat-sensitive paper.
- These printers are commonly used in calculators and fax machines; and although they are inexpensive and print relatively fast, they produce low resolution print jobs.
- Thermal printers are of the following kinds/types:

   (i) Direct thermal printing or **thermal printing** is a digital printing process which produces a printed image by selectively heating coated thermo-chromic paper, or thermal paper as it is commonly known, when the paper passes over the thermal print head. The coating turns black in the areas where it is heated, producing an image.

   (ii) A **thermal wax-transfer printer** melts a wax-based ink onto paper in tiny dots, using dithering to create different colors.

   (iii) A **thermal dye-transfer printer**, sometimes called a dye sublimation printer, heats ribbons containing dye and then diffuses the dye into specially coated paper.

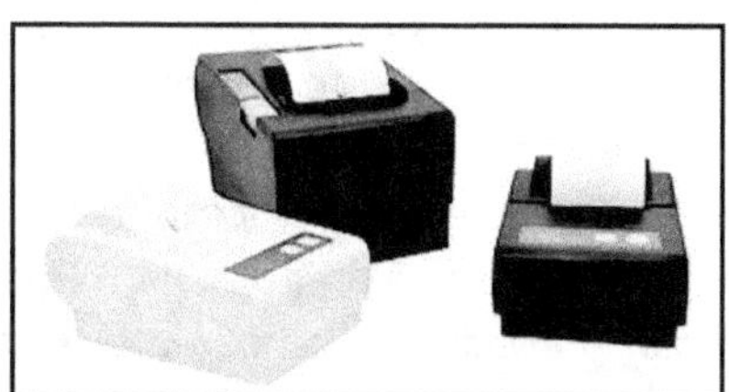

Fig. 1.41: Thermal Printers

Advantages	Disadvantages
(i) Low initial cost.	(i) Slow print speed.
(ii) Quiet operation.	(ii) Expensive, special paper or ribbon (is required)

3. Plotter:

- A plotter is a device that draws images on paper after receiving a command from a computer.
- A plotter is a special output device used to produce hardcopies of graphs and designs on the paper.
- A plotter is typically used to print large-format graphs or maps such as construction maps, engineering drawings and big posters.
- Plotters are divided into two types drum plotter and flatbed plotter.

Fig. 1.42: Drum Plotter

   (i) Drum Plotter: A drum plotter is also known as roller plotter. Drum plotter consists of a drum or roller on which a paper is placed and the drum rotates back and forth to produce the graph on the paper. Drum plotters are used to produce continuous output, such as plotting earthquake activity.

   (ii) Flatbed Plotter: A flatbed plotter is also known as table plotter. Flatbed plotter plots on paper that is spread and fixed over a rectangular flatbed table.

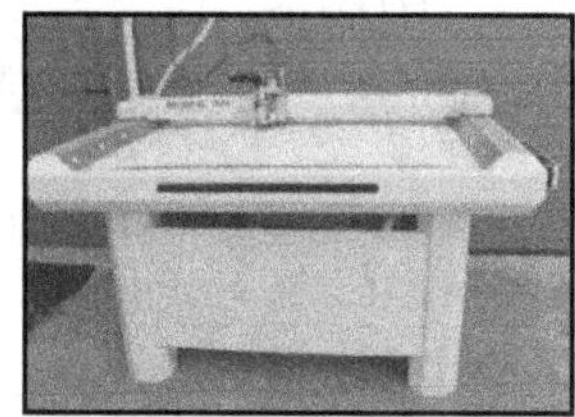

Fig. 1.43: Flatbed Plotter

- Today, mechanical plotters have been replaced by thermal, electrostatic and inkjet plotters. These systems are faster and cheaper. They also produce large size drawings.

Advantages	Disadvantages
(i) Plotters can work on very large sheets of paper while maintaining high resolution.	(i) Plotters are slower in printing speed.
(ii) Plotters allow the same pattern to be drawn thousands of times without any image degradation.	(ii) They do not produce very high quality printouts. (iii) Plotters are also much more expensive than a printer.

1.5 COMPUTER LANGUAGES

- To write a program, a programmer should know a programming language. A computer language or programming language is a language used for writing a program.
- A program can be defined as "a series of step-by-step specific instructions that tells the computer what to do". **OR** A program is, "a set of instructions given to a computer to performing a particular task".
- The art of writing a program is called programming. A set of programs written by specialists called as programmers.
- A programming language can be defined as, "a set of symbols and the rules governing their use that are used in constructing programs".
- Programming can be defined as, "an art of performing computational tasks and converting these tasks into machine readable form".
- Programmers write instructions in various programming languages. Some are directly understandable by the computer while others require some compilation or translation. Computer languages can be broadly classified into three categories i.e., machine language, assembly language and high-level language.

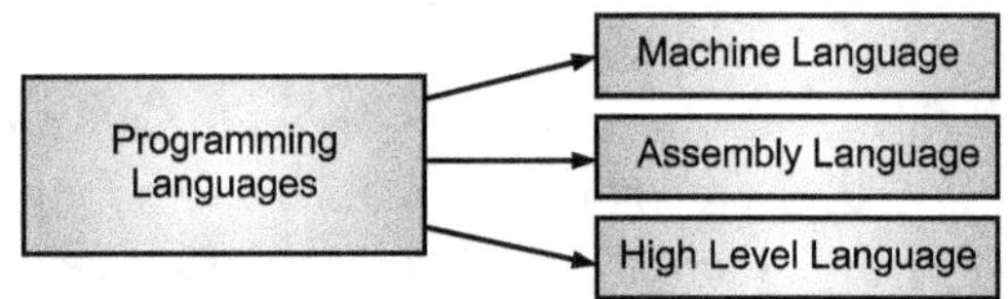

Fig. 1.44: Types of Programming Languages

- Computer programming languages may be classified in three major categories as shown in Fig. 1.44. Let use see above programming languages in detail.

1. Machine Level Language:

- The lowest level of computer programming languages made up of binary bits is called as machine language. Machine language is the only language the computer can understand directly.
- The basic operating principle reveals that they work on only binary information. Therefore, the information must be supplied in the form of binary bits 0's and 1's. To perform the different kinds of operations the different groups of binary bits are needed.
- A machine language program consists of a sequence of bits that are all zeros (0's) and ones (1's). Each combination of zeros and ones is an instruction to the computer about something.
- Machine language can be defined as, "the language with set of instructions or machine codes formulated using groups of binary bits (0's and 1's) is called as machine language".
- **Example of Machine Level Language:**
      00001101   111010000

Advantages of Machine Level Language	Disadvantages of Machine Level Language
(i) It makes fast and efficient use of the computer.	(i) These languages are machine dependent.
(ii) It is directly understood by the computer, so no translator is required.	(ii) There is hard to find errors in a program.

2. Assembly Language:

- Assembly language is a symbolic representation of machine language.
- The language which substitutes letters and symbols for the numbers in the machine language program is called an Assembly Language or Symbolic Language.
- A program written in symbolic language that uses symbols instead of numbers is called assembly code or symbolic code i.e., MNEMONICS.
- The computer instructions are written in easily understandable short words which are called MNEMONICS. For example, MOV stands for moving the data, SUB stands for subtraction, ADD stands for addition, etc. All instructions are written in capital letters.
- A program in assembly language needs software known as an assembler which converts the program consisting of mnemonics and data into machine language.
- The assembly language can be defined as, "the computer programming language that uses MNEMONICS".

Advantages of Assembly Language	Disadvantages of Assembly Language
(i) Assembly level language programs can run much faster and use less memory and other resources.	(i) Assembly level language generated applications are slower and time consuming to develop.
(ii) Assembly language is easier to understand and use because mnemonic is used instead of numeric op-codes and suitable names are used for data.	(ii) Assembly language programs are machine dependent i.e., assembly language programs are differ from computer to computer.
(iii) In assembly language programmers can easily track and locate errors and correct them.	(iii) A machine cannot execute an assembly language program directly. It needs to be first translated into machine language code. This is done by an assembler.
(iv) Assembly language programs are easier for user to modify than machine language programs.	

3.   High Level Language:

- During 1960s computers started to gain popularity and it became necessary to develop languages that were more like natural languages such as English so that a common user could use the computer efficiently.

- High level languages are similar to English language. Programs written using these languages can be machine independent.

- A single high-level statement can substitute several instructions in machine or assembly language.

- In high-level language, programs are written in a sequence of statements to solve a problem. For example: sum = x + y.

- The high level languages can be defined as, "the computer programming languages consisting of a set of pre-defined words, symbols, punctuation marks and mathematical operations. Such elements of the language can be combined to form an executable statement that directs the computer to perform particular task or operation".

- Languages such as COBOL, FORTRAN, BASIC, and C are considered as high-level languages.

Advantages	Disadvantages
(i) High Level Language (HLL) programs are machine independent i.e., they can be used on different platforms with very little or no change at all.	(i) Because the automatic features of high level languages always occur and are not under the control of the programmer, they are less flexible than assembly languages.
(ii) Program written in high level languages are easier to maintain and write than assembly or machine language programs.	(ii) The programs written in high level languages take more time to run and require more main storage space.
(iii) These languages are very similar to the languages normally used us in our day to day life. Hence, they are easy to learn and use.	(iii) They are not directly executed by the machine and therefore always require a translator to convert them into machine codes.

Differentiation between Machine, Assembly and High Level Languages:

Sr. No.	Machine Language	Assembly Language	High Level Language
1.	It is in the form of 0's and 1's.	It is in the form of symbolic phrases called as MNEMONIC like ADD, MOV etc.	It is in the form of English words.
2.	Machine dependent language.	Machine dependent language.	Machine independent language.

contd. ...

3.	It does not need any language translators like assembler or compiler.	A translator 'assembler' is required to convert assembly language into machine code.	A translator like 'compiler or interpreter' is required to translate high level language into machine level languages.
4.	It is difficult to programming.	It is difficult to programming.	Easy to programming.
5.	Speed is high.	Speed is high.	Slow speed.
6.	Hard and difficult to understand.	Hard and difficult to understand.	Easy and simple to understand.
7.	Not user friendly.	Not user friendly.	User friendly.

1.6 COMPUTER MEMORY

- Computers are used not only for processing of data and/or information for immediate use, but also for storing of large volume of data for future use.
- The computer memory is just like a human brain. Memory is used to store the data and instructions on temporary or permanent basis.
- The primary goal of computer memory is to store data. The computers memory stores data, instructions required during the processing of data and output results.
- Fig. 1.45 shows all the three categories of computer memory.
- The storage location where the data are held temporarily is referred to as the primary memory while the storage location where the programs and data are stored permanently for future use is referred to as the secondary memory.

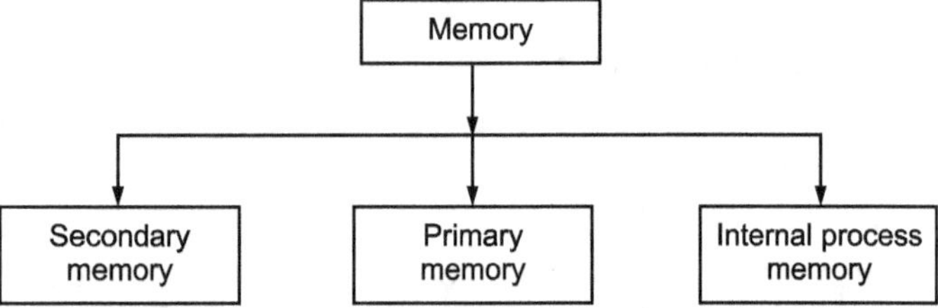

Fig. 1.45: Memory Categories (Types) in Computer

- The primary memory is generally known as "memory" and the secondary memory as "storage".
- Computers also use a third type of storage location or memory known as the internal process memory and this memory is placed either inside the CPU or near the CPU.
- Fig. 1.45 shows following categories of computer memory:
  1. **Primary Memory:** It is also known as main memory. It includes two types, namely, Random Access Memory (RAM) and Read Only Memory (ROM). The data stored in RAM are lost when the power is switched OFF and therefore, it is known as volatile memory. However, the data stored in ROM stay permanently even after the power is switched OFF and therefore ROM is known as non-volatile memory.
  2. **Secondary Memory:** It is also known as auxiliary memory. It includes magnetic disks and magnetic tapes. These storage devices have much larger storage capacity than the primary memory. Information or data stored in such devices remains permanent (until we remove it).
  3. **Internal Process Memory:** Usually it includes cache memory and registers both of which store data temporarily and are accessible directly by the CPU. This memory is placed inside or near the CPU for the fast access of data.

Differentiation Primary Memory and Secondary Memory:

Sr. No.	Primary Memory	Secondary Memory
1.	The primary memory (main memory) is used to store the data and instructions temporarily.	The secondary memory (auxiliary memory) is used to store the data and instructions permanently.
2.	It is directly accessed by the CPU for the purpose of storing and retrieving information.	It is not directly accessible by the CPU.

contd. ...

3.	It is accessed with the help of address bus and data bus.	It is accessed by using input/output channels.
4.	It does not hold the data when the power is turned OFF, i.e., it is volatile in nature.	It holds the data even when the power is turned OFF, i.e., it is non-volatile in nature.
5.	It is much faster than secondary memory.	It is comparatively slower.
6.	It is more costly as compared to secondary memory.	Secondary memory devices are less costly.
7.	Examples: RAM, ROM etc.	Examples: Magnetic tape, optical disks etc.

Primary Storage/Memory:

1. RAM:

- o RAM is read/write memory. RAM is used to store data and instructions during the operation of computer.
- o RAM is volatile in nature, i.e. data stored in it is lost when switch OFF the computer or if there is a power failure.
- o This memory is accessible from any memory location anytime. One can switch to one place to another place in memory randomly.
- o RAM is of two types i.e. Static RAM (SRAM) and Dynamic RAM (DRAM).

(i) SRAM:

- ▪ SRAM (Static Random Access Memory) is a type of semi-conductor memory where the word static indicates that, it does not need to be periodically refreshed, as SRAM uses bi-stablelatching circuitry to store each bit.
- ▪ SRAM is volatile in the conventional sense that data is eventually lost when the memory is not powered.

(ii) DRAM:

- ▪ DRAM (Dynamic Random Access Memory) stores each bit of data in a separate capacitor within an integrated circuit.
- ▪ Since, real capacitors leak charge, the information eventually fades unless the capacitor charge is refreshed periodically.

Comparison between SRAM and DRAM:

Sr. No.	Static RAM (SRAM)	Dynamic RAM (DRAM)
1.	Each static RAM cell is a flip-flop.	A dynamic RAM cell consists of a MOSFET and a capacitor.
2.	SRAM has less number of memory cells/unit area.	DRAM has more number of memory cells/unit area.
3.	SRAM has more number of components per cell.	DRAM has only two components per cell.
4.	It does not require refreshing.	It require refreshing.
5.	Faster memories.	Slower memories.
6.	Power consumption is less.	More power consumption.

Advantages of RAM	Disadvantages of RAM
(i) RAM uses much less power than disk drives. (ii) RAM is the fastest storage medium outside of the CPU.	(i) A power failure will cause irrecoverable data loss. (ii) RAM's cost per bit is high. (iii) RAM has limited memory space.

2. ROM:

- o   ROM is the memory from which one can only read but cannot write on it.
- o   ROM is a non-volatile primary memory. It does not lose its content when the power is switched OFF.
- o   ROM is also known as firmware and it is an integrated circuit programmed with specific data when it is manufactured.
- o   A ROM stores an instruction as are required to start computer when electricity is first turned ON, this operation is referred to as bootstrap.

Types of ROM:

(a) MROM (Masked ROM):

- ▪   The very first ROMs were hard-wired devices that contained a pre-programmed set of data or instructions. These kinds of ROMs are known as masked ROMs.
- ▪   MROM is inexpensive ROM.

(b) PROM (Programmable Read Only Memory):

- ▪   PROM is read-only memory that can be modified only once by a user. The user buys a blank PROM and enters the desired contents using a PROM programmer.
- ▪   Inside the PROM chip there are small fuses which are burnt open during programming.
- ▪   PROM can be programmed only once and is not erasable.

(c) EPROM (Erasable and Programmable Read Only Memory):

- ▪   An EPROM is a type of memory chip that retains its data when its power supply is switched OFF.
- ▪   The EPROM can be erased by exposing it to ultra-violet light for a duration of up to 40 minutes. Usually, a EPROM eraser achieves this function.

(d) EEPROM (Electrically Erasable and Programmable Read Only Memory):

- ▪   EEPROM is a type of non-volatile memory used in computers and other electronic devices to store small amounts of data that must be saved when power is OFF.
- ▪   The EEPROM is programmed and erased electrically. It can be erased and reprogrammed about ten thousand times. Both erasing and programming take about 4 to 10 ms (milli second).
- ▪   In EEPROM, any location can be selectively erased and programmed. EEPROMs can be erased one byte at a time, rather than erasing the entire chip. Hence, the process of re-programming is flexible but slow.

Advantages of ROM	Disadvantages of ROM
(i) Non-volatile in nature.	(i) More power consumption.
(ii) Cheaper and more reliable than RAMs.	(ii) Increase in size with increase in number of input variables.

Comparison between RAM and ROM:

Sr. No.	RAM	ROM
1.	RAM stands for Random Access Memory.	ROM stands for Read Only Memory.
2.	It is a read/write memory.	It is a read only memory.
3.	It is volatile in nature.	It is a non-volatile in nature.
4.	Faster in speed.	Slower in speed.
5.	Data is erased as soon as power supply is turned OFF.	Data remains stored even after power supply has been turned OFF.

Secondary Storage/Secondary Memory Devices:

- •   Secondary storage (memory) devices facilitate storing of data and instructions permanently.

- Fig. 1.46 shows classification of secondary storage devices.

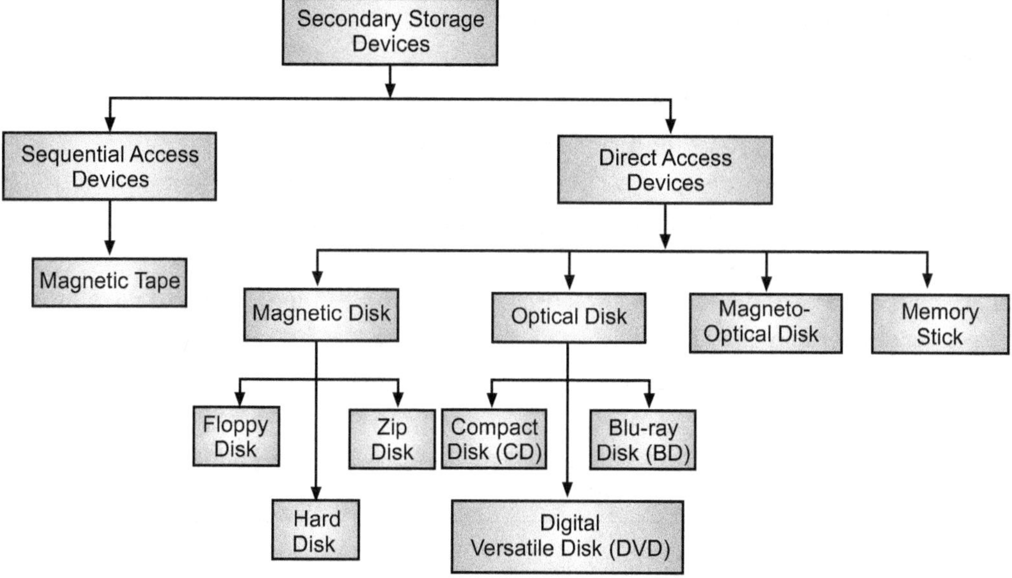

Fig. 1.46: Classification of Secondary Storage (Memory) Devices

1. Floppy Disks :

- A floppy disk is a secondary storage device used to store data permanently.
- A floppy disk is a round, flat piece of Mylar plastic coated with ferric oxide (a rust like substance containing tiny particles capable of holding a magnetic field) and encased in a protective plastic cover (disk jacket).

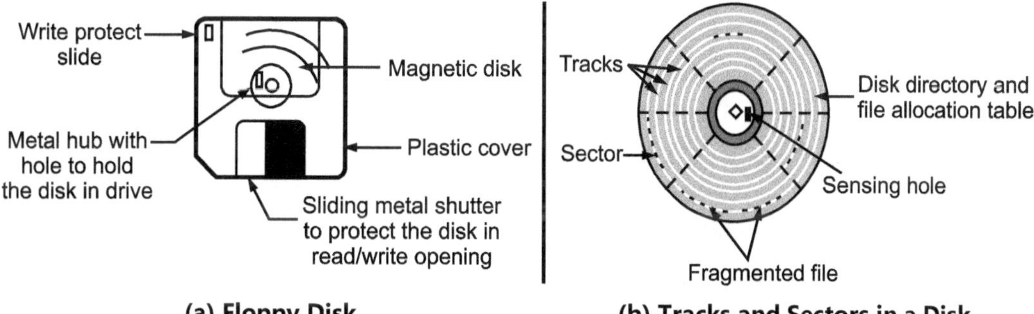

(a) Floppy Disk **(b) Tracks and Sectors in a Disk**

Fig. 1.47: Structure of Floppy Disk

- A floppy disk is a removable disk and is read and written by a Floppy Disk Drive (FDD).
- Data are stored in a floppy disk in concentric circles known as tracks. Tracks are divided into many storage locations called sectors (Fig. 1.47 (b)).
- Since, the tracks are circles, there is a sensing hole that identifies the physical beginning of the track. Tracks and sectors on a disk are identified by the disk drive through formatting.
- FDD is a device that performs the basic operation on a floppy disk, including rotating the disk and reading and writing data onto it.
- The disk drive's read/write head alters the magnetic orientation of the particles, where orientation in one-direction represents '1' and orientation in the other represents '0'.
- Traditionally, floppy disks were used an Personal Computers (PCs) to distribute software, transfer data between computers, and create small backups.

- Earlier, 5¼ inch floppy disks were used. Later, a new format of 3½ inch floppy disk came into existence, which has larger storage capacity (1.44 MB) and supports faster data transfer than 5¼ inch floppy disks.

Working (Operation) of Floppy Disk:

- To read and write data onto a floppy disk FDD is used as shown in Fig. 1.48.

- The floppy disk drive is made up of a box with a slot (having a drive gate) into which user inserts the disk. When user inserts a disk into the floppy disk drive, the drive grabs the disk and spins it inside its plastic jacket.

- The FDD has a read/write head which reads/writes data on to the disk. The disk rotates at 360 rpm while reading or writing data on to it.

- The FDD drives circuit board receives instructions for reading or writing the data from/to disk through the floppy disk controller.

- A motor located beneath the disk spins a shaft that engages a notch on the hub of the disk, causing the disk to spin.

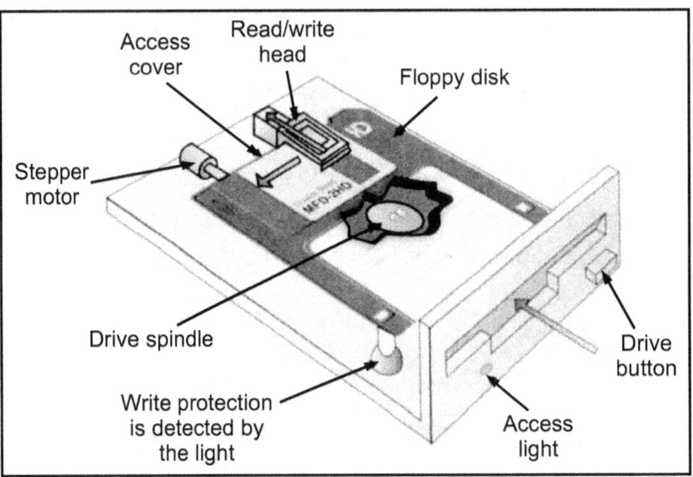

Fig. 1.48: Floppy Disk Drive (FDD)

Advantages of Floppy Disks	Disadvantages of Floppy Disks
(i) A floppy disk is portable and inexpensive. (ii) Data on a floppy disk can be accessed randomly.	(i) Floppy disk is not durable (due to dust and dirt) and can be destroyed by magnetic field. (ii) Storage capacity of floppy disk is limited (only 1.44 MB).

2. Hard Disk:

- The hard disk, also called the hard drive or fixed disk, is the primary storage unit of the computer.

- It is a non-volatile, random access digital magnetic data storage device.

- Hard disks consist of stacks of flat, circular plates made of aluminium or glass and coated on both surfaces with magnetic material, (See Fig. 1.49 (a)).

- A Hard Disk (HD) consists of one or more platters divided into concentric tracks and sectors. It is mounted on a central spindle, like a stack. It can be read by a read/write head that pivots across the rotating disks.

- The data is stored on the platters covered with magnetic coating as shown in Fig. 1.49 (b).

- The recording surfaces are divided into concentric circles called tracks, and each track is subdivided into invisible segments called sectors. Each sector can hold a specific number of characters (say, 512 bytes).

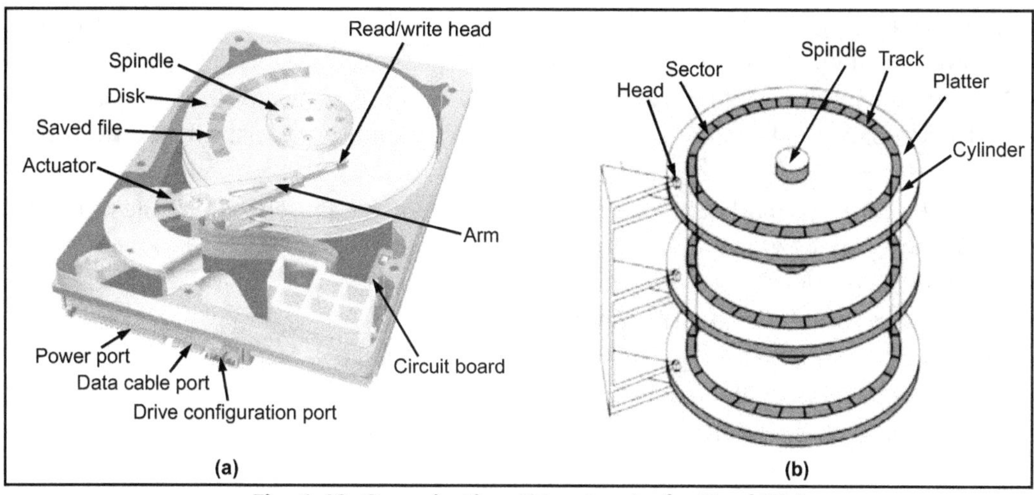

Fig. 1.49: Organisation (Structure) of a Hard Disk

- External hard disk is a type of portable disk drive, which is connected to the computer using a USB cable.

Advantages of HDs	Disadvantages of HDs
(i) They have large storage capacity (upto 500 GB). (ii) Stores and retrieves on data much faster than a floppy disk or CD-ROM. (iii) Data is not lost when switch OFF the computer.	(i) Regular 'head' crashes can damage the surface of the disk, leading to loss of data in that sector. (ii) The disk is fixed inside the computer and cannot easily be transferred to another computer.

Comparison between Floppy Disk and Hard Disk:

Sr. No.	Floppy Disk	Hard Disk
1.	Floppy disks are also known as floppies or micro disks.	Hard disks are also known as fixed disks.
2.	The computer takes more time to read from a floppy disk.	The computer takes less time to read from hard disk.
3.	More prone to damage by heat, dust and improper handling as it is made of a flexible material.	Less prone to damage as it is within the system unit.
4.	Can be used to store 1.44 MB of data.	Used to store upto 500 GB of data.
5.	It is cheaper.	It is costly.

3. Optical Disks:

- An optical disk is any computer disk that uses optical storage techniques and technology to read and write data.
- It is a computer storage disk that stores data digitally and uses laser beams (transmitted from a laser head mounted on an optical disk drive) to read and write data.
- Compact Disk (CD) and Digital Versatile Disk (DVD) are commonly used optical disks.

(i) CD:

- o CDs are circular discs that are 4.75 in (12 cm) in diameter. CDs can hold up to 700 MB of data.
- o The CD standard was proposed by Sony and Philips in 1980.
- o The data on a CD is stored as small notches on the disc and is read by a laser from an optical drive. The drives translate the notches (which represent 1's and 0's) into usable data.
- o A compact disc is a small, portable, round medium disk for electronically recording, storing, and playing back audio, video, text, and other information in digital form.

Fig. 1.50: A CD

- o CD is made of a polycarbonate substance and is coated with a metallic film, usually an aluminum alloy. This aluminum film is the portion of the disc that the CD-ROM drive reads for information.
- o The aluminum film is then covered by a plastic polycarbonate coating that protects the underlying data. A label will usually be placed on the top of the disc and data is read from the bottom of the CD.
    - **(a) CD-R (CD Recordable):** It is similar to a CD-ROM. The user can write to the disk only once.
    - **(b) CD-RW (CD Rewritable):** It is similar to a CD-ROM. The user can erase and rewrite to the disk multiple times.

CR-ROM (Compact Disc Read-Only Memory):

- o The CD-ROM drives in computers that read the data on discs are almost identical to audio CD players.
- o CD-ROM drives have speeds ranging from 1x all the way up to 72x, meaning it reads the CD roughly 72 times faster than the 1x version.

Fig. 1.51: CD-ROM Drive

(ii) DVD:

- o DVD is an optical disk storage media format that can be used for data storage.
- o DVD also formerly known as Digital Video Disk. A standard DVD disk store up to 4.7 GB of data.

Fig. 1.52: DVDs

Fig. 1.53: DVD ROM Drive

- o Digital Video Disk-Read Only Memory (DVD-ROM) is an optical storage device used to store digital video or computer data. DVDs look like CDs in shape and physical size.
    - **(a) DVD-R (DVD Recordable):** It is similar to a DVD-ROM. The user can write to the disk only once. Only one-sided disks can be used.
    - **(b) DVD-RW (DVD Rewritable):** It is similar to a DVD-ROM. The user can erase and rewrite to the disk multiple times. Only one-sided disks can be used.

Advantages of Optical Disks	Disadvantages of Optical Disks
(i) Portable because of small size and lightweight. (ii) It possesses large capacity to store data/ information in the form of multimedia, graphics, audio and video files. (iii) It holds more data recording density as com-pared to other storage media therefore, it has low cost per bit of storage.	(i) It possesses slow data access speed as compared to magnetic disks. (ii) The drive mechanism of optical disk is more complicated than the floppy disks.

(iii) High Definition (HD) Optical Disks:

- o High-definition optical disks in computer are designed to store high-definition videos and to provide significantly greater storage capacity compared to DVDs.
- o Blu-ray Disc (or BD) is an optical disc storage medium that can have a capacity of upto 50 GB.
- o Blu-ray disks are available in two sizes i.e. Standard (12 cm) and Mini (8 cm).
- o Blu-ray disks are also available in following formats:
    - **(a) BD-R** is a disk on which you can record data only once.
    - **(b) BD-RW** is similar to BD-R disk but the difference is that it is rewritable. This means data can be erased and recorded number of times on the same disk.

(c) BD-ROM comes with pre-recorded content that can only be read.

(d) BD-RE is also a rewritable disk but is used only for high definition audio/video and television recording.

(iv) Magneto-Optical Disk (MO Disk):

o A magneto-optical disk is a rewritable disk that makes use of both magnetic disk and optical technologies. It is similar to a magnetic diskette except for its larger size.

o Magneto-optical disks are seldom manufactured and used due to the advent of flash drives and DVD/CD drives, which are less expensive and have better writing time and reliability.

Comparison between CD and DVD:

Sr. No.	Parameters	CD	DVD
1.	Stands for	Compact Disc.	Digital Versatile Disc.
2.	Purpose	CDs are made with the purpose of holding audio files as well as program files.	DVDs are made with the purpose of holding video files, movies, substantial amount of programs.
3.	Media type	Optical disc.	Optical disc.
4.	Capacity	Typically upto 700 MB (upto 80 minutes audio).	DVD can range from 4.7 GB to 17.08 GB.
5.	Types	CD-R, CD-RW etc.	DVD-RW, DVD+RW, etc.

4. Pen Drive:

- A pen drive is a removable storage device that is frequently used nowadays to transfer audio, video, and data files from one computer to another.

- Pen drive also called as flash drive.

- A pen drive consists of a small printed circuit board, which is fitted inside a plastic, metal, or rubber casing to protect it.

- The USB connector which is present at one end of pen drive is protected by either a removable cap or pulling it back in the casing.

- Fig. 1.54 shows a pen or flash drive.

- Pen drive is a high storage (ranging from 1 GB to 32 GB) capacity device and is physically small enough to fit into a pocket.

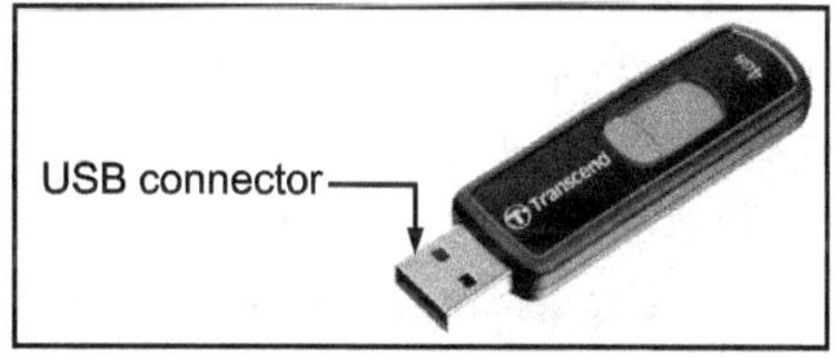

Fig. 1.54: Pen Drive

Advantages	Disadvantages
(i) Smaller in size and faster read and write speed than HDDs.	(i) Most flash drives do not have a write-protection mechanism.
(ii) Cheaper than traditional drives in small storage capacities.	(ii) Smaller size drives make them easier to lose.

5. Magnetic Tape:

- Magnetic tape is one of the most popular storage medium for large data that sequentially accessed and processed.

- Magnetic tapes are the plastic tapes with magnetic coating that are used for storing the data, such as text, audio or video.

Organization of Magnetic Tape:

- Magnetic tape is a medium for magnetic recording, made of a thin, magnetizable coating on a long, narrow strip of plastic film.

- Devices that record and play back audio and video using magnetic tape are tape recorders and video tape recorders.

- A device that stores computer data on magnetic tape is a tape drive (tape unit, streamer).
- The magnetic tape is divided into vertical columns i.e., frames and horizontal rows i.e., channels or tracks.
- In magnetic tape data is stored in a string of frames with one character per frame and each frame spans multiple tracks (usually 7 or 9 tracks).
- For this reason a single bit is stored in each track, that is, one byte per frame. The remaining track (7th or 9th) stores the parity bit, (See Fig. 1.55).
- When a byte is written to the magnetic tape, the number of 1s in the byte is counted, the parity bit is then used to make number of 1s even (even parity) or odd (odd parity).
- When the magnetic tape is read again, the parity bit is checked to see if any bit has been lost. In case of odd parity, there must be an odd number of 1s represented for each character and an even number of 1s in case of even parity.

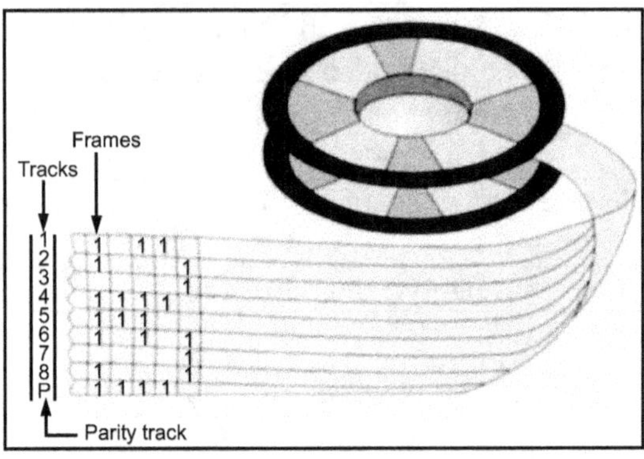

Fig. 1.55: Representing Data in Magnetic Tape

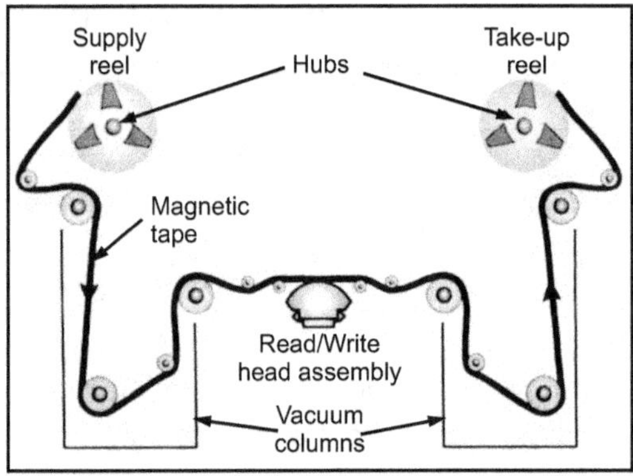

Fig. 1.56: Basic Tape Drive Mechanism

- Fig. 1.56 shows the basic tape drive mechanism.
- Magnetic tape drive uses two reels, supply reel and take-up reel. Both reels are mounted on the hubs and the rape moves from the supply reel to the take-up reel.
- The magnetic oxide coated side of the tape passes directly over the read/write head assembly, thus making contact with the heads.
- As the tape passes under the read/write head, the data can be either read and transferred to the primary memory or read from primary memory and written onto the tape.

Advantages of Magnetic Tapes	Disadvantages of Magnetic Tapes
(i) Low cost.	(i) Low data transmission speed due to sequential access.
(ii) Large storage capacity.	
(iii) Magnetic tapes are portable because they are compact in size and lightweight.	(ii) Not suitable for random access.
	(iii) Updation, such as insertion or deletion is difficult.

1.7 OPERATING SYSTEMS

- An operating system processes system data and user input, and responds by allocating and managing tasks and internal system resources as a service to users and programs of the system.
- An OS is an intermediary between users and computer hardware. It provides users an environment in which a user can execute programs conveniently and efficiently.
- Popular modern operating systems include Android, iOS, Unix, Windows, Linux, Mac, VMS, etc.
- An operating system is the collection of the various programs that take control over the operations of the computer system.

- An operating system is an interface between a computer user and computer hardware, (See Fig. 1.57).

Objectives of an Operating System:

1. To make a computer system convenient to use in an efficient manner.
2. To hide the details of the hardware resources from the users.
3. To provide users a convenient interface to use the computer system.
4. To manage the resources of a computer system.

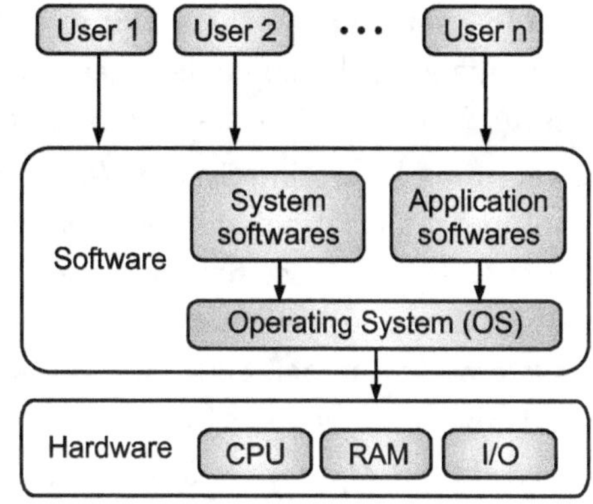

Fig. 1.57: Interaction among Hardware, OS, Software and Users

Definition of Operating System (OS):

- An operating system is defined as, "a program that acts as an interface between the user and the computer hardware and controls the execution of all kinds of programs".		**OR**
- An operating system is defined as, "a collection of system programs that together controls the operation of a computer system".

Functions of Operating System:

1. **Memory Management:** OS keeps tracks of primary memory i.e. what part of it is in use by whom, what part is not in use etc., and allocates the memory when a process or program requests it.
2. **Processor Management:** OS allocates the processor (CPU) to a process and deallocates processor when it is no longer required.
3. **Device Management:** It keeps track of all devices. This is also called I/O controller that decides which process gets the device, when, and for how much time.
4. **File Management:** It allocates and de-allocates the resources and decides who gets the resources.
5. **Security:** It prevents unauthorized access to programs and data by means of passwords and similar other techniques.
6. **Job Accounting:** OS keeps track of time and resources used by various jobs and/or users.
7. **Control over System Performance:** OS records delays between request for a service and from the system.
8. **Co-ordination between other Softwares and Users:** Co-ordination and assignment of compilers, interpreters, assemblers and other software to the various users of the computer systems.
9. **Execution of Program:** The OS is responsible for executing various programs whether user programs or system programs, i.e., special programs required for the machine functioning.
10. **Detection of Errors:** The OS is also responsible for detecting any type of error that occurs and then handling it.

Types of OS:

- The broad categories of OS are described below:
  1. **Single-user, Single-tasking OS:** As the name implies, this operating system is designed to manage the computer so that one user can effectively do one thing at a time. The palm OS for palm handheld computers is a good example of a modern single-user, single-task operating system.
  2. **Single-user, Multi-tasking OS:** This type of OS which allows a single user to execute two or more tasks at a time. Single-user, multi-tasking is the type of operating system most people use on their desktop and laptop computers today. Both Windows 98 and the Macintosh OS are examples of operating systems that will let a single user have several programs in operation at a time.

3. **Multi-user, Multi-tasking OS:** A multi-user operating system allows many different users to take advantage of the computer's resources simultaneously. Unix, Linux, VMS and mainframe operating systems such as MVS, are examples of multi-user operating systems.

4. **Real-time Operating System:** Real-Time Operating System (RTOS) is managing the resources of the computer so that a particular operation executes in precisely the same amount of time every time it occurs. The response time of a real-time operating system is very quick.The operating system which guarantees the maximum time for these operations are commonly referred to as **hard real-time systems**, while operating systems that can only guarantee a maximum of the time are referred to as **soft real-time systems**.Examples of real-time OS are QNX, RTLINUX etc.

5. **Distributed Operating System:** Distributed operating system is a type of multi-user operating system, where processing is distributed, i.e. each user workstation is assigned a task to be processed. The processed results are later compiled together on a central computer.

6. **Time Sharing OS:** In time sharing OS a number of simultaneous users are there. Each user is given a trivial amount of time (a quantum/time slice) in which he/she processes interactively or conversationally.

7. **Multi-programming OS:** Multi-programming is a term given to a system that may have several (multiple) processes in the 'state of execution' at the same time.

8. **Multi-processing OS:** In a multi-processing OS, a single CPU has more than one processor. All these processors may or may not be equally powerful and may or may not prefer same operation.

9. **Batch-processing OS:** In this type of OS, there is no direct interaction between user and the computer. The user has to submit a job (written on cards or tape) to a computer operator. Then computer operator places a batch of several jobs on an input device. Jobs are batched together by type of languages and requirement. Then a special program, the monitor, manages the execution of each program in the batch. The monitor is always in the main memory and available for execution.

10. **Multi-tasking OS:** It allows executing more than one task at the same time. An operating system that is capable of allowing multiple software processes to run at the same time. Some examples of multi-tasking operating systems are Unix and Windows 2000 Windows XP, Windows 7, Windows Vista etc.

11. **Multi-threading OS:** It allows different parts of a single program to run concurrently. Multithreading operating systems allow different parts of a software program to run concurrently. Operating systems that would fall into this category are Linux, Unix and Windows 2000.

Working with Windows 7 Operating System:

- Microsoft Windows or simply 'Windows' is a meta-family of graphical operating systems marketed, developed, and sold by Microsoft Corporation.

- Microsoft introduced an operating environment named Windows on November 20, 1985, as a graphical operating system shell for MS-DOS in response to the growing interest in Graphical User Interfaces (GUIs).

- Windows 7 is an operating system that Microsoft has produced for use on personal computers.

- The major upgrade versions of Windows 7 are:

  1. **Home Premium** is the most popular version of Windows 7 and will likely suit the needs of most users.

  2. The **Professional** version may appeal to owners of small to medium-sized businesses because it has extras like Windows XP mode and networking backup features.

  3. **Ultimate** is the most powerful version, with added security features like Bitlocker and the flexibility of use in 35 languages.

Desktop of Windows 7 OS:

- Desktop is the working area on Windows operating system where we place and arrange application and document windows to work efficiently.
- When we turn ON a computer, the desktop is displayed first. The desktop contains icons of useful applications and shortcuts of frequently required documents.
- We use wallpapers on desktop background to make desktops attractive. Wallpapers are the images displayed on the desktop background.

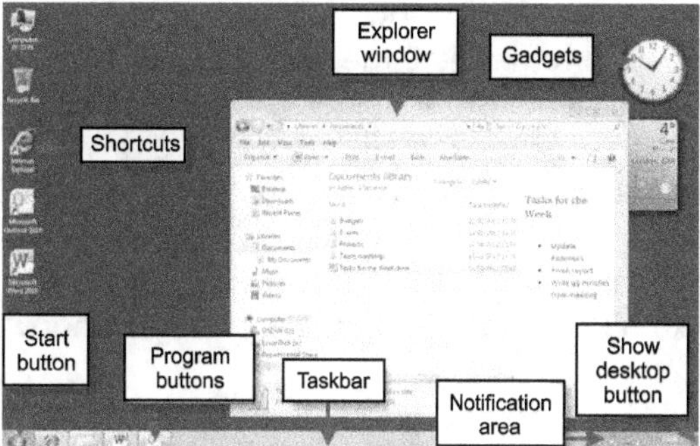

Fig. 1.58: Desktop of Windows 7 OS

- Following table shows components of Windows 7 desktop:

Sr. No.	Windows 7 Features	Description
1.	Start Button	The Start Button is located on the taskbar at left down corner of screen.
2.	Program Buttons	Each open program appears as a button - even when multiple items for a program are open - but we can choose to change this view. We can pin programs to the taskbar. Appear Next to Start button.
3.	Taskbar	The horizontal line at the bottom of our screen is the taskbar.
4.	Show Desktop Button	This button will minimise all open windows to display the desktop and located at right down corner of screen.
5.	Explorer Window	In the Explorer Window we can manage our files and folders. We can use the back and forward buttons to navigate around.
6.	Gadgets	Gadgets can be placed anywhere on the desktop. They offer information - such as the time or the weather - at a quick glance.
7.	Notification Area	The notification area displays the time, the date and any program related icons. We can click an icon to display a window of options.
8.	Shortcuts	Shortcuts allow us to open a program without having to search for it in the Start Menu.

Working with Files and Folders:

- Understanding how to work with files and folders is an important part of using the computer.
- A file is a collection of data that is stored on disk and that can be manipulated as a single unit by its name.
- A file can be defined as, "a collection of logically related records". Files are usually represented by an icon, (See Fig. 1.59).
- An icon is a small image displayed on the screen which represent an application program, file or folder and are displayed on the desktop.
- A folder is a container for storing programs and files. Windows uses folders to organize files. We can put files inside a folder, just like we would put documents inside a real folder.
- Folders are containers that can contain icons, programs, data or other folders (sub-folders).
- A folder can hold different types of files, such as text, spreadsheets, and presentations.
- We can view and organize files and folders using a built-in application known as Windows Explorer in Windows 7.
- The Explorer windows are powerful easy-to-use tools for working with files consistently across Windows 7.

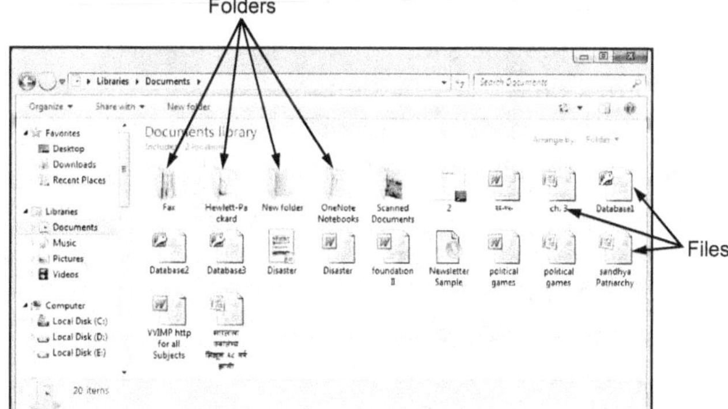

Fig. 1.59: Windows Explorer with Files and Folders

- Windows Explorer is the file management application in Windows. Windows Explorer can be used to navigate the hard drive and display the contents of the folders and subfolders we use to organize our files on the hard drive.
- Windows Explorer is automatically launched any time we open a folder in Windows XP. We can also right-click on a folder and select Explore to open that folder using Windows Explorer.
- Fig. 1.60 shows a typical window and each of its parts.

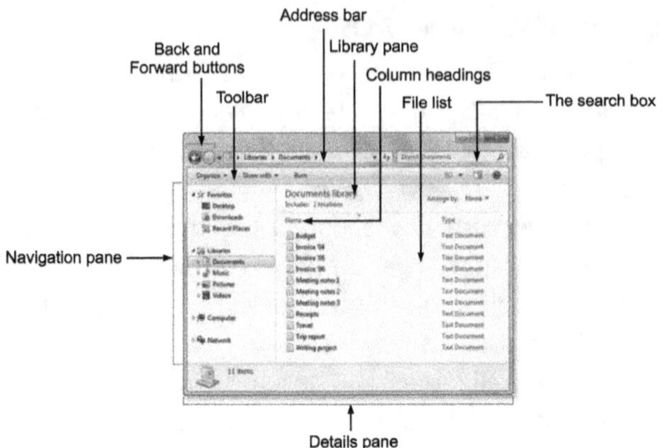

Fig. 1.60: Window Explorer with its Elements

- Various part of Windows Explorer are explained below:

Part	Description
Navigation pane	The Navigation Pane provides quick links to various folders and locations on the computer.
Back and Forward buttons	Use the Back button and the Forward button to navigate to other folders or libraries we have already opened without closing the current window.
Toolbar	Use the toolbar to perform common tasks, such as changing the appearance of the files and folders.
Address bar	The Address Bar allows us to navigate up and down a series of windows by double-clicking on a folder. Use the address bar to navigate to a different folder or library or to go back to a previous one.
Library pane	The library pane appears only when we are in a library (such as the Documents library). Use the library pane to customize the library or to arrange the files by different properties.
File list	This is where the contents of the current folder or library are displayed. If we type in the search box to find a file, only the files that match the current view (including files in subfolders) will appear.
Search box	Type a word or phrase in the search box to look for an item in the current folder or library. The search begins as soon as we begin typing - so if we type "B," for example, all the files with names starting with the letter B will appear in the file list.
Details pane	Use the details pane to see the most common properties associated with the selected file. File properties are information about a file, such as the author, the date we last changed the file, and any descriptive tags we might have added to the file.
Preview pane	Use the preview pane to see the contents of most files. If we select an e-mail message, text file, or picture, for example, we can see its contents without opening it in a program. If we do not see the preview pane, click the Preview pane button in the toolbar to turn it on.

Creating New Files and Folders:

- File management is organizing and keeping track of files and folders, helping we stay organized, so information is easily located.
- We can create a file or folder just about anywhere on the computer in following way:
  If we are creating a new file or folder on the desktop, or in a folder, right-click on an empty area, point to New, and click Folder or File icons as shown in Fig. 1.61.

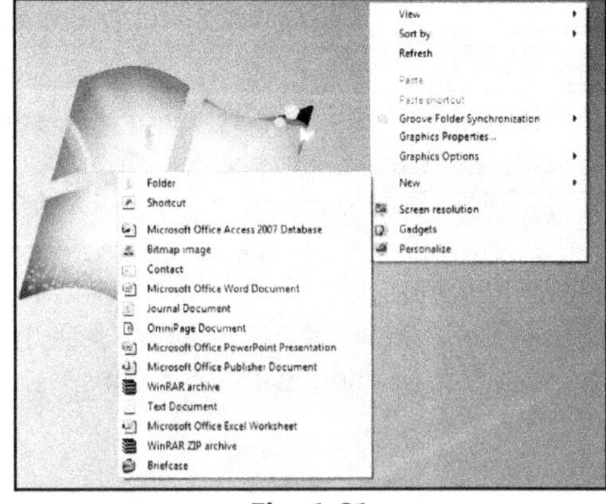

Fig. 1.61

1. For New File Creation:

- In the Windows Explorer select a file then Right click on it and then select New option which creates a new file.

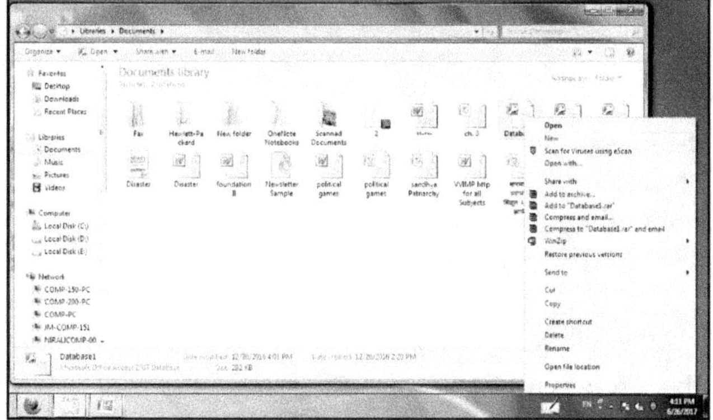

Fig. 1.62: Creation of a New File

2. For New Folder Creation:

- Folders are really helpful to organize files. We can create new folders and sub folders so that we can arrange files logically.
- In the Windows Explorer click New folder button as shown in Fig. 1.63 (a) which creates new folder as shown in Fig. 1.63 (b). We can also create a new folder using Ctrl+Shift+N.

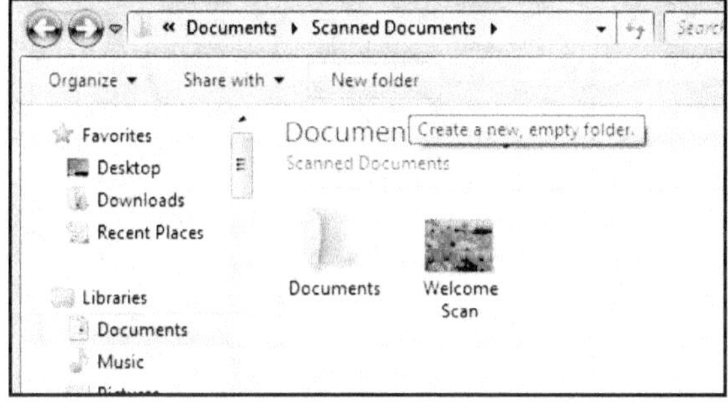

(a)

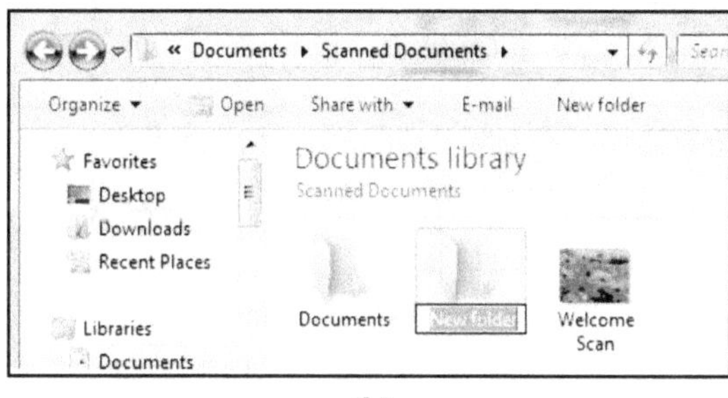

(b)

Fig. 1.63: Creation of a New Folder

Copying and Pasting Files and Folders:

- Sometimes, we will need to move a file from one folder to another, or copy a file from one folder to another, leaving the file in the first location and placing a copy of it in the second.
- We can move or copy a file or folder using a variety of methods. The procedure as follows:

  Step 1 : Open the drive or folder containing the file or folder we want to copy, (See Fig. 1.64).

  Step 2 : Select the files or folders we want to copy, (See Fig. 1.65).

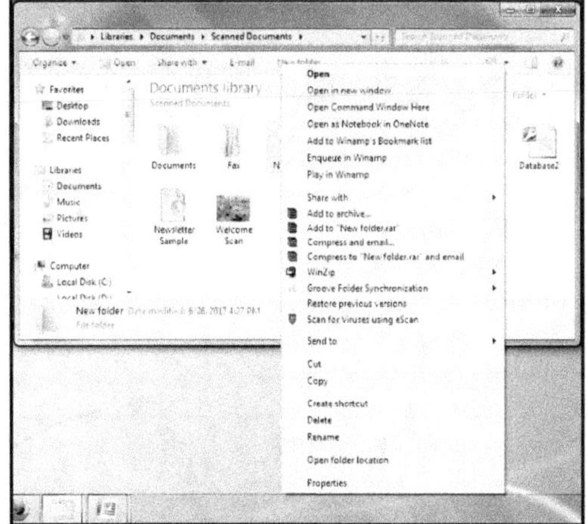
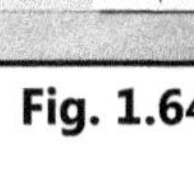

Fig. 1.64

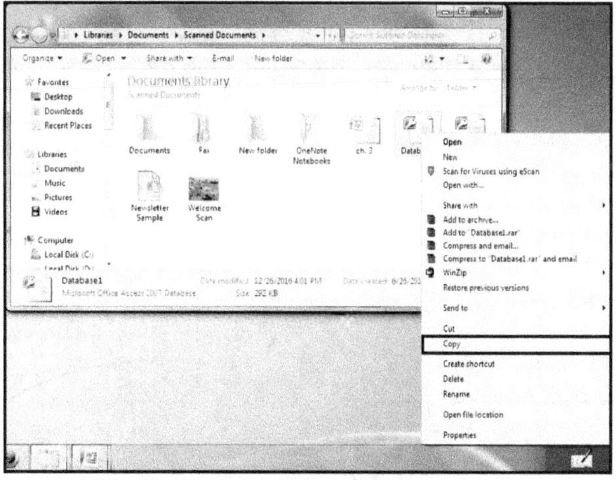

Fig. 1.65: Select Files/Folder to Copy

Step 3 : Display the destination folder where we want to copy the files or folder.

Step 4 : Then click Paste. Which display the copied file or folder on the Window.

Rename a File or Folder:

- We can also change the name of any file or folder.
- A unique name will make it easier to remember what type of information is saved in the file or folder.
- For **renaming a file** follow the following steps:

  Step 1 : Right Click the file which will be rename, (See Fig. 1.67).

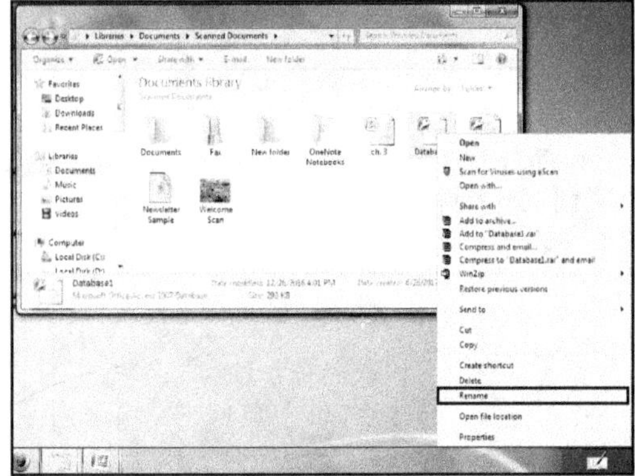

Fig. 1.67

  Step 3 : Type the desired name using the keyboard and press Enter. The name will be changed, (See Fig. 1.69).

- For **renaming a folder** follow the following steps:

  Step 1 : Right Click the folder which is renamed, (See Fig. 1.70).

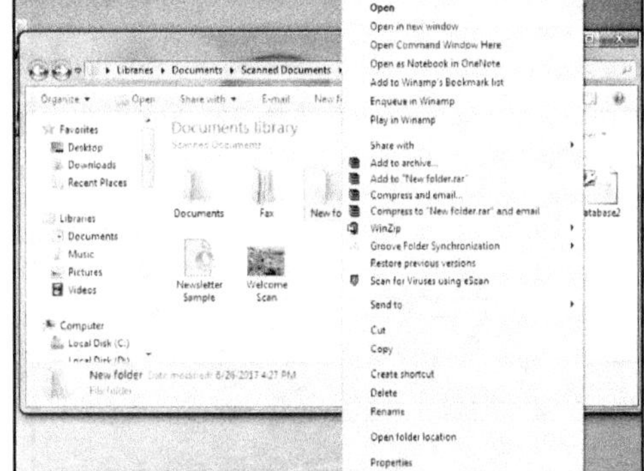

Fig. 1.70

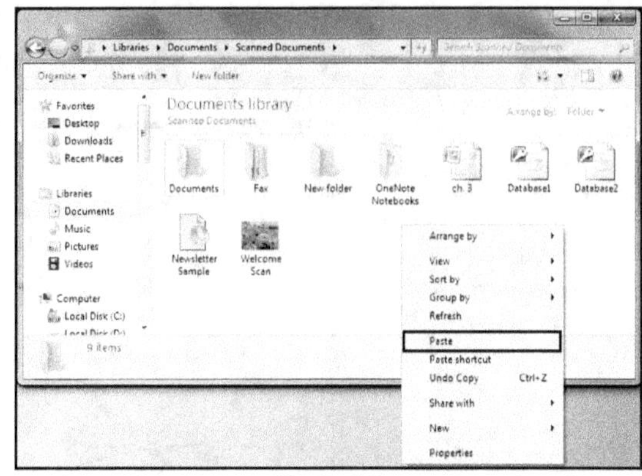

Fig. 1.66: Paste File/Folder to Destination

Step 2 : An editable text field will appear, (Se Fig. 1.68).

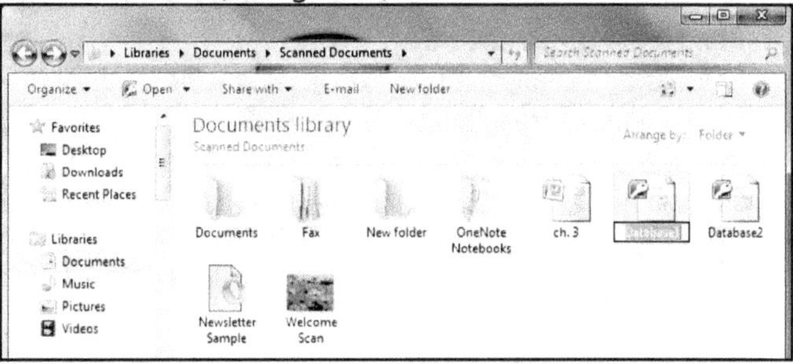

Fig. 1.68

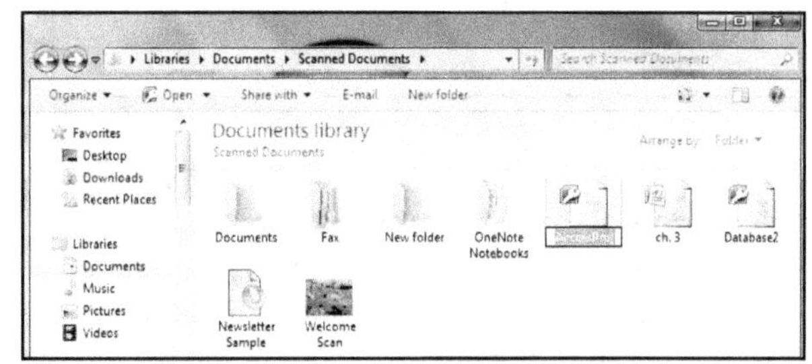

Fig. 1.69

Step 2 : An editable text field will appear, (See Fig. 1.71).

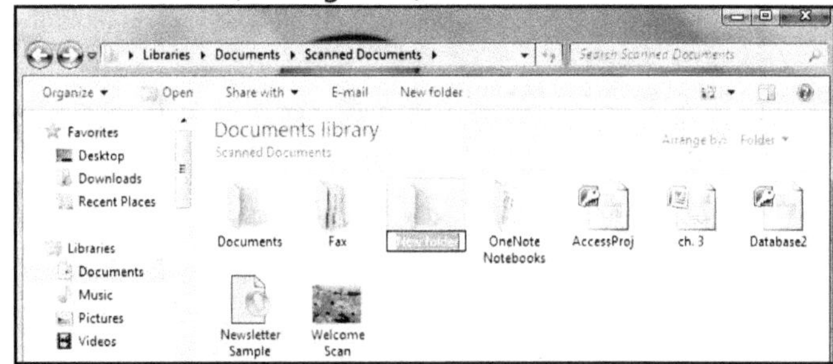

Fig. 1.71

Step 3 : Type the desired name using the keyboard and press Enter. The name will be changed, (See Fig. 1.72).

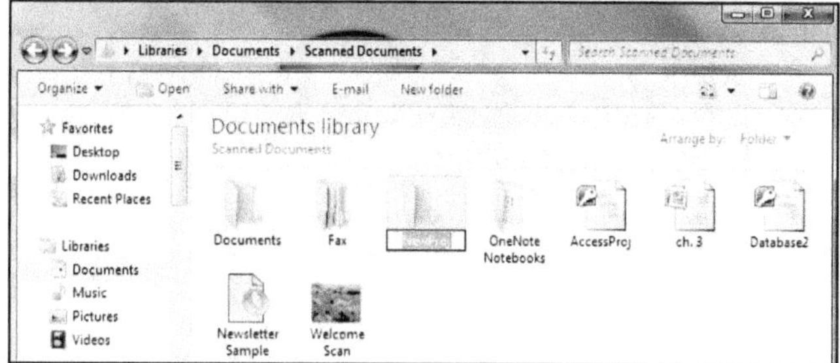

Fig. 1.72

Deleting Files and Folders:

- Windows Explorer lets we to delete files and folders easily.
- Select the files or folders that we want to delete.
- Press the Delete key on keyboard or right click the files then select Delete from shortcut menu. Click on Yes if we are sure to wish to delete the files.

Searching Files and Folders:

- If we save a file and then cannot remember what we named it or where it was saved, we can search for the file using the Windows Search feature.
- Windows allows us to locate files based on file name, a portion of a name, date, and/or the text included in a file. To conduct a search, we can:

1. Using Search in the Start Menu:

- Searching locally inside a folder is easy to do with the Search bar, but doesn't help we very much if we do not know what folder to look in! Therefore, Windows includes a broad search tool in the Start menu, just above the Start button.
- Type the first few letters of the file name or program (See Fig. 1.73) we are looking for and Windows will display some of the best matches for we as we type based on files that have been indexed.

2. Using a Search Box:

- Every folder contains a search box, usually right at the top of the window. To use a Search Box, just type in the name of a file we are looking for.
- While the computer is searching, we will see a green progress indicator behind the address bar showing the approximate overall progress of the search, (See Fig. 1.74).
- The Search Results dialog box comes up, (See Fig. 1.75).

Fig. 1.73: Searching File/Folder

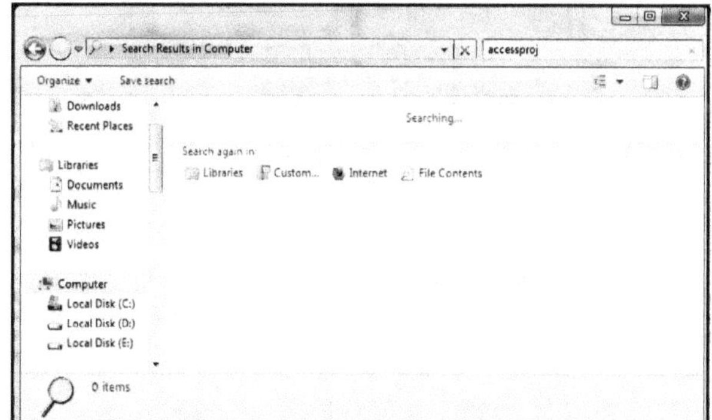

Fig. 1.74: Search Progress Indicator

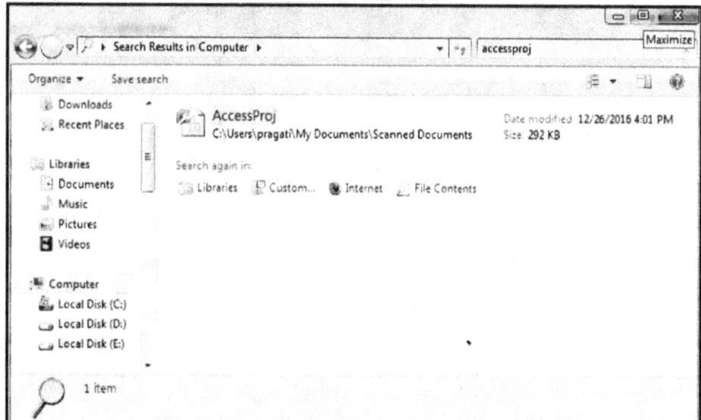

Fig. 1.75: Search Results

Windows 7 Accessories:

- Windows 7 operating system ships with some handy applications known as Windows accessories.
- Calculator, Notepad, Paint, WordPad etc., are some of the most frequently used accessories in Windows 7 operating system.
- To use Accessories in Windows 7 click on Start Button → All Programs → Accessories. Which displays various Accessories as shown in Fig. 1.77.

Fig. 1.76　　　　**Fig. 1.77**

- Let use see common accessories in detail:

1. Calculator:

- Windows Calculator is a calculating application included in all the versions of Windows 7.

- It can be used to perform simple calculation, scientific calculation and Programming calculation.

- To open calculator follow following steps:
  - Choose Start → Programs → Accessories → Calculator to start Calculator application. **OR**
  - Alternately we can open Run dialog box (Start → Run) dialog box then type calc and press enter.

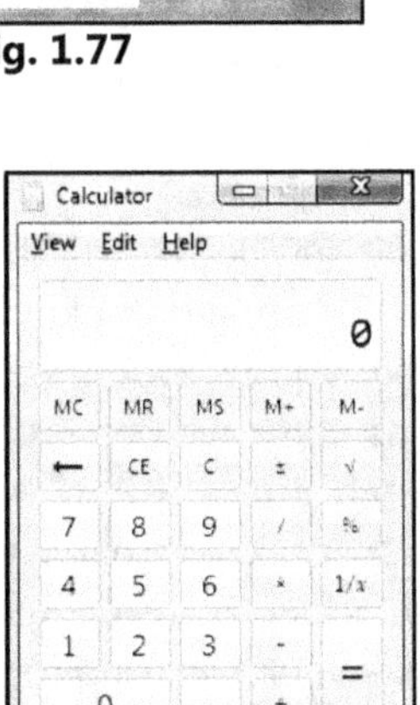

Fig. 1.78: Calculator Window

2. Paint:

- Paint or Paintbrush is a simple graphics painting program that has been included with Microsoft Windows 7. It is often referred to as MS Paint or Microsoft Paint.

- The Paint program opens and saves files as Windows bitmap (24-bit, 256 color, 16 color, and monochrome) .BMP, JPEG, GIF.

- The program can be in color mode or two-color black-and-white, but there is no grayscale mode. For its simplicity, it rapidly became one of the most used applications in the early versions of Windows introducing many to painting on a computer for the first time and still has strong associations with the immediate usability of the old Windows workspace.

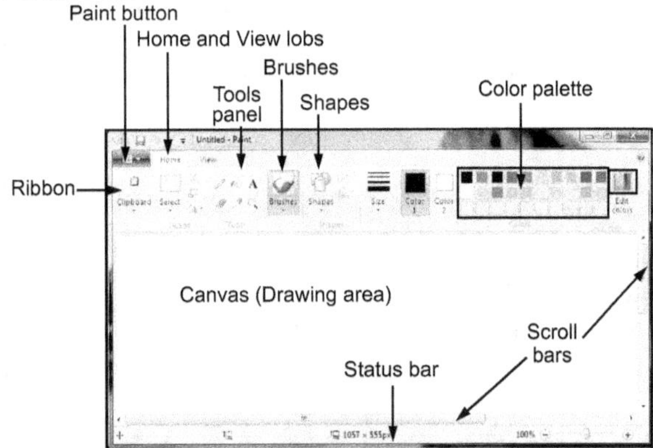

Fig. 1.79: Paint Window

3. Snipping Tool:

- Snipping Tool is another useful of Windows accessories included in Windows 7.
- The program is a screen-capture tool that allows taking screenshots (called snips) of an open window, rectangular areas, a free-form area, or the entire screen. Snips can then be annotated using a mouse or a tablet, saved as an image file (PNG, GIF, or JPEG file) or an HTML page, or e-mailed.

- The Snipping Tool captures all or part of the computer display screen as a picture. We can save the picture and attach it to an e-mail or paste the picture into a document.
- Click the Start button, type snip, and click the Snipping Tool from the search results. The screen fades slightly, and the Snipping Tool toolbar appears, (See Fig. 1.80).

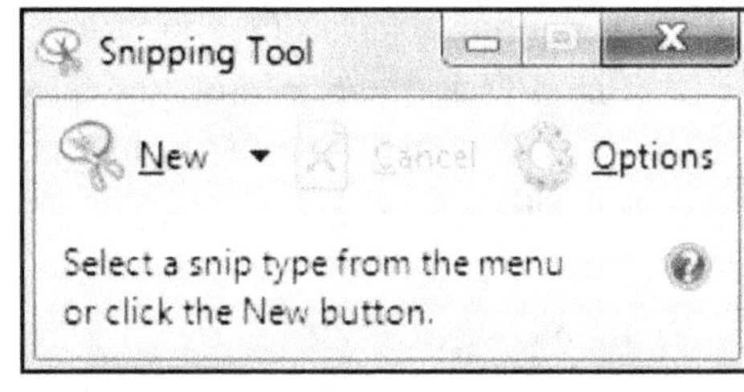

Fig. 1.80

4. Character Map:

- Character Map is a utility included with Microsoft Windows 7 operating systems and is used to view the characters in any installed font, to check what keyboard input (Alt code) is used to enter those characters, and to copy characters to the clipboard in lieu of typing them.
- The tool is usually useful for entering special characters. It can be opened via the command line or Run Command dialog using the 'charmap' command.

Fig. 1.81: Character Map

5. Notepad:

- Notepad is a common text-only (plain text) editor.
- The resulting files typically saved with the .txt file extension have no format tags or styles, making the program suitable for editing system files that are to be used in a DOS environment.

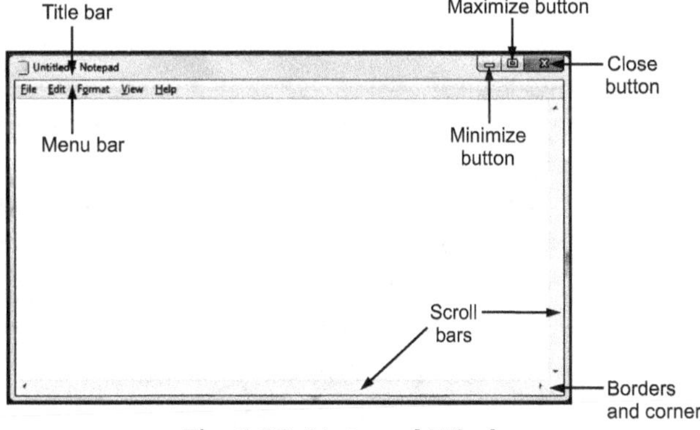

Fig. 1.82: Notepad Window

6. WordPad:

- Microsoft WordPad is a free rich text editor included with Microsoft Windows 7.
- WordPad can format and print text, but lacks intermediate features such as a spell checker, thesaurus, and support for tables.

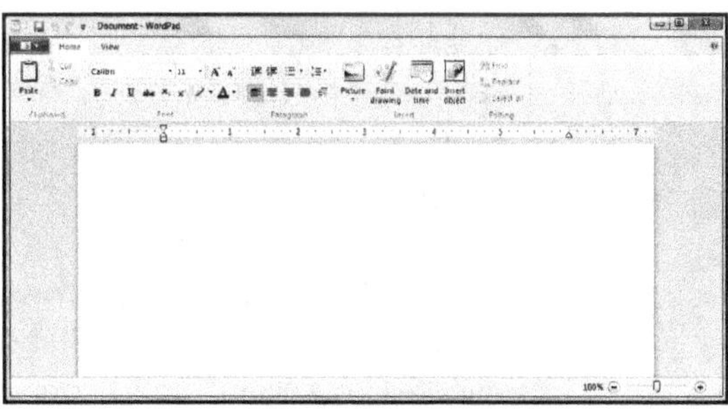

Fig. 1.83: WordPad Window

7. System Tools:

- For a smooth performance of a system, periodic maintenance is necessary. Tools like Disk Check, Disk Derangement, etc., helps in system maintenance. Windows includes some of the system utilities such as Disk Cleanup, Disk Defragment, System Restore, Control Panel and so on.

(i) Disk Cleanup:

- o In Windows 7, Disk Cleanup frees disk space by cleaning areas that gather unneeded files.
- o Using Disk Cleanup regularly, along with ScanDisk, DEFRAG, and up-to-date virus patterns, will help keep the computer running smoothly.

To run Disk Cleanup:

- o From the Start menu, select Programs or All Programs, then Accessories, then System Tools, and then Disk Cleanup.
- o Select the drive on which we would like to clear disk space (usually C: drive), and click OK.
- o Select from the four areas that Disk Cleanup will check and clean up:

  - **Temporary Internet Files:** Files stored on the local drive that allow web pages to load more quickly.
  - **Downloaded Program Files:** A storage location for programs downloaded when we visit specific web sites that use ActiveX and Java applets.
  - **Recycle Bin:** Files that have been marked as deleted.
  - **Temporary Files:** Items in the TEMP folder.

Fig. 1.84

(ii) Disk Defragmenter:

- o Disk Defragmenter consolidates files so that each is saved in contiguous physical space on the hard drive.
- o When the computer crashes or freezes, the drive becomes fragmented. This means the files may be broken up and stored in different locations in the computer. This causes programs and files to run more slowly.

  To run Disk Defragmenter:

  - Close all programs.
  - Choose Start menu → All Programs → Accessories → System Tools → Disk Defragmenter.
  - Select the drive we want to clean and click OK.
  - Click Analyze. This may take several minutes.
  - Click Close or Defragment. If the drive needs defragmented and we choose Defragment, this may take several minutes to several hours, depending on the number of files to consolidate.

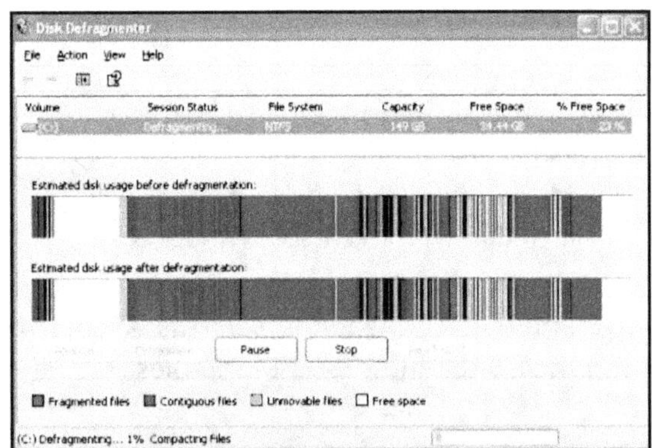

Fig. 1.85

(iii) System Restore:

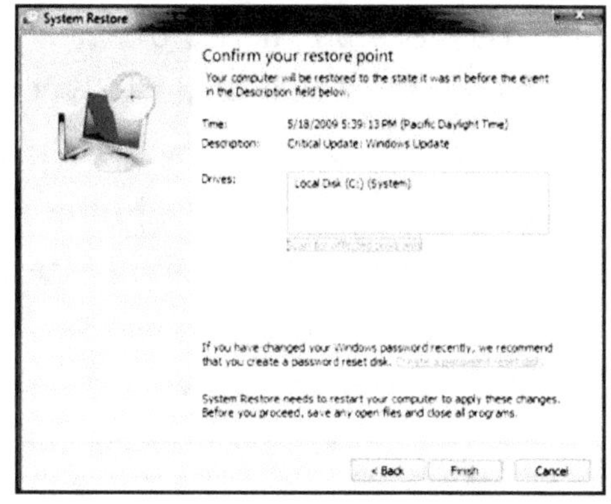

Fig. 1.86

- o System restore is a feature of Windows 7 that allows us to roll back the system to the same configuration it had at an earlier time. If this feature is turned on (which it is by default) Windows will take periodic snapshots (called restore points) of the installation.

- o It can use the information gathered in these snapshots to restore the system to a previous configuration. It is a good idea to create a restore point before the make major changes to the system, such as before installing new hardware or doing any major updates.

To run System Restore:

- Go to Start → Programs → Accessories → System Tools → System Restore.

Set Wall Paper in Windows 7:

- Microsoft Windows 7 users can turn off or change their wallpaper and screen saver and adjust its properties.
- We can easily change the desktop background in Windows 7 the steps are given below:

Step 1 : Right-click a blank part of the desktop and choose Personalize (See Fig. 1.87). The Control Panel's Personalization pane appears, (See Fig. 1.88).

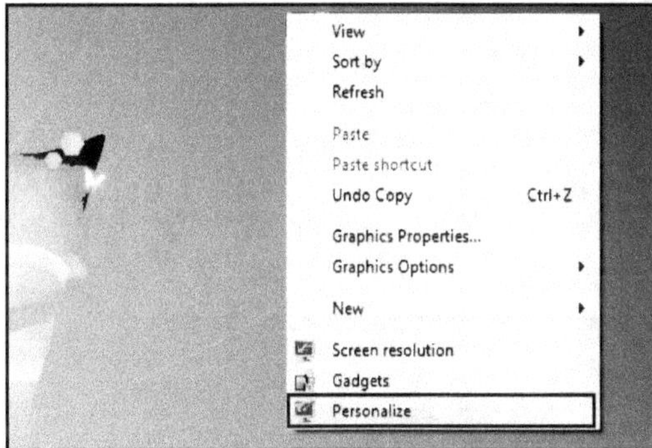

Fig. 1.87

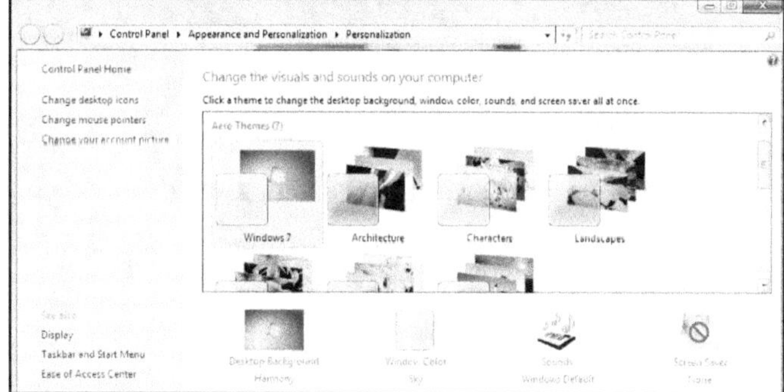

Fig. 1.88

Step 2 : Click the Desktop Background option along the window's bottom left corner, (See Fig. 1.89).

Step 3 : Click any of the pictures, and Windows 7 quickly places it onto your desktop's background. Click the Browse button and click a file from inside the personal Pictures folder.

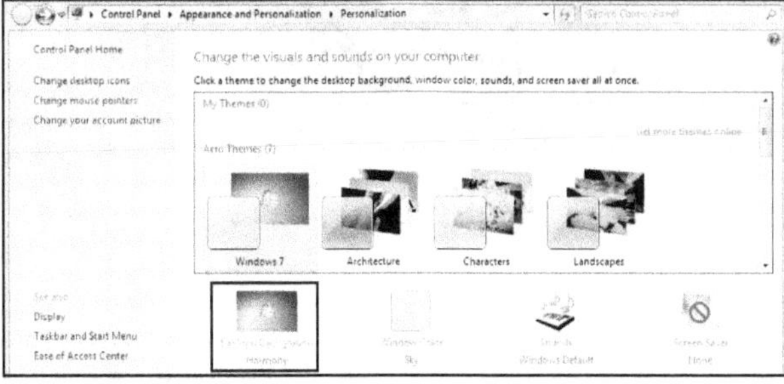

Fig. 1.89

Fig. 1.90

Step 4 : Click Save Changes (See Fig. 1.90) and exit the Desktop Background window. Which displays changed wallpaper on the desktop.

Set Screen Saver in Windows 7:

- A screen Saver is a moving picture or pattern that appears on a computer screen when the mouse or keyboard has not been used for a specified period of time.

- To change screen saver follow the following steps:

   Step 1 : Right-click a blank part of the desktop and choose Personalize (See Fig. 1.87). The Control Panel's Personalization pane appears, (See Fig. 188).

   Step 2 : Click the Screen Saver option along the window's bottom right corner, (See Fig. 1.91). It will displays a dialog box as shown in Fig. 1.92.

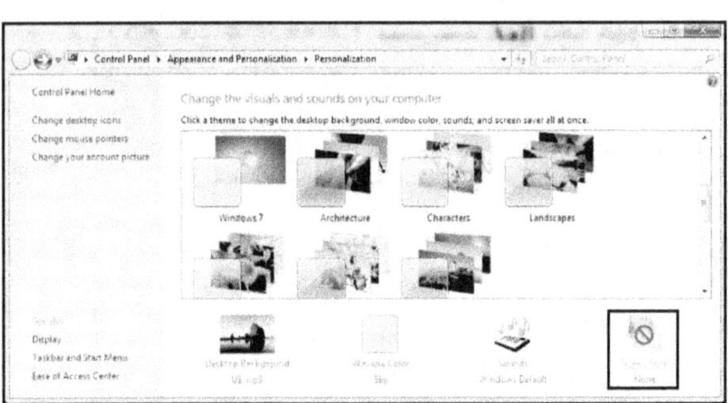

Fig. 1.91

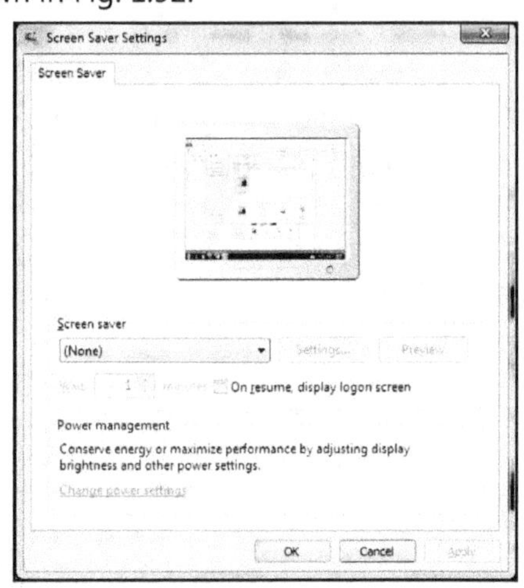

Fig. 1.92

   Step 3 : Under Screen saver, in the drop-down list, click the screen saver we want to use, (See Fig. 1.93). Click Preview to see what the chosen screen saver will look like, (Fig. 1.94).

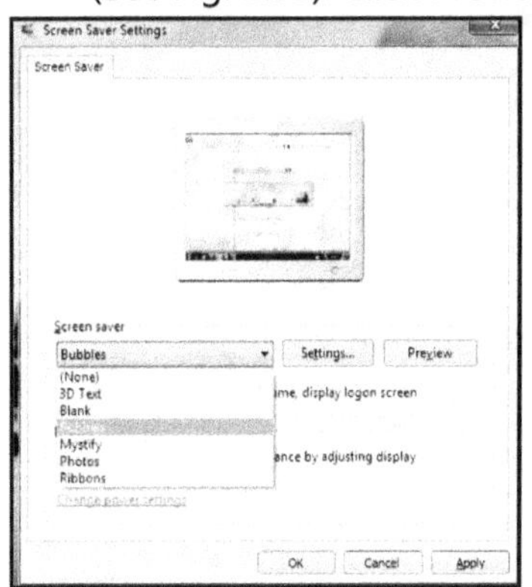

Fig. 1.93

Fig. 1.94

   Step 4 : Click OK, which changes screen saver an desktop.

Set the Date and Time in Windows 7:

- Windows 7 normally keeps good track of the current date and time, but often with a new computer, you have to change the Windows time zone to match the location so it can maintain time accurately.

- Follow the following steps to set date and time in Windows 7:

   Step 1 : Right-click the Date/Time display on the far right end of the taskbar, and click Adjust Date/Time from the shortcut menu that appears, (See Fig. 1.95).

Step 2 : The Date and Time dialog box appears, (See Fig. 1.96). Then click the Change date and time button.

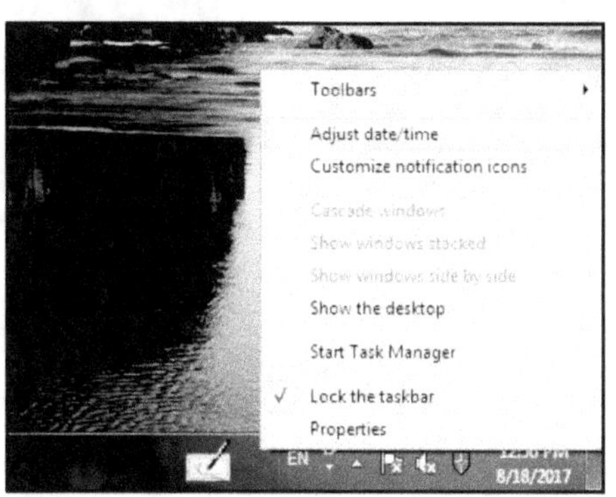

Fig. 1.95

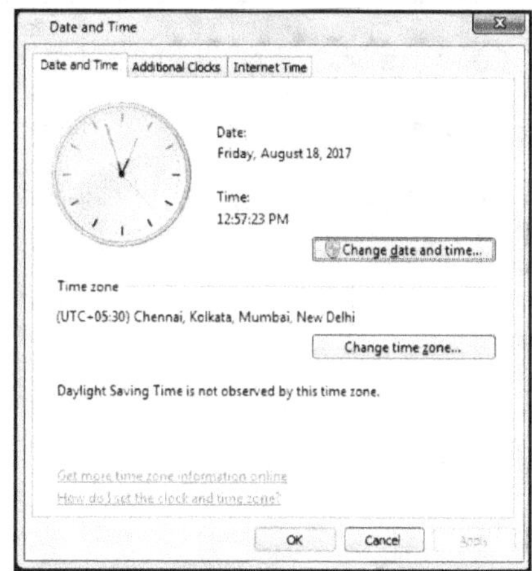

Fig. 1.96

Step 3 : Click in the Time field and enter a new time, or click the up and down arrows next to that field to change the time. If we wish, click a new date in the calendar to set the date.

Step 4 : When we are finished with adjusting the date and time settings, click OK. We are returned to the Date and Time dialog box.

Step 5 : To change the time zone, click the Change Time Zone button, choose an option from the Time Zone drop-down list, and click OK.

Step 6 : Click OK. The new settings are applied, and the dialog box closes.

Practice Questions

1. What is computer?
2. Define the term computer.
3. With the help of diagram describe working of the computer.
4. Enlist characteristics/features of the computers.
5. What is meant by computer hardware and software?
6. With the help of diagram explain the block diagram of computer.
7. Explain the terms computer hardware and software diagrammatically.
8. Compare application software and system software.
9. Explain generations of the computers in detail.
10. Explain the classification of computers according to their purpose, data handling and functionality.
11. Differentiate between computer hardware and software.
12. What is program, programming and programming language? Define them.
13. What is computer memory? Compare RAM and ROM.

14. Enlist types of the programming languages with their advantages and disadvantages.

15. What is meant by OS? Define it. Enlist objectives of OS?

16. Compare Machine, assembly and high level languages.

17. What are the types of OS? Explain them with example.

18. How to search for files and folders in Windows?

■■■

Chapter 2

Introduction to MS Office

Contents

2.1 INTRODUCTION TO MS OFFICE

- Microsoft Office is an office suite of desktop applications (or services) developed by Microsoft. A suite is a group of applications that are designed to work together with particular User Interface (UI).
- Microsoft Office 2013 provides a comprehensive toolkit for tackling day-to-day productivity and communication tasks for business or personal purposes. In this chapter we study Microsoft Office 2013 with its applications and how to use them.
- Microsoft Office 2013 includes a word processor (Word), a spreadsheet program (Excel), a presentation graphics program (PowerPoint), e-mail program (Outlook), a database program (Access), a notetaking program (OneNote), a desktop publishing program (Publisher) and so on.

2.2 MS WORD

- Microsoft Word is word processing application in Microsoft Office 2013. Word processing means the process of manipulation of text using computers.
- Word processing includes entering (typing), editing, formatting and manipulation of text on a computer screen and printing of the processed text.
- A computer program/application/software exclusively used for word-processing is called a Word processor.
- Microsoft Word is a word processor designed by Microsoft. It was first released in 1983 under the name Multi-Tool Word for Xenix systems.

- The documents created in MS Word are characterized by the .doc (binary file format for Word 97–Word 2003) or .docx (default XML-based file format for Word 2016, Word 2013, Word 2010 and Word 2007) file extension.
- Microsoft Word is used to create, format, edit, save and print electronic documents.

2.2.1 Starting MS Word

- To start MS Word 2013 follow the following steps:

   Step 1 : Click the Start button on the Windows taskbar. The Start menu opens, (See Fig. 2.1).

   Step 2 : Click All Programs, (See Fig. 2.1).

   Step 3 : Click the Microsoft Office, (See Fig. 2.2).

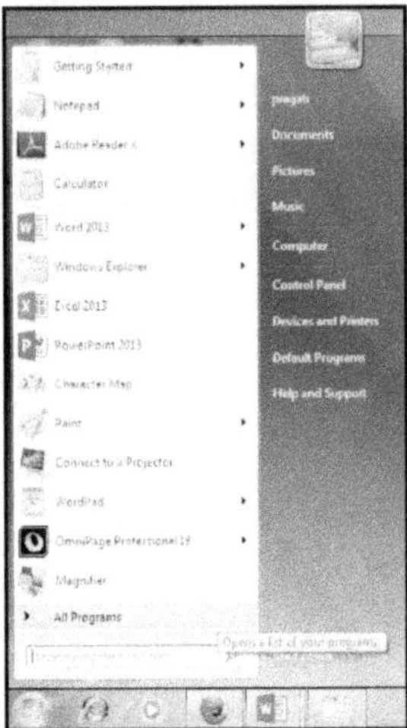

Fig. 2.1

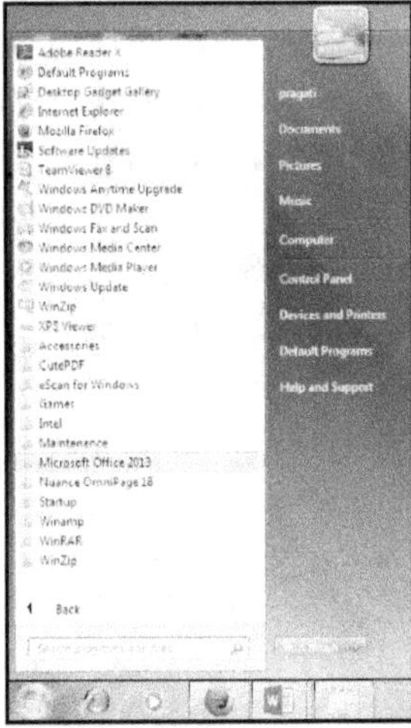

Fig. 2.2

   Step 4 : Click Word 2013, (See Fig. 2.3), it opens backstage view (See Fig. 2.4) with various options for saving, opening a file, printing, and sharing the document. Then click Blank document option which display a new Word document as shown in Fig. 2.5.

Fig. 2.4

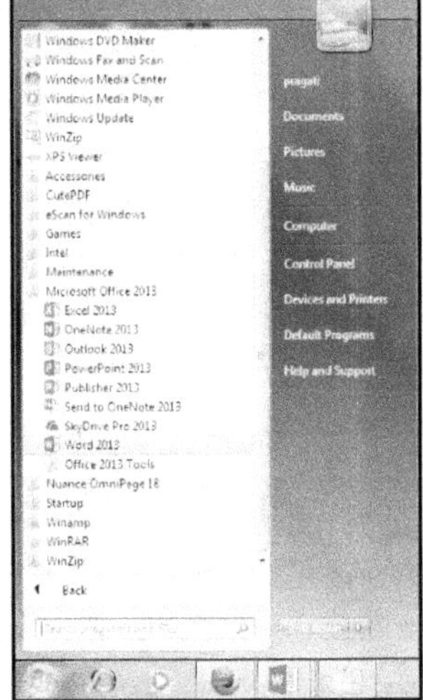

Fig. 2.3

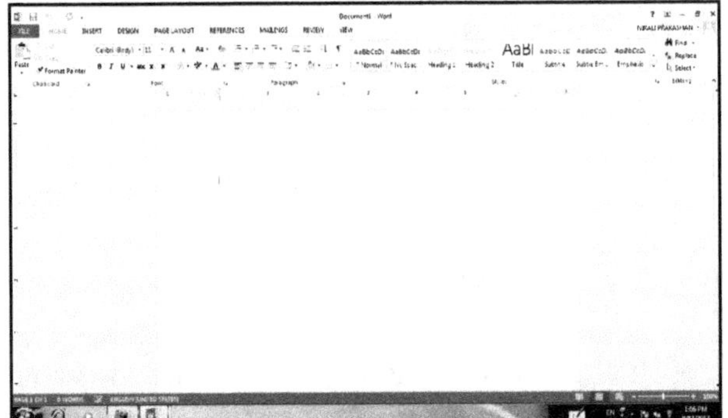

Fig. 2.5

2.2.2 Screen Components of MS Word

- Fig. 2.6 shows Word 2013 User Interface (UI) with basic screen elements/components.

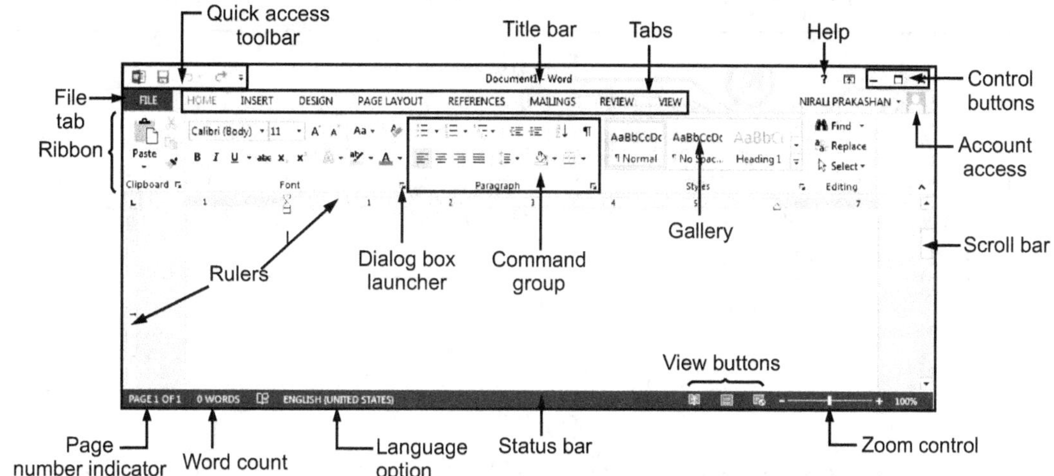

Fig. 2.6: Screen Components of Word 2013 Window

- Fig. 2.6 shows following screen elements of MS Word 2007 window:

1. **Quick Access Toolbar:** The quick access toolbar lets us to access common commands no matter which tab is selected. By default, it includes the Save, Undo, and Redo commands. We can also add other commands depending on our preference.

2. **Title Bar:** This lies in the middle and at the top of the window. Title bar shows the program and document titles. It shows name of active (current) working file.

3. **Account Access:** From here, we can access the Microsoft account information, view this profile, and switch accounts.

4. **Ribbon:** The ribbon contains all of the commands we will need to perform common tasks in Word. The Ribbon is composed of three parts i.e., Tabs, Groups, and Commands as explained below:

   (i) Tabs: Microsoft Word is a powerful program that is used to create many different types of documents, including articles, letters, books, contracts, marketing documents, and much more. Microsoft Word has hundreds of commands for working with documents. To make it easier for users to find the specific commands they are looking for, commands are organized onto tabs. The main tables are shown in Fig. 2.7 and explained below:

   - **HOME:** The HOME tab includes commands for formatting documents.
   - **INSERT:** Use the INSERT tab to insert pages, tables, pictures, links, headers and footers, custom text and symbols, and more.
   - **DESIGN**: Use the DESIGN tab to set document formatting and page backgrounds.
   - **PAGE LAYOUT:** Use the PAGE LAYOUT tab to change the margins, add columns, change the page orientation, and more.
   - **REFERENCES:** Use the REFERENCES tab to add a table of contents, add footnotes, add a bibliography, and more.
   - **MAILINGS:** Use the MAILINGS tab to create labels, start a mail merge, and more.
   - **REVIEW:** Use the REVIEW tab to check spelling and grammar, track and accept or reject changes, compare documents, and more.
   - **VIEW:** Use the VIEW tab to change the document view, show the Ruler or navigation pane, zoom in or out, and more.

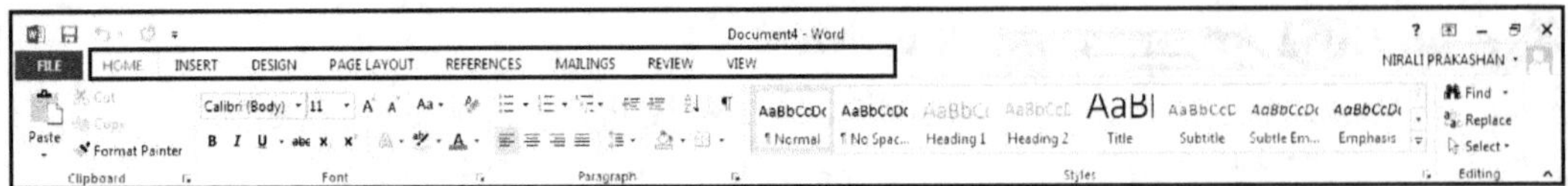

Fig. 2.7

In addition to the main tabs, there are numerous tool tabs which include less commonly used commands. Some of the most commonly used tool tabs are Smartart, Chart, Drawing, Picture, Table and Header and Footer. That they will appear when we select commands that have related tool tabs. For example, when we insert a table, two table-specific tool tabs (DESIGN and LAYOUT) will appear as shown in Fig. 2.8.

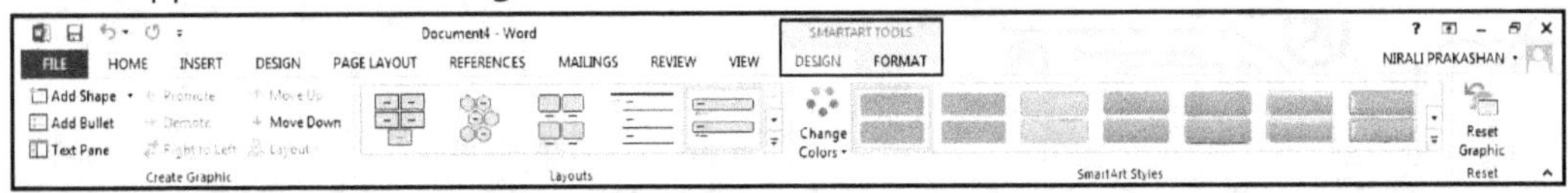

Fig. 2.8

(ii) Command Group: Each group contains a series of different commands. To further organize the many commands available in Microsoft Word, commands are organized in groups on each tab. Each group contains three or more related commands.

(iii) Commands: Commands are controls that enable we to accomplish specific tasks, such as adding a list, inserting a picture, or adding page numbers.

(iv) Dialog Box Launcher: This appears as very small arrow in the lower-right corner of many groups on the Ribbon. Clicking this button opens a dialog box (or task pane) that provides more options about the group.

5. **Ruler:** The Ruler is located at the top and to the left of the document. It makes it easier to make alignment and spacing adjustments. Word has two rulers; a horizontal ruler and a vertical ruler. The horizontal ruler appears just beneath the Ribbon and is used to set margins and tab stops. The vertical ruler appears on the left edge of the Word window and is used to gauge the vertical position of elements on the page.

6. **Help:** The Help icon can be used to get word related help anytime we like. This provides description on various topics related to word.

7. **FILE Tab:** It replaces the Office button from Word 2007/2010. We can click it to open the Backstage view.

8. **Scroll Bar:** Click, hold, and drag the scroll bar to scroll up and down through the pages of the document.

9. **Document Area/Workspace:** This is the area where we type. The flashing/blinking vertical bar is called the insertion point and it represents the location where text will appear when we type.

10. **Zoom Control:** Zoom control lets us to zoom in for a closer look at the text in the document. The zoom control consists of a slider that can slide left or right to zoom in or out; we can click the + (plus) button to increase or − (minus) button to decrease the zoom factor.

11. **Document Views:** Word 2013 has a variety of viewing options that change how the document is displayed. The group of three buttons located to the left of the Zoom control, near the bottom of the screen, lets us to switch through the Word's various document views. There are three ways to view a document. Simply click to select the desired view:

(i) **Read Mode** displays the document in full-screen mode.

(ii) **Print Layout** is selected by default. It shows the document as it would appear if it were printed.

(iii) **Webpage Layout** shows how the document would look as a webpage.

12. Status Bar: It is located at the bottom of Word, shows basic information about the document and enables us to change viewing settings. From left to right, this bar contains the total number of pages and words in the document, language, etc. The page number indicator helps us to keep track of the number of pages the document contains. Word Count displays the number of words in the document. If we see 🔖, Word found no errors. 🔖 indicates Word has found spelling or grammatical errors in the document.

13. Control Buttons: Word windows controls button contains:

    (i) Minimize button, which reduces the windows and places it onto the taskbar.

    (ii) Maximize button, which displays the full screen view of the word window.

    (iii) Restore button, which puts the window back to its previous size.

    (iv) Close button, automatically close the word application.

To save a document:

- It is important to save the document whenever we start a new word document or make changes to an existing one. Saving early and often can prevent the work from being lost.

- We will also need to pay close attention to where we save the document so it will be easy to find later.

    Step 1 : Locate and select the Save command on the Quick Access toolbar, (See Fig. 2.9).

    Step 2 : If we are saving the file for the first time, the Save As pane will appear in Backstage view. Double-click Browse to save the file to the computer, (See Fig. 2.10).

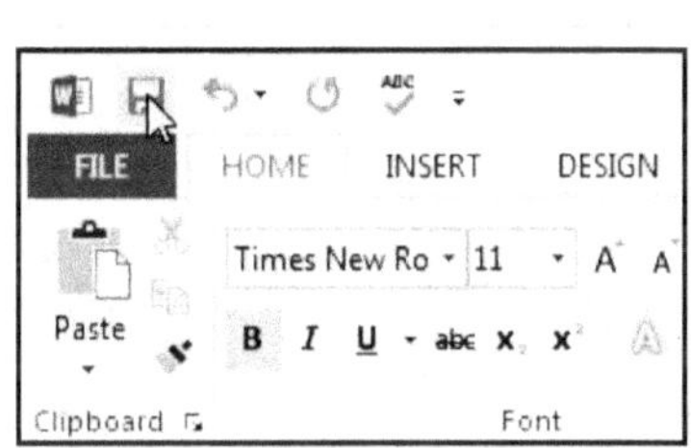

Fig. 2.9

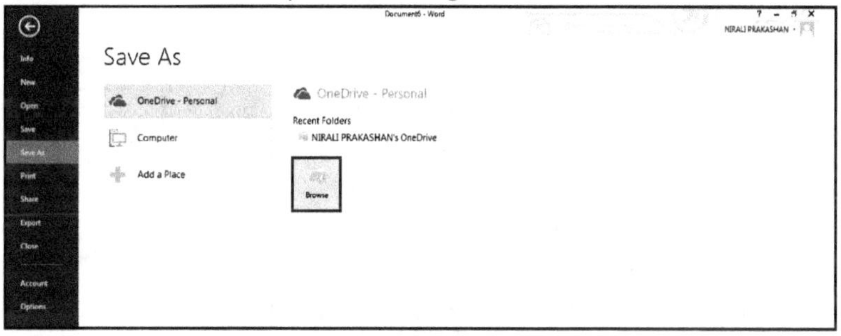

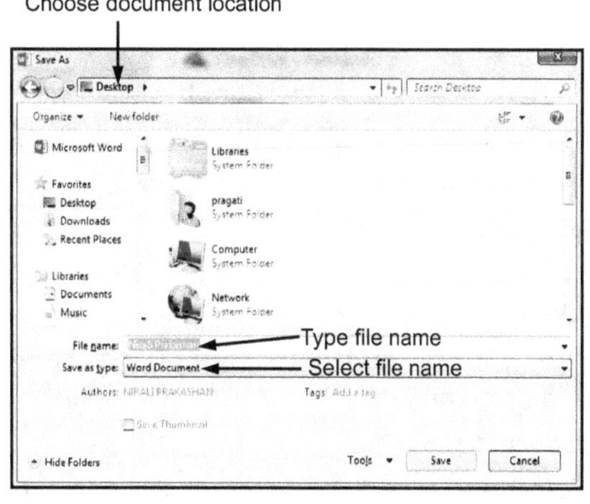

Fig. 2.10

Selecting Save As opens up a dialog box (See Fig. 2.11) in which we can see:

1. The **document location**, or where on the computer Word will save the document. We can select a new location by clicking on the arrows.

2. The **file name** is highlighted as Microsoft Word expects we to choose the own name for the document. Simply begin typing to do so.

3. The **file type** of word 2013 is ".docx", which is the default file type for Microsoft Word 2013 documents. When final, we can choose to save the document as another type, such as a pdf, simply by choosing "PDF (*.pdf)".

Fig. 2.11

Step 3 : After we have entered these fields, simply click Save (See Fig. 2.11) to save the document.

2.2.3 Elementary Working with MS Word

- In this section, we will study the basic and advanced operations in MS Word 2013.

Working with Text:

- Document area is the area where we type the text. The flashing vertical bar is called the insertion point (See Fig. 2.12) and it represents the location where text will appear when we type.

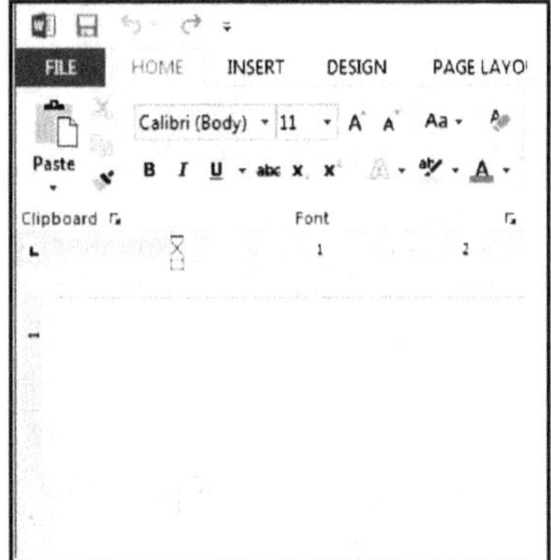

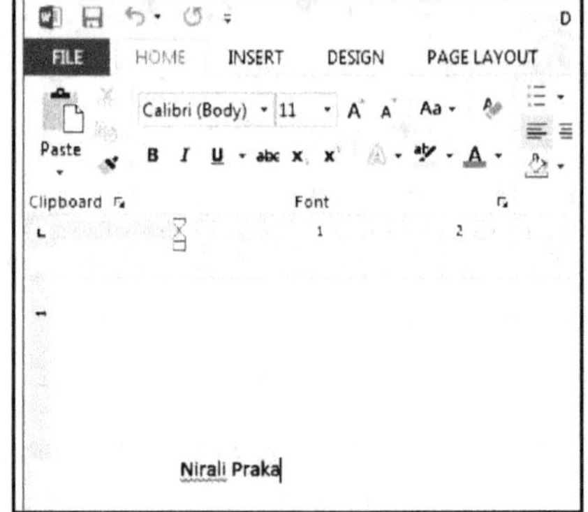

Fig. 2.12   **Fig. 2.13**

- **Adding spaces:** For adding spaces press the spacebar to add spaces after a word or in between text.
- **New paragraph line:** For new line press Enter on the keyboard to move the insertion point to the next paragraph line.

To Select Text:

- Before applying formatting to text, we will first need to select it.

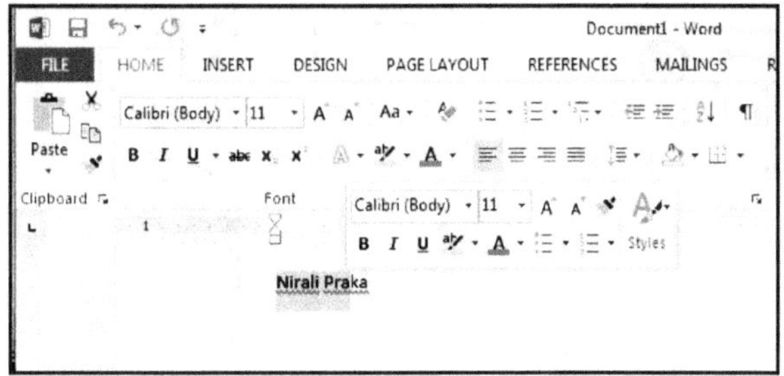

Fig. 2.14

Step 1 : Place the insertion point next to the text we want to select.

Step 2 : Click the mouse, and while holding it down drag the mouse over the text to select it.

Step 3 : Release the mouse button. We have selected the text. A highlighted box will appear over the selected text, (See Fig. 2.14).

To Delete Text:

- There are several ways to delete or remove text in Word as explained below:
  - To delete text to the left of the insertion point, press the Backspace key on the keyboard.   **OR**
  - To delete text to the right of the insertion point, press the Delete key on the keyboard.   **OR**
  - Select the text we want to remove, then press the Delete key.

Copying and Moving Text:

- Word allows us to copy text that is already in the document and paste it to other areas of the document, which can save time.
- If there is text we want to move from one area of the document to another, we can cut and paste or drag and drop the text.

To copy and paste text:

- Copying text creates a duplicate of the text. To copy and paste text follow the following types:

  Step 1 : Select the text we want to copy. Click the Copy command on the HOME tab. We can also right-click the selected text and select Copy, (See Fig. 2.15).

Step 2 : Place the insertion point where we want the text to appear. Click the Paste command on the HOME tab, (See Fig. 2.16).

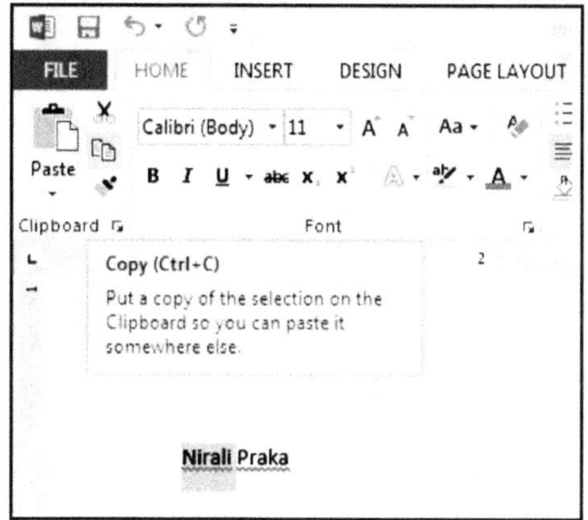

Fig. 2.15

Fig. 2.16

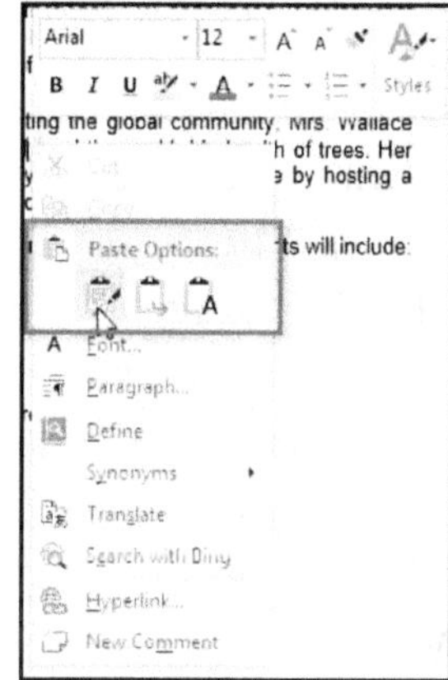

Fig. 2.17

- We can also cut, copy, and paste by right-clicking the document and choosing the desired action from the drop-down menu. When we use this method to paste, we can choose from three options that determine how the text will be formatted, Keep Source Formatting (K), Merge Formatting (M), and Keep Text Only (T).

- We can hover the mouse over each icon to see what it will look like before we select it, (See Fig. 2.17).

To Find and Replace Text:

- When we are working with longer documents, it can be difficult and time consuming to locate a specific word or phrase.

- Word can automatically search the document using the Find feature, and it allows us to quickly change words or phrases using Replace.

To find Text:

- The Find command enables we to locate specific text in the document.In this example, we have to the Find command to locate all references to the word 'Nirali'.

    Step 1 : From the HOME tab, click the Find command, (See Fig. 2.18).

    Step 2 : The navigation pane will appear on the left side of the screen, (See Fig. 2.19).

Fig. 2.18

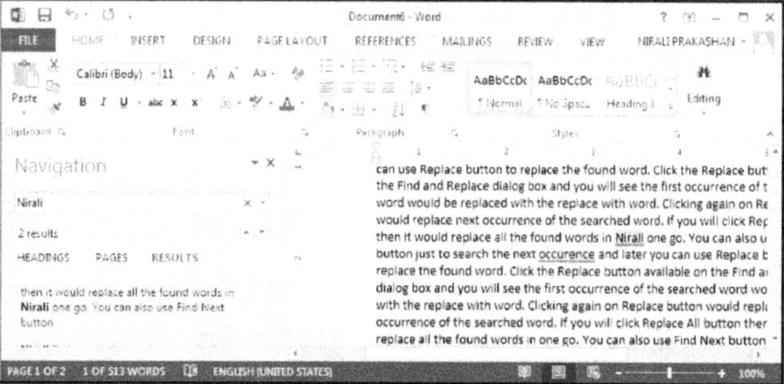

Fig. 2.19

Step 3 :　Type the text we want to find in the field at the top of the navigation pane. In this example, we will type the Nirali. If the text is found in the document, it will be highlighted in yellow, and a preview of the results will appear in the navigation pane.

To Replace Text:

- At times, we may discover that we have made a mistake repeatedly throughout the document like misspelling a name or that we need to exchange a particular word or phrase for another.

- We can use Word's Find and Replace feature to quickly make revisions. In this example, we will use Find and Replace to change the title of a magazine so it is abbreviated. The steps are:

Step 1 :　From the HOME tab, click the Replace command.

Fig. 2.20

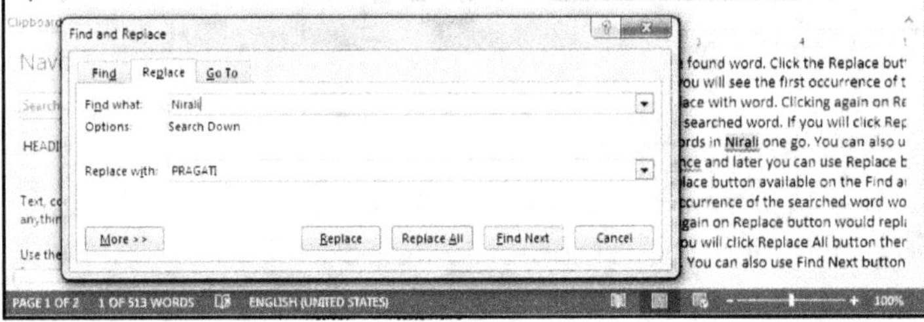

Fig. 2.21

Step 2 :　The Find and Replace dialog box will appear, (See Fig. 2.21). Type the text we want to find in the Find what: field.

Step 3 :　Type the text we want to replace it with in the Replace with: field. Then click Find Next.

Step 4 :　Word will find the first instance of the text and highlight it in gray, (See Fig. 2.21). Review the text to make sure we want to replace it, if the text does not need to be replaced Click Find Next button as shown in Fig. 2.22.

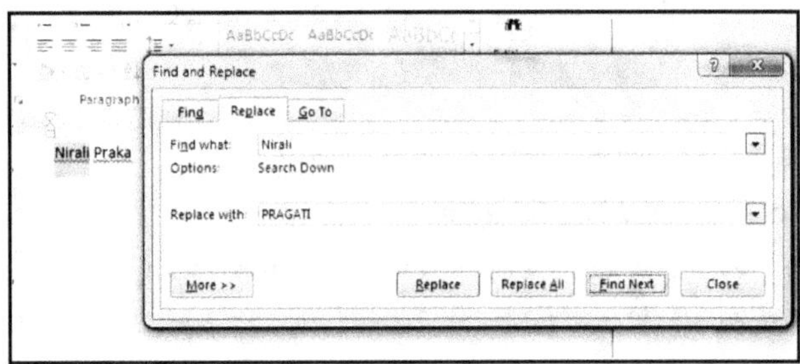

Fig. 2.22

Step 5 :　Word will jump to the next instance of the text. If we want to replace it, select one of the replace options:

(i) Replace will replace individual instances of text. In this example, we will choose this option.

(ii) Replace All will replace every instance of the text throughout the document. The selected text will be replaced.

- The replace text is shown in Fig. 2.24.

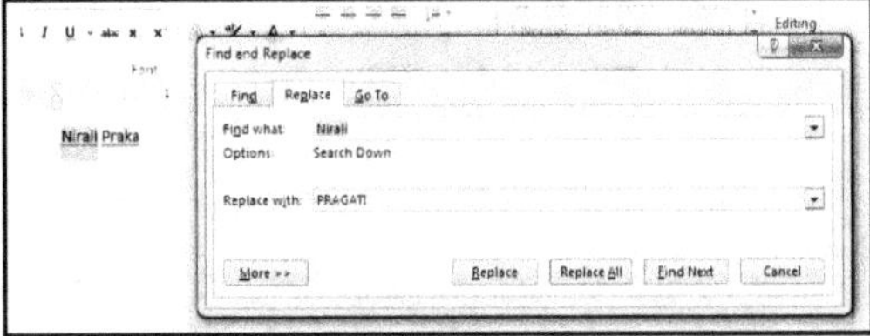

Fig. 2.23

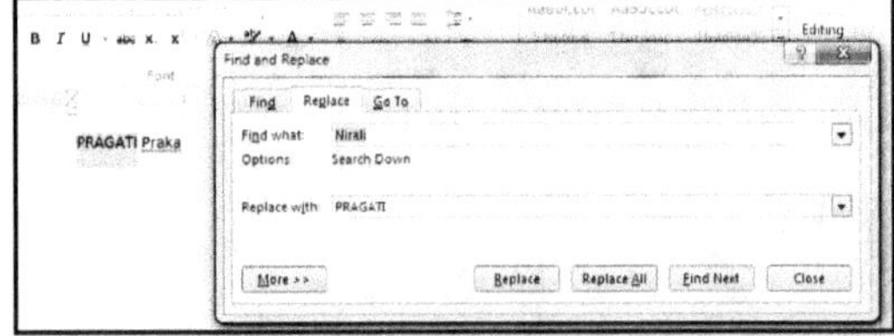

Fig. 2.24

Text Formatting in Word:

- Formatted text can draw the reader's attention to specific parts of a document and emphasize important information.

- In Word, we have several options for adjusting the font of the text, including size, color, and inserting special symbols.

- We can also adjust the alignment of the text to change how it is displayed on the page.

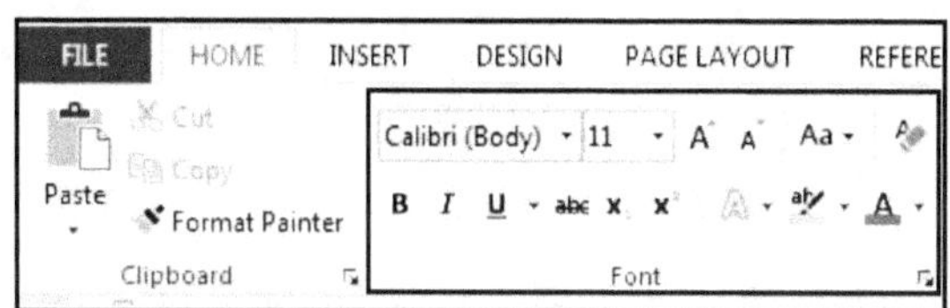

Fig. 2.25

To change the font:

- By default, the font of each new document is set to Calibri (body) and default size is 11 points.

- However, Word provides many other fonts we can use to customize text and titles.

 Step 1 : Select the text we want to modify, (See Fig. 2.26).

 Step 2 : On the HOME tab, click the drop-down arrow next to the Font box. A menu of font styles will appear, (See Fig. 2.27).

 Step 3 : Move the mouse over the various font styles. A live preview of the font will appear in the document. Select the font style we want to use.

 Step 4 : The font will change in the document.

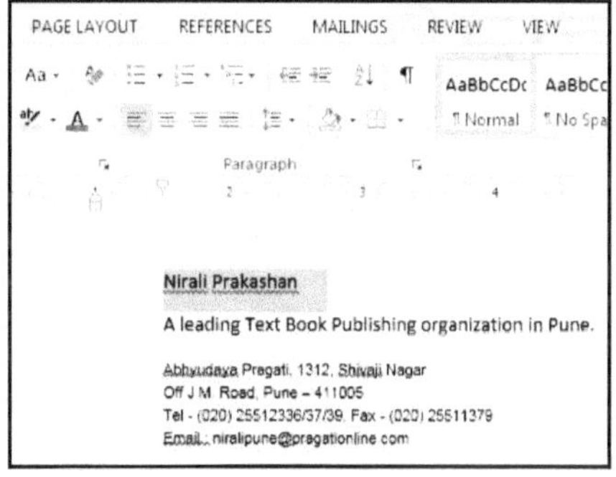

Fig. 2.26

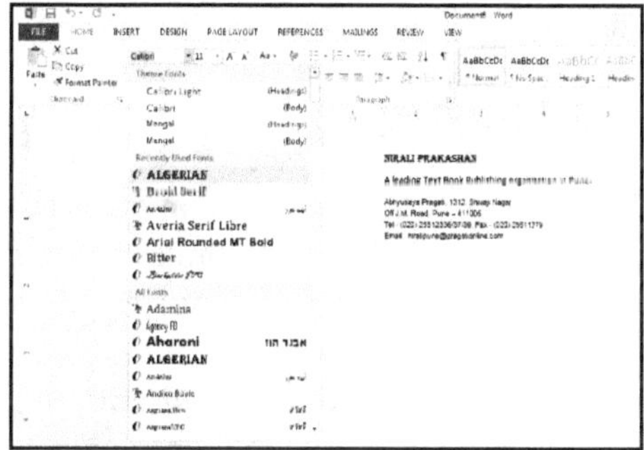

Fig. 2.27

To change the font size:

 Step 1 : Select the text we want to modify.

 Step 2 : Select the desired font size formatting option:

 (i) Font size drop-down arrow: On the HOME tab, click the Font size drop-down arrow. A menu of font sizes will appear. When we move the mouse over the various font sizes, a live preview of the font size will appear in the document, (See Fig. 2.28).

 (ii) Grow and shrink font commands: Click the Grow Font or Shrink Font commands, (See Fig. 2.29) to change the font size.

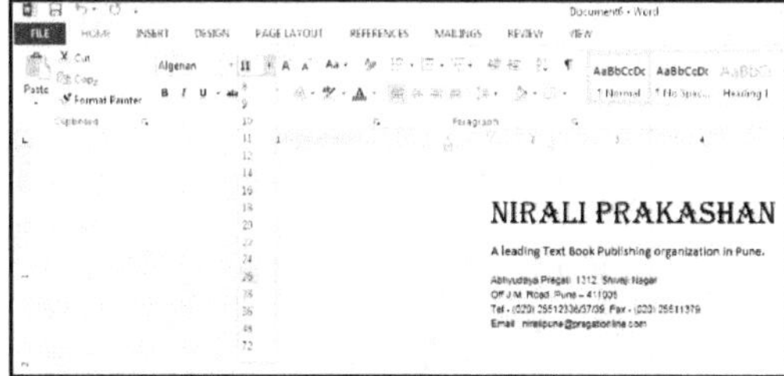

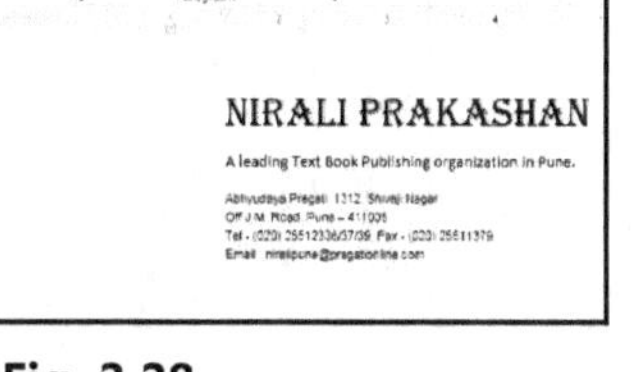

Fig. 2.28

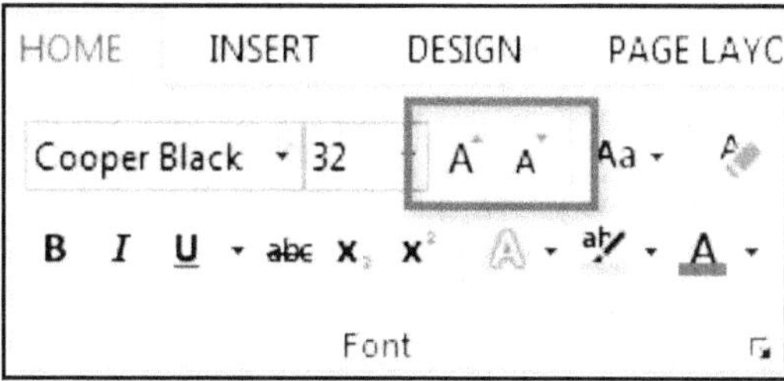

Fig. 2.29

To change the font color:

Step 1 : Select the text we want to modify, (See Fig. 2.30).

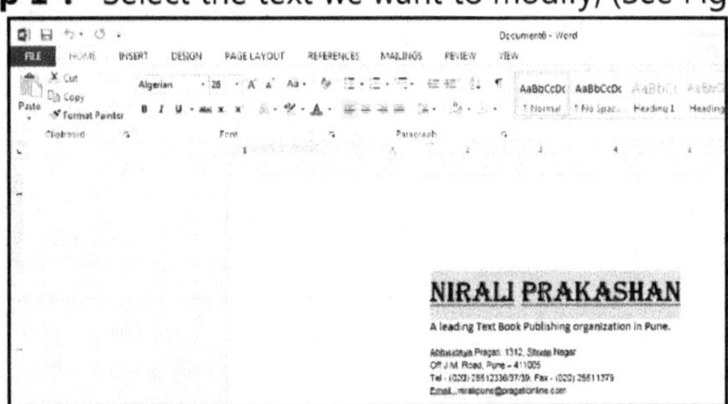

Fig. 2.30

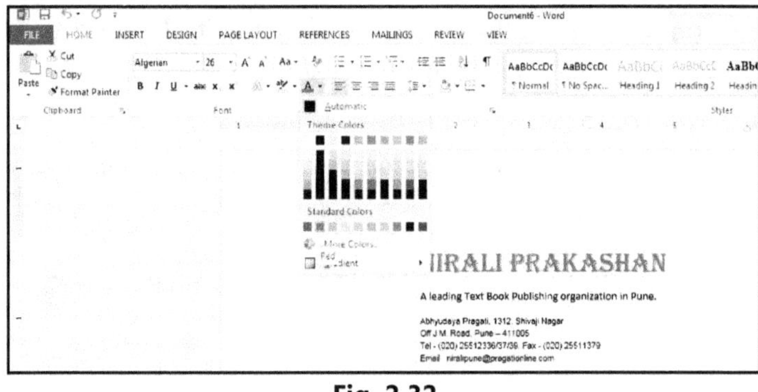

Fig. 2.31

Step 2 : On the HOME tab, click the Font Color drop-down arrow. The Font Color menu appears, (See Fig. 2.31).

Step 3 : Move the mouse over the various font colors. A live preview of the color will appear in the document.

Step 4 : Select the font color we want to use (we use Red). The font color will change in the document, (See Fig. 2.32).

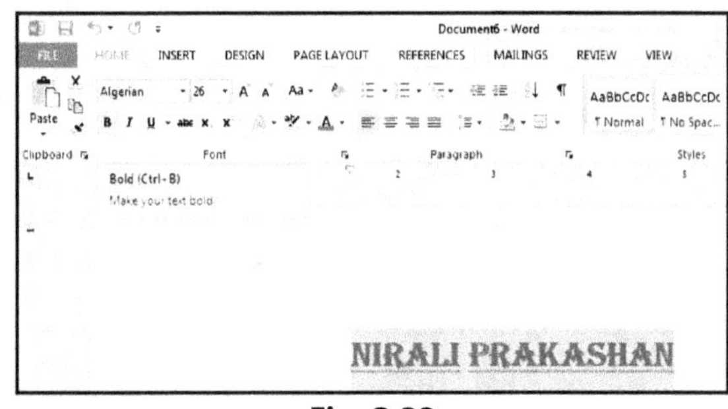

Fig. 2.32

To use the Bold, Italic and Underline Commands:

- The Bold, Italic, and Underline commands can be used to help draw attention to important words or phrases.

  Step 1 : Select the text we want to modify.

  Step 2 : On the HOME tab, click the Bold (**B**), Italic (*I*), or Underline (U) command in the Font group. In this example, we will click Bold.

Fig. 2.33

To Change the Text Case:

- When we need to quickly change text case, we can use the Change Case command instead of deleting and retyping text. The steps are:

  Step 1 : Select the text we want to modify, (See Fig. 2.34).

  Step 2 : On the HOME tab, click the Change Case command in the Font group.

  Step 3 : A drop-down menu will appear, (See Fig. 2.35). Select the desired case option from the menu.

  Step 4 : The text case will be changed in the document, (See Fig. 2.36).

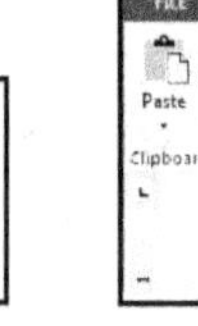

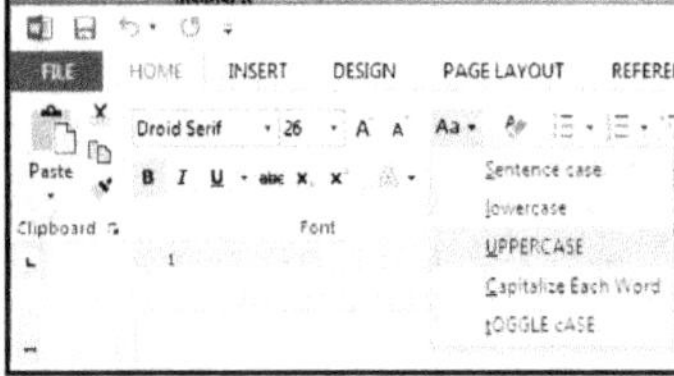

Fig. 2.34   Fig. 2.35

Fig. 2.36

To change text alignment:

- By default, Word aligns text to the left margin in new documents. However, there may be times when we want to adjust text alignment to the center or right. The steps are:

  Step 1 : Select the text we want to modify.

  Step 2 : On the HOME tab, select one of the this alignment options from the Paragraph group, (See Fig. 2.37).

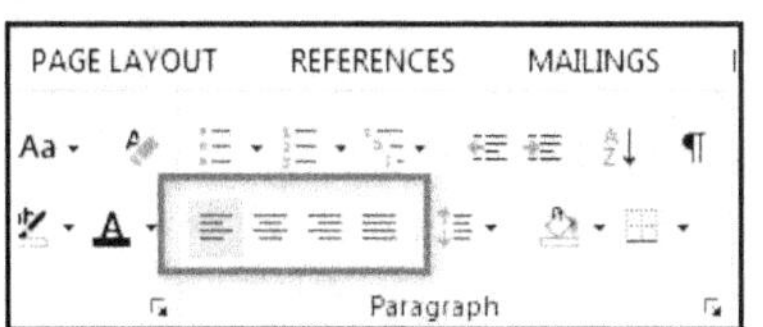

Fig. 2.37   Fig. 2.38

- **The options for aligning text in word are:**

  (i) **Left:** Aligns all of the selected text to the left margin.

  (ii) **Center:** Aligns text an equal distance from the left and right margins.

  (iii) **Right:** Aligns all of the selected text to the right margin.

  (iv) **Justify:** Aligns text equally to the right and left margins; used in many books, newsletters and newspapers.

To Indenting Text:

- Indenting text adds structure to the document by allowing us to separate information. Whether we would like to move a single line or an entire paragraph, we can use the tab selector and the horizontal ruler to set tabs and indents.

- In many types of documents, we may want to indent only the first line of each paragraph. This helps to visually separate paragraphs from one another. It is also possible to indent every line except for the first line, which is known as a hanging indent.

Fig. 2.39   Fig. 2.40

To indent using the Tab Key:

- A quick way to indent is to use the Tab key. This will create a first-line indent of 1/2 inch. The steps are:

  Step 1 : Place the insertion point at the very beginning of the paragraph we want to indent, (See Fig. 2.41).

  Step 2 : Press the Tab key. On the ruler, we should see the first-line indent marker move to the right by 1/2 inch. The first line of the paragraph will be indented, (See Fig. 2.42).

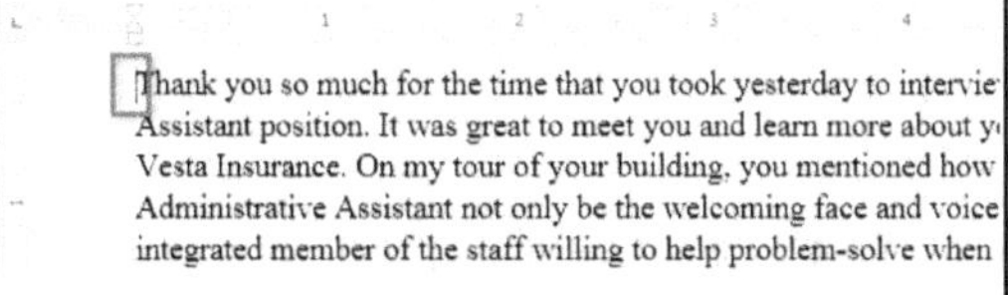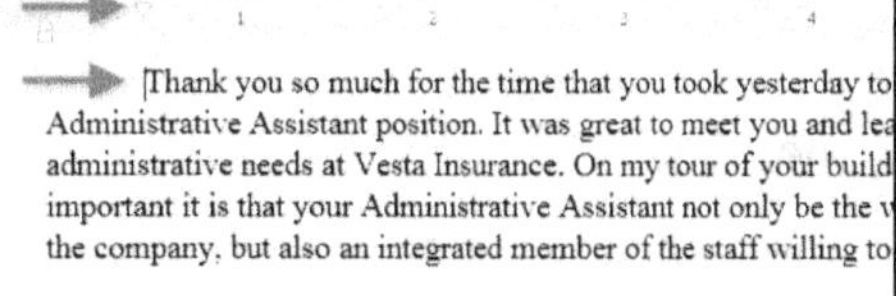

Fig. 2.41   Fig. 2.42

Indent Markers:

- In some cases, we may want to have more control over indents. Word provides indent markers that allow we to indent paragraphs to the location we want.

- The indent markers are located to the left of the horizontal ruler, and they provide several indenting options:

  1. **First-line Indent Marker** ▽: Adjusts the first-line indent.

  2. **Hanging Indent Marker** △: Adjusts the hanging indent.

  3. **Left Indent Marker** ▢: Moves both the first-line indent and hanging indent markers at the same time (this will indent all lines in a paragraph).

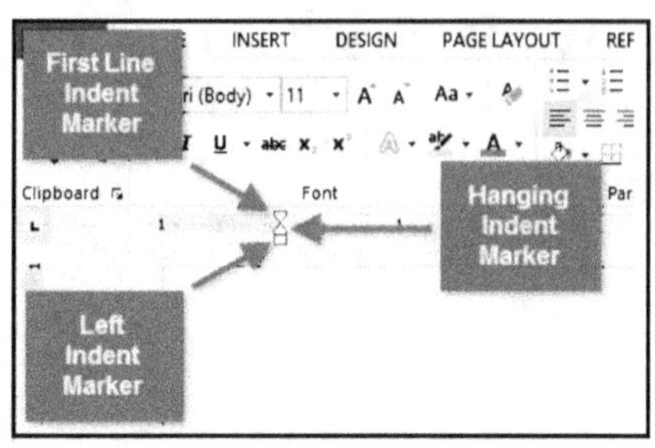

Fig. 2.43

To indent using the Indent commands:

- If we want to indent multiple lines of text or all lines of a paragraph, we can use the Indent commands.

  Step 1 : Select the text we want to indent, (See Fig. 2.44).

  Step 2 : On the HOME tab, click the desired Indent command, (See Fig. 2.45).

    o **Increase Indent**: This increases the indent by increments of 1/2 inch. In this example, we will increase the indent.

    o **Decrease Indent**: This decreases the indent by increments of 1/2 inch.

  Step 3 : The text will indent, (See Fig. 2.46).

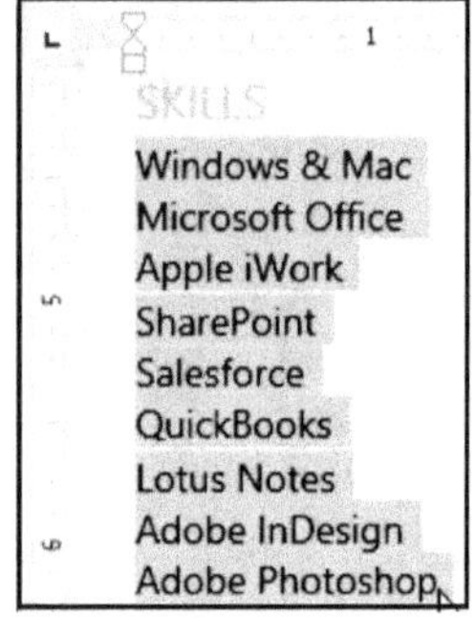

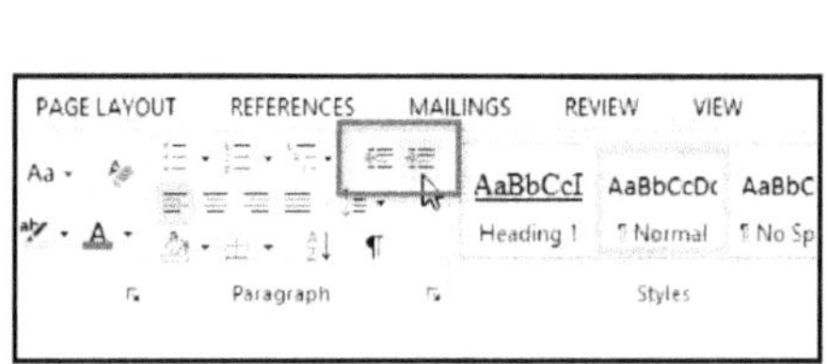

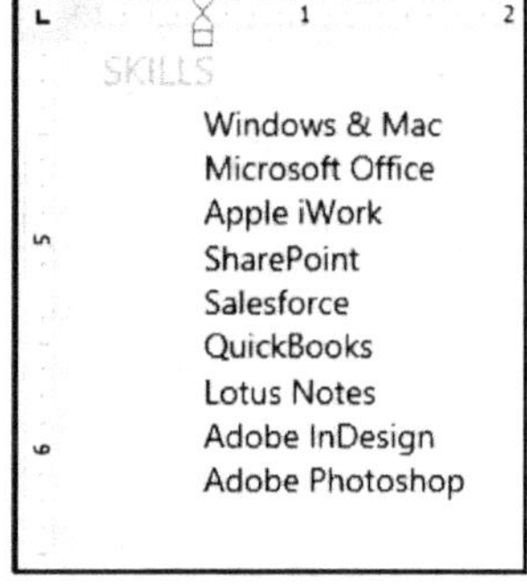

Fig. 2.44   **Fig. 2.45**   **Fig. 2.46**

Tabs:

- Using tabs gives we more control over the placement of text.

- By default, every time we press the Tab key, the insertion point will move 1/2 inch to the right.

- Adding tab stops to the Ruler allows us to change the size of the tabs, and Word even allows us to apply more than one tab stop to a single line.

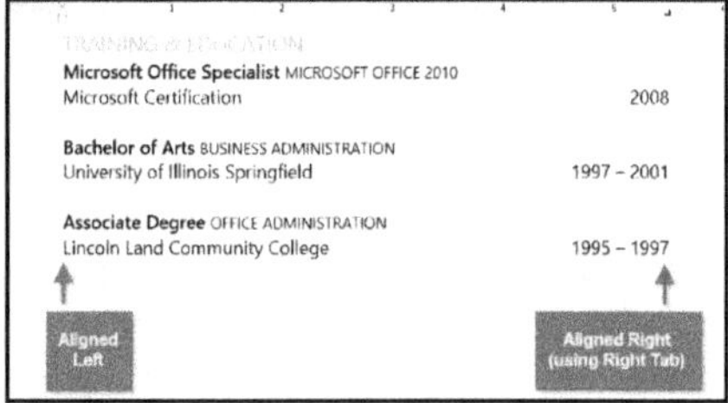

Fig. 2.47

- For example, on a resume we could left align the beginning of a line and right align the end of the line by adding a Right Tab, (See Fig. 2.47).

Tab Selector:

- The tab selector is located above the vertical ruler on the left, (See Fig. 2.48). Hover the mouse over the tab selector to see the name of the active tab stop.
- Types of tab stops include in word are listed below:
    o **Left Tab** ⌊: Left-aligns the text at the tab stop.
    o **Center Tab** ⊥: Centers the text around the tab stop.
    o **Right Tab** ⌋: Right-aligns the text at the tab stop.
    o **Decimal Tab** ⊥: Aligns decimal numbers using the decimal point.
    o **Bar Tab** ⎢: Draws a vertical line on the document.

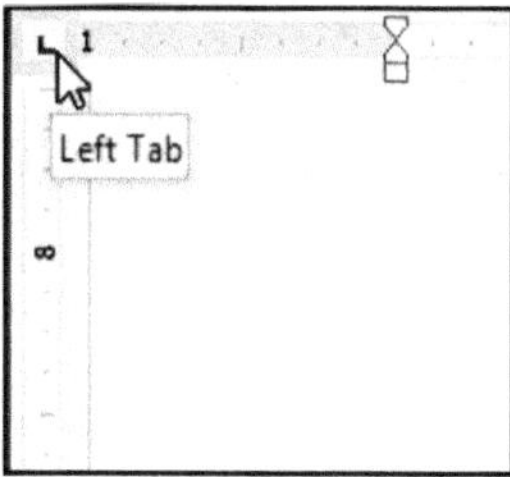

Fig. 2.48

Use of Symbols:

- Sometimes we may find that we need to add a symbol to the text, such as the Copyright symbol ©. Word offers a collection of symbols for currency, languages, mathematics, and more according to use.

To insert a symbol:

Step 1 : Place the insertion point in the location where we want to insert a symbol, (See Fig. 2.49).

Step 2 : On the INSERT tab, click the Symbol drop-down arrow. A menu of symbols will appear, (See Fig. 2.50). Select the symbol.

Fig. 2.49

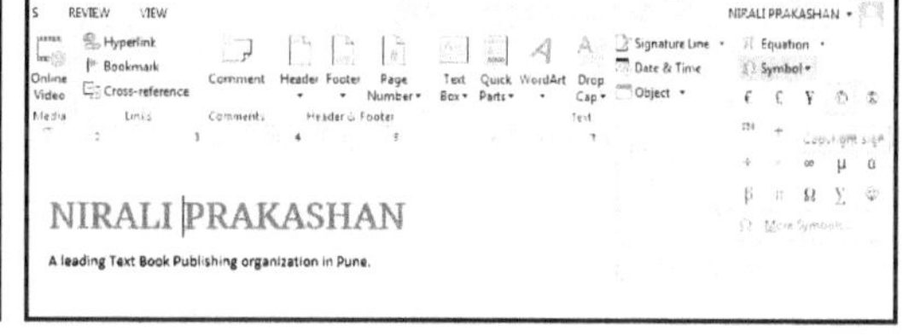

Fig. 2.50

Step 3 : If we do not see the symbol we are looking for, click More Symbols... to open the Symbol dialog box (See Fig. 2.51). Locate and select the desired symbol, then click Insert. The symbol will appear in the document as shown in Fig. 2.52.

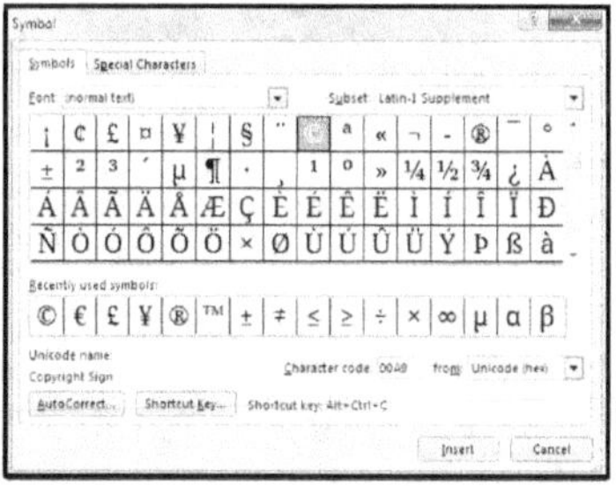

Fig. 2.51

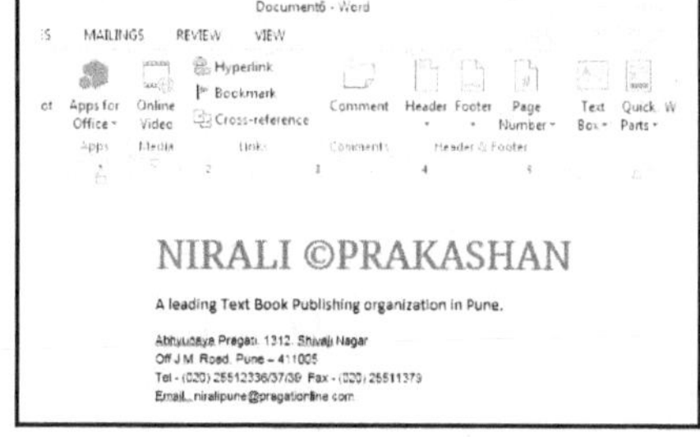

Fig. 2.52

Line Spacing:

- Line spacing is the space between each line in a paragraph. Microsoft Word allows us to customize the line spacing to be single spaced (one line high), double spaced (two lines high), or any other amount we want.
- The default spacing in Word 2013 is 1.08 lines, which is slightly larger than single spaced.
- In the Fig. 2.53, we can compare different types of line spacing. From left to right, these images show the default line spacing, single spacing, and double spacing.

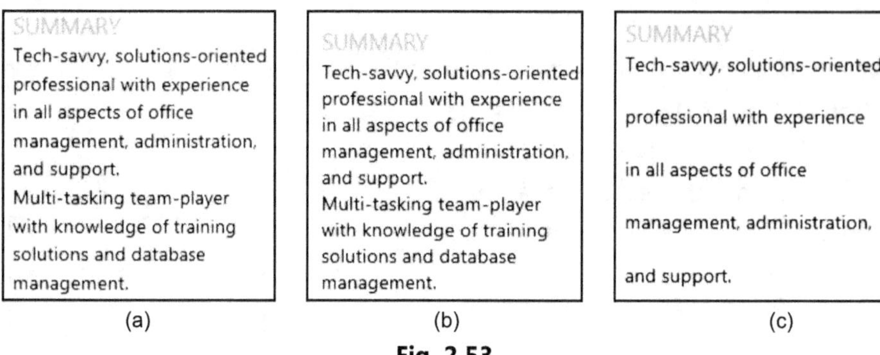

(a)   (b)   (c)

Fig. 2.53

To format line spacing:

Step 1 : Select the text we want to format, (See Fig. 2.54).

Step 2 : On the HOME tab, click the Line and Paragraph Spacing command. A drop-down menu will appear, (See Fig. 2.55).

Step 3 : Move the mouse over the various options. A live preview of the line spacing will appear in the document. Select the line spacing we want to use.

Step 4 : The line spacing will change in the document, (See Fig. 2.56).

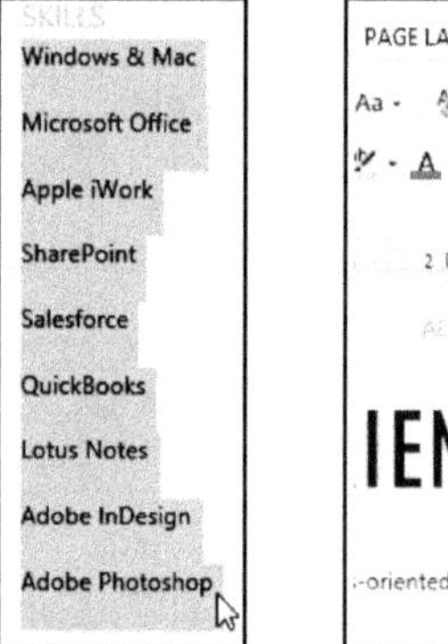

Fig. 2.54

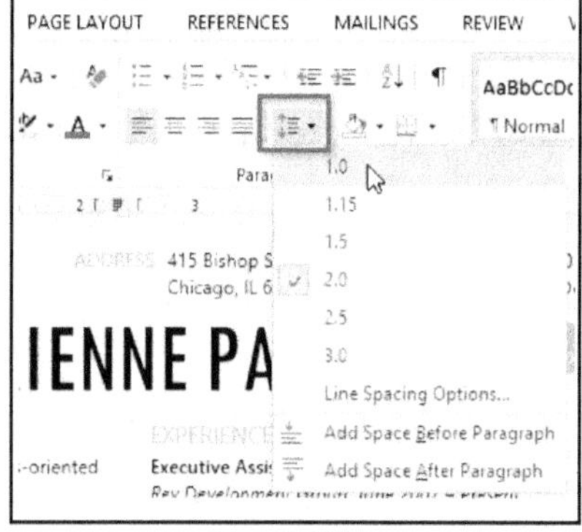

Fig. 2.55

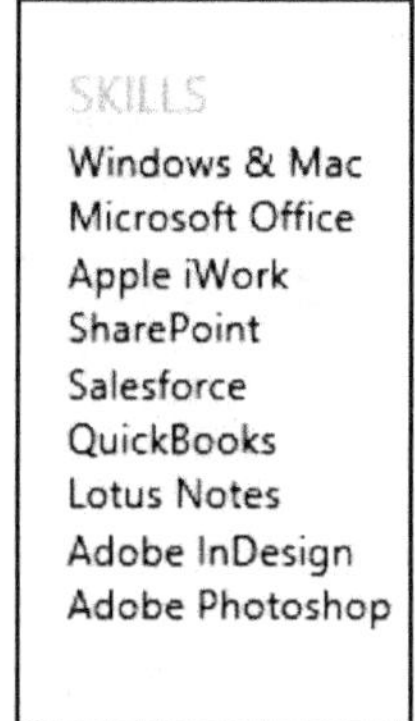

Fig. 2.56

Paragraph Spacing:

- By default, when we press the Enter key Word 2013 moves the insertion point down a little farther than one line on the page. This automatically creates space between paragraphs.

- Just as we can format spacing between lines in the document, we can adjust spacing before and after paragraphs. This is useful for separating paragraphs, headings, and subheadings.

To format paragraph spacing:

- In this example, we will increase the space before a paragraph to separate it from a heading. This will make this document easier to read. The steps are:

Step 1 : Select the paragraph or paragraphs we want to format, (See Fig. 2.57).

Step 2 : On the HOME tab, click the Line and Paragraph Spacing command. Hover the mouse over Add Space Before Paragraph or Remove Space After Paragraph from the drop-down menu. A live preview of the paragraph spacing will appear in the document.

Fig. 2.57

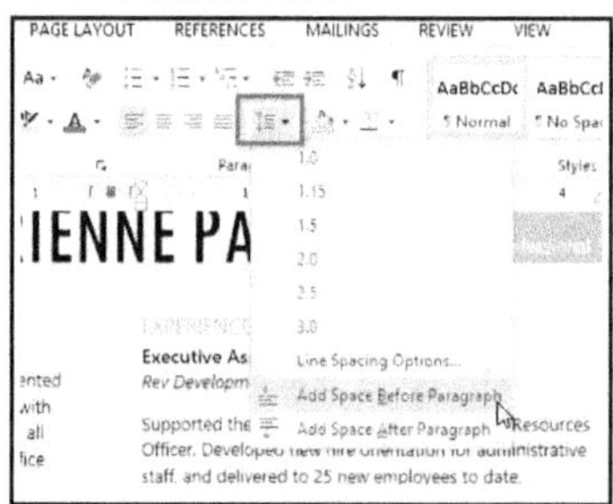

Fig. 2.58

Step 3 :   Select the paragraph spacing we want to use. In this example, we will select Add Space Before Paragraph.

Step 4 :   The paragraph spacing will change in the document, (See Fig. 2.59).

- From the drop-down menu, we can also select Line Spacing Options to open the Paragraph dialog box. From here, we can control how much space there is before and after the paragraph, (See Fig. 2.60).

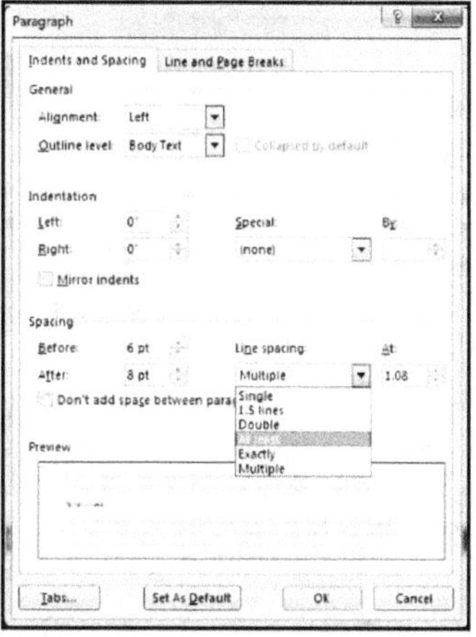

Fig. 2.59   **Fig. 2.60**

Lists:

- Bulleted and numbered lists can be used in word documents to outline, arrange, and emphasize text.

To create a bulleted list:

Step 1 :   Select the text we want to format as a list, (See Fig. 2.61).

Step 2 :   On the HOME tab, click the drop-down arrow next to the Bullets command (See Fig. 2.62). A menu of bullet styles will appear, (See Fig. 2.63).

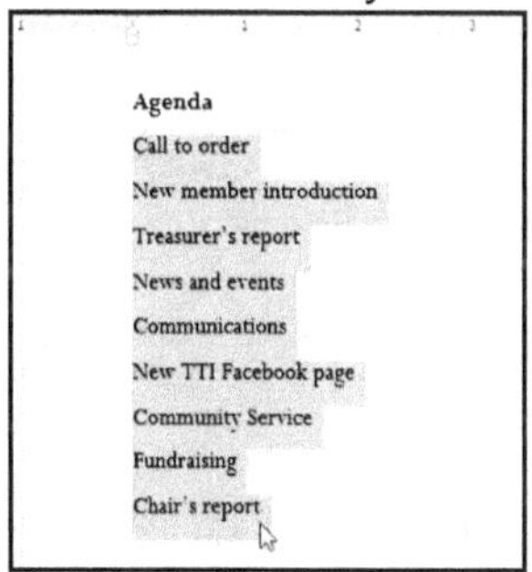

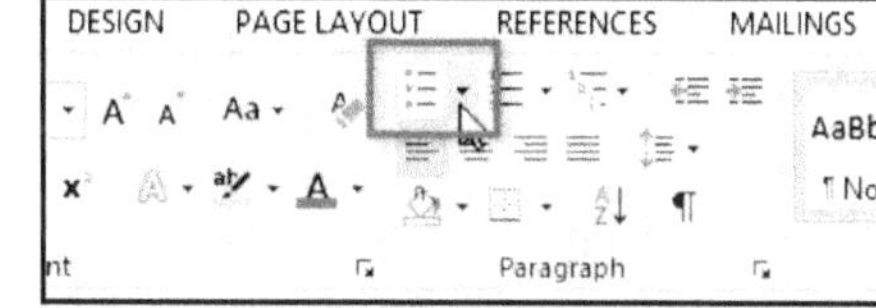

Fig. 2.61   **Fig. 2.62**

Step 3 :   Move the mouse over the various bullet styles. A live preview of the bullet style will appear in the document. Select the bullet style we want to use.

Step 4 :   The text will be formatted as a bulleted list, (See Fig. 2.64).

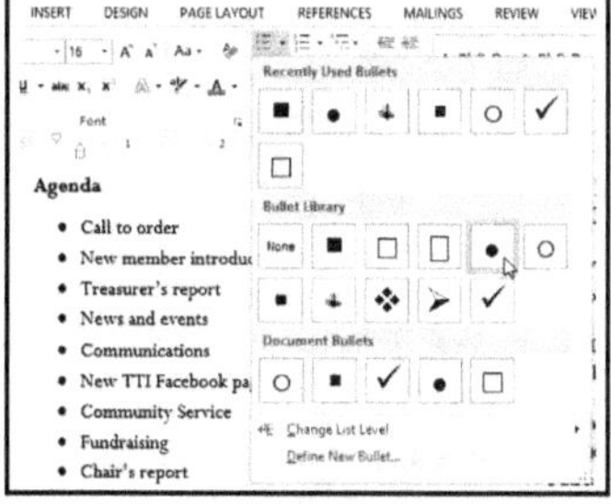

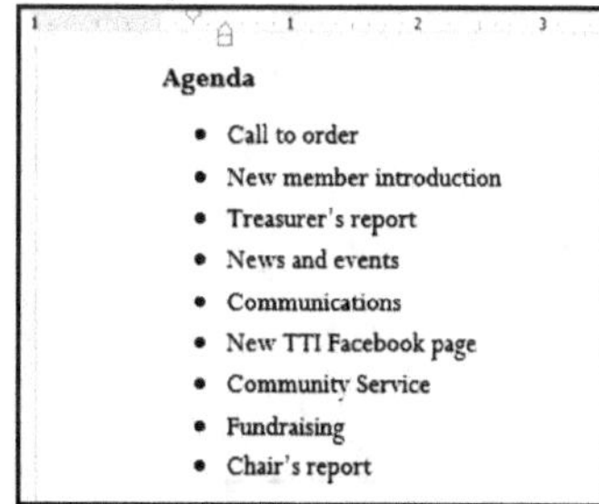

Fig. 2.63   **Fig. 2.64**

To create a numbered list:

- When we need to organize text into a numbered list, Word offers several numbering options.
- We can format the list with numbers, letters, or Roman numerals.

Step 1 : Select the text we want to format as a list, (See Fig. 2.65).

Step 2 : On the HOME tab, (See Fig. 2.66). click the drop-down arrow next to the Numbering command (See Fig. 2.67). A menu of numbering styles will appear, (See Fig. 2.68).

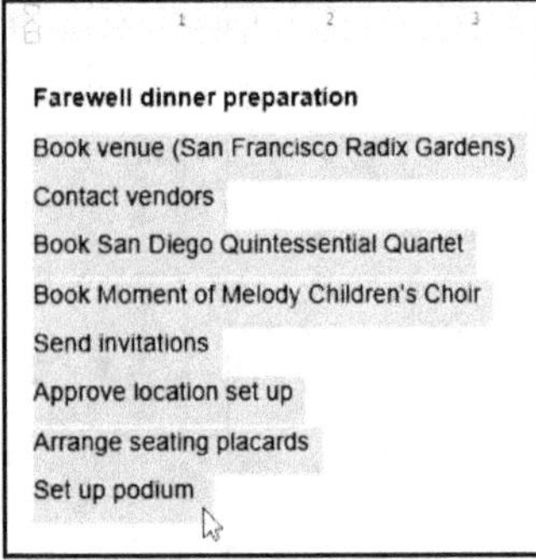

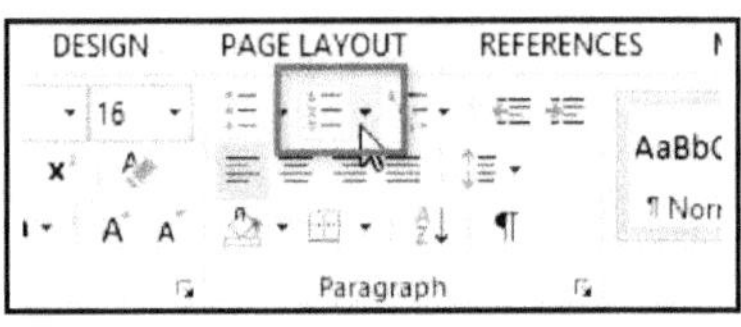

Fig. 2.65   **Fig. 2.66**   **Fig. 2.67**

Step 3 : Move the mouse over the various numbering styles. A live preview of the numbering style will appear in the document. Select the numbering style we want to use.

Step 4 : The text will format as a numbered list, (See Fig. 2.69).

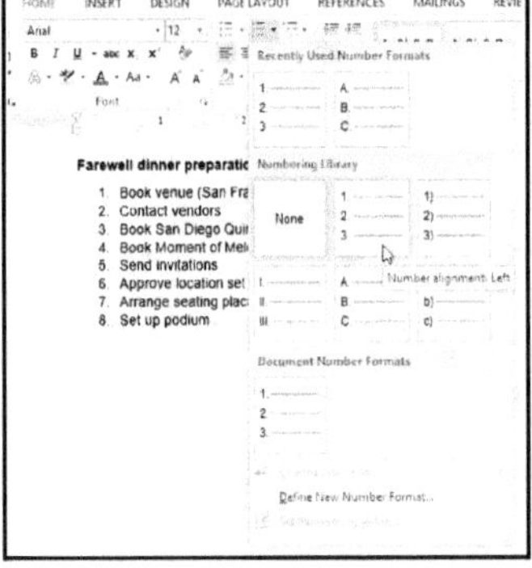

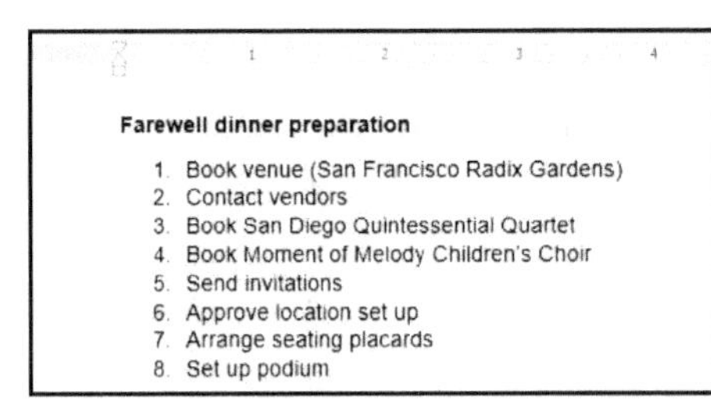

Fig. 2.68   **Fig. 2.69**

Shapes:

- We can add a variety of shapes to the document, including arrows, callouts, squares, stars, and flowchart shapes.

To insert a shape:

Step 1 : Select the INSERT tab, then click the Shapes command (See Fig. 2.70). A drop-down menu of shapes will appear.

Step 2 : Select the desired shape.

Step 3 : Click, hold, and drag in the desired location to add the shape to the document, (See Fig. 2.71).

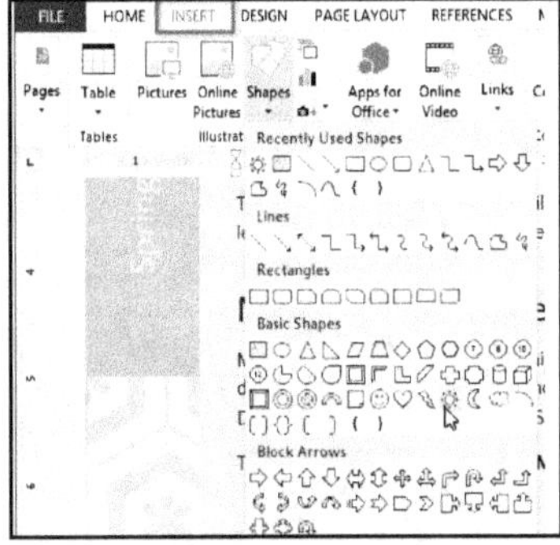

Fig. 2.70   **Fig. 2.71**

Borders and Shading:

- We can add borders and shading to paragraphs and entire pages. To create a border around a paragraph or paragraphs using following steps:

  Step 1 : Select the area of text where we want the border or shading, (See Fig. 2.73).

  Step 2 : Click the Borders Button on the Paragraph Group on the HOME Tab, (See Fig. 2.72).

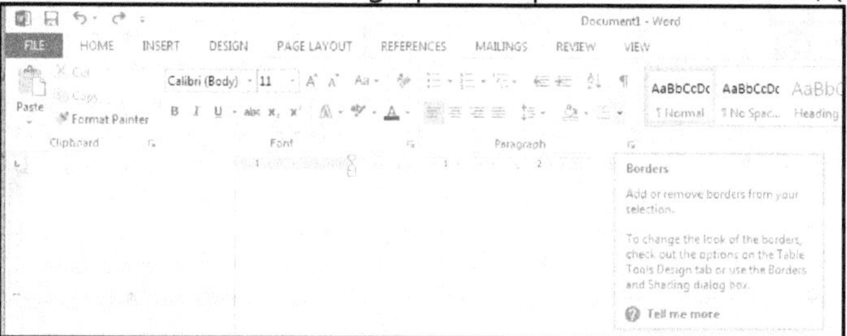

Fig. 2.72

Step 3 : Choose the Border and Shading which displays a dialog box as shown in Fig. 2.73.

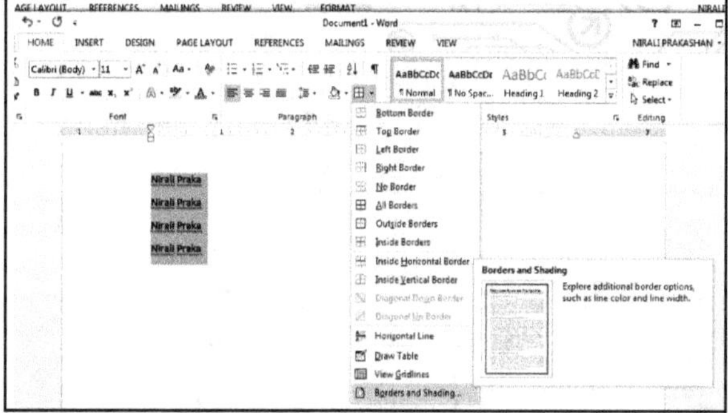

Fig. 2.73

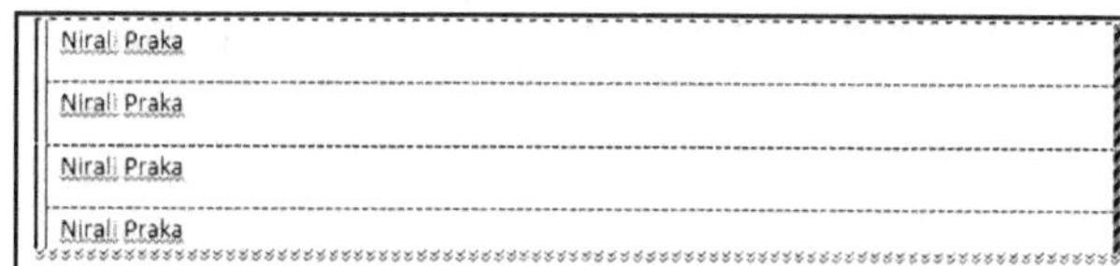

Fig. 2.74

Step 4 : Select the various options according to use, (See Fig. 2.74).

Hyperlinks:

- Hyperlinks or links allows us to click on text and go to another Web page or website.
- Creating links in a Word document allows us to put in a URL that readers can click on to visit a web page.
- Adding hyperlinks to text can provide access to websites and e-mail addresses directly from the document.
- There are a few ways to insert a hyperlink into the document. Depending on how we want the link to appear, we can use Word's automatic link formatting or convert text into a link.

To Insert Hyperlink:

Step 1 : Select the text we want to format as a hyperlink, (See Fig. 2.75).

Step 2 : Select the INSERT tab, then click the Hyperlink command, (See Fig. 2.76). The Insert Hyperlink dialog box will appear, (See Fig. 2.77).

Step 3 : The selected text will appear in the Text to display: field at the top. We can change this text if we want.

Step 4 : In the Address: field, type the address we want to link to, then click OK, (See Fig. 2.77).

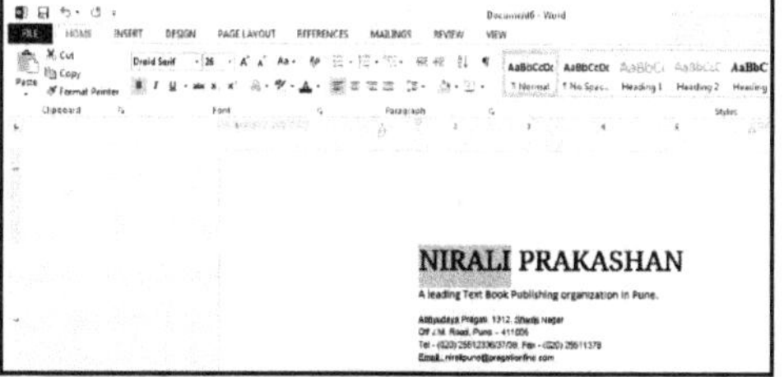

Fig. 2.75

Fig. 2.76

- The hyperlink created hyperlink is shown in Fig. 2.78.

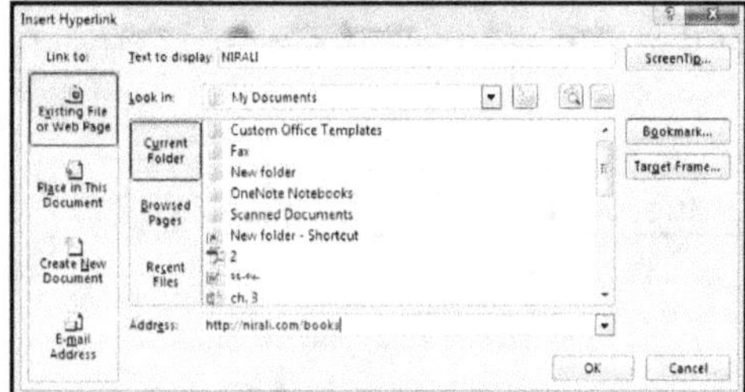 

Fig. 2.77	Fig. 2.78

Pictures:

- Adding pictures to the document can be a great way to illustrate important information or add decorative accents to existing text. Used pictures can improve the overall appearance of the document.

To insert a picture from a file:

- If we have a specific image in mind, we can insert a picture from a file. In this example, we will insert a picture saved locally on the computer.

    Step 1 : Place the insertion point where we want the image to appear.

    Step 2 : Select the INSERT tab on the Ribbon, then click the Pictures command, (Se Fig. 2.79).

    Step 3 : The Insert Picture dialog box will appear, (Se Fig. 2.80). Select the desired image file, then click Insert, (Se Fig. 2.80). The image will appear in the document (See Fig. 2.81).

 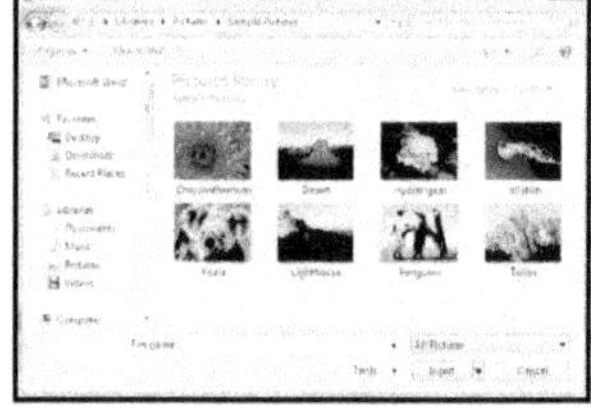 

Fig. 2.79	Fig. 2.80	Fig. 2.81

To insert an online picture (in Word 2007/2010, this option is called as Clip Art):

    Step 1 : Place the insertion point where we want the image to appear, (See Fig. 2.82).

    Step 2 : The INSERT pictures dialog box will appear, (See Fig. 2.83).

    Step 3 : Choose Bing Image Search or the OneDrive, in this example, we will use Bing Image Search.

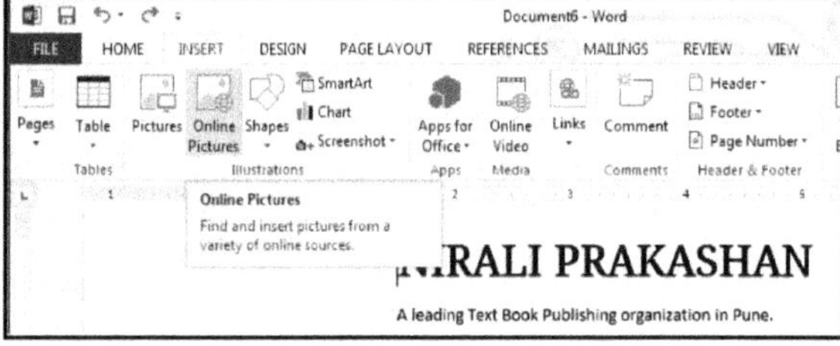 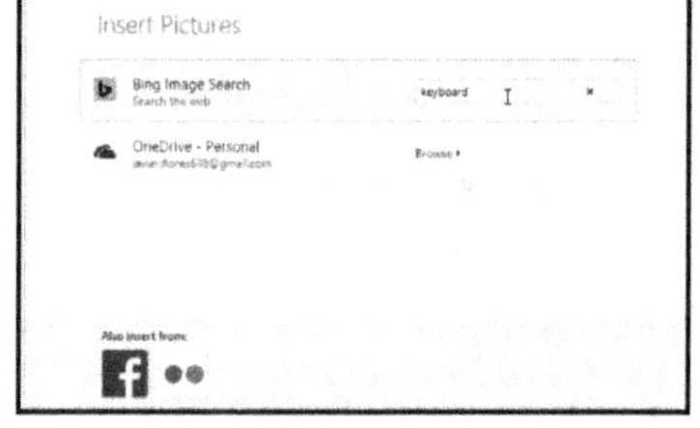

Fig. 2.82	Fig. 2.83

    Step 4 : Press the Enter key. The search results will appear in the dialog box, (See Fig. 2.84). Then select the desired image, then click Insert.

    Step 5 : The image will appear in the document, (See Fig. 2.85).

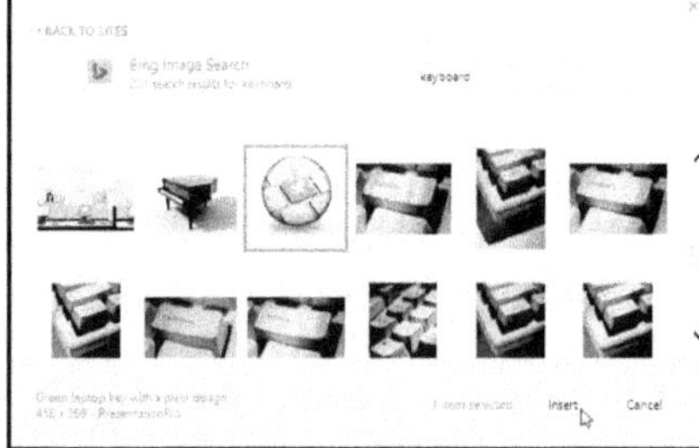  

Fig. 2.84	Fig. 2.85

WordArt:

- Text boxes can be useful for drawing attention to specific text. They can also be helpful when we need to move text around in the document.

- Word allows us to format text boxes and the text within them as WordArt. To insert WordArt follow the following steps:

  Step 1 : Select the text we want to convert, then click the INSERT tab, (See Fig. 2.86).

  Step 2 : Click the WordArt drop-down arrow in the Text group. A drop-down menu of WordArt styles will appear. Select the style we want to use, (See Fig. 2.87).

  Step 3 : Word will automatically create a text box for the text, and the text will appear in the selected style, (See Fig. 2.88).

Fig. 2.86

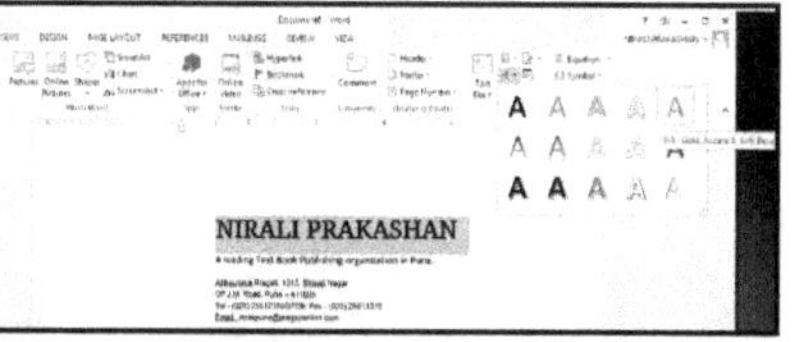
Fig. 2.87

Fig. 2.88

Headers and Footers:

- The header is a section of the document that appears in the top margin, while the footer is a section of the document that appears in the bottom margin.

- Headers and footers generally contain additional information which can help keep longer documents organized and make them easier to read. Text entered in the header or footer will appear on each page of the document.

To create a Header or Footer:

- In this example, we want to display the chapter name at the top of each page, so we will place it in the header.

  Step 1 : Double-click anywhere on the top or bottom margin of the document. In this example, we will double-click the top margin. We can also click a Header command then select Blank header, (See Fig. 2.89).

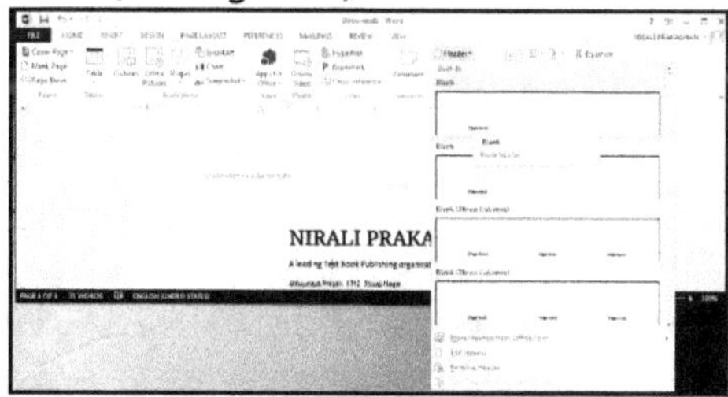
Fig. 2.89

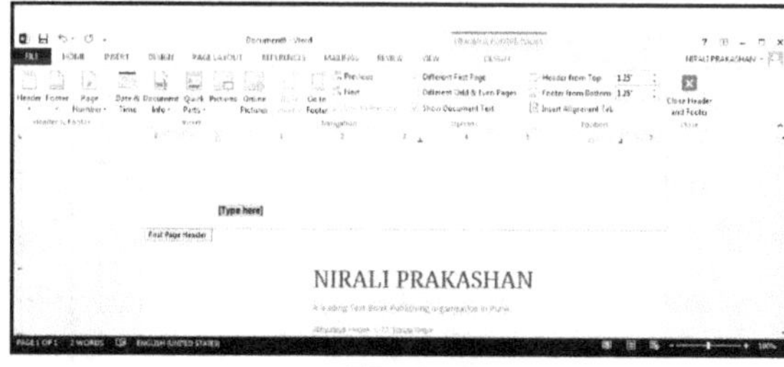
Fig. 2.90

  Step 2 : Type the desired information into the header or footer. In this example, we will type the chapter name, (See Fig. 2.91).

  Step 3 : When we are finished, click Close Header and Footer. Alternatively, we can press the Esc key, See Fig. 2.92).

Fig. 2.91

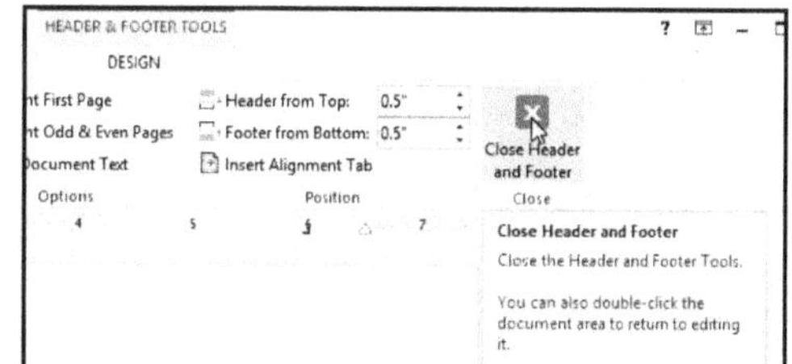
Fig. 2.92

Step 4 : The header text will appear as shown in Fig. 2.93.

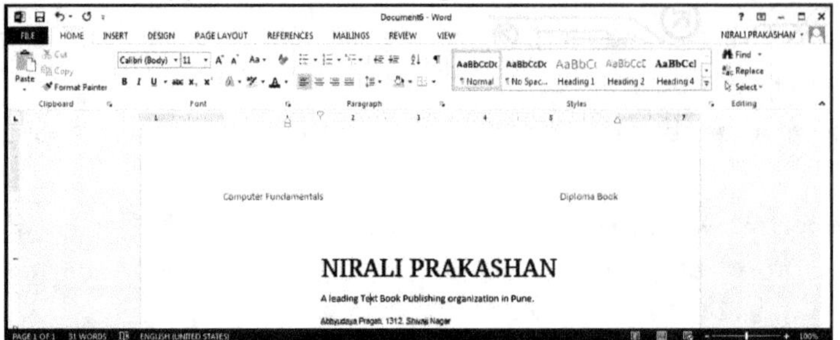

Fig. 2.93

Remove the Header:

* If we want to remove all information contained in the header, click the Header command and select Remove Header from the menu that appears.

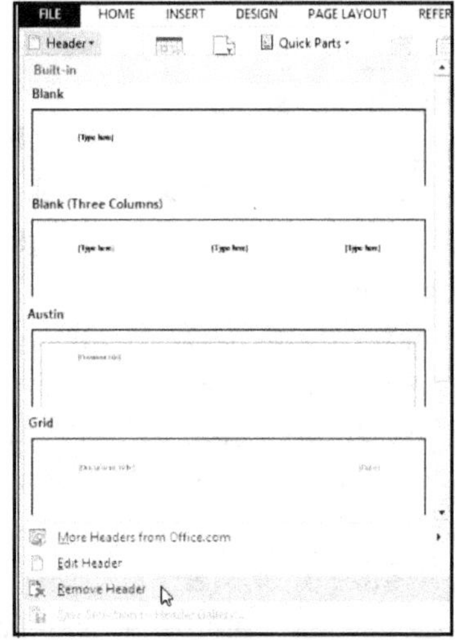

Fig. 2.94

To Insert Footer:

Step 1 : Click on INSERT tab select Footer command as shown in Fig. 2.95. Which displays various options as shown in Fig. 2.96, select Blank footer.

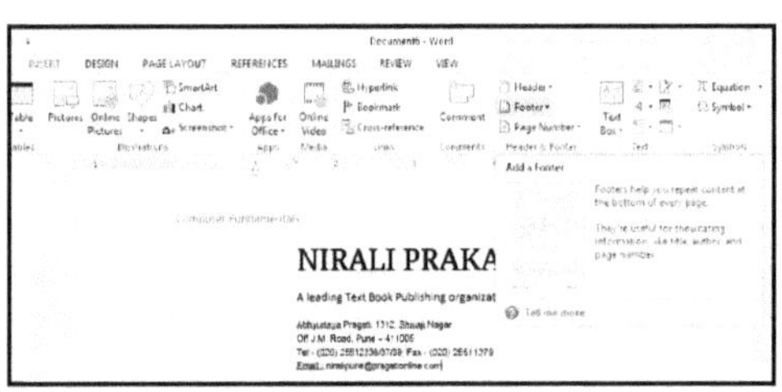

Fig. 2.95

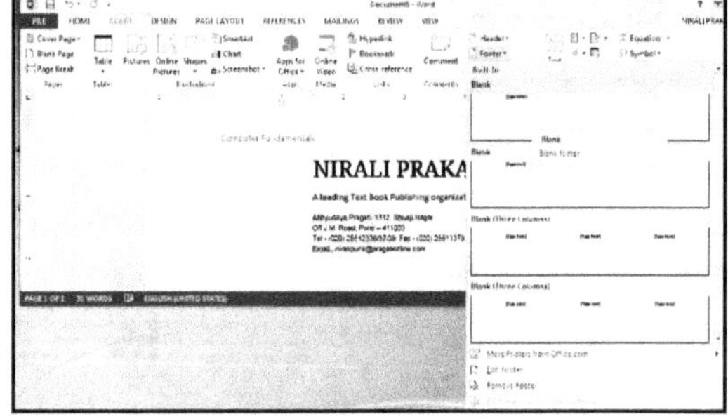

Fig. 2.96

Step 2 : Type the desired information into the header or footer. The footer display at the bottom of the document as shown in Fig. 2.98.

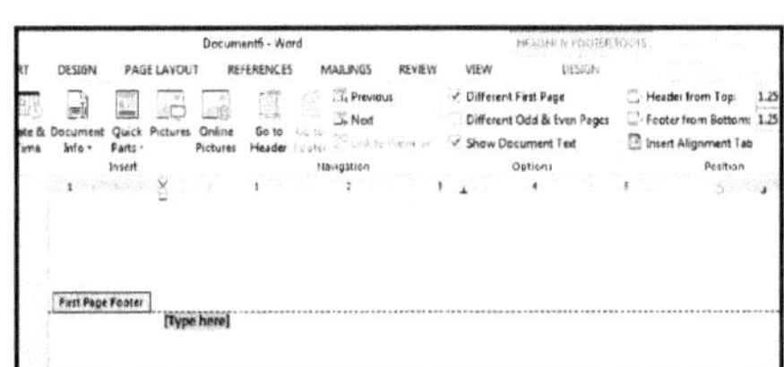

Fig. 2.97

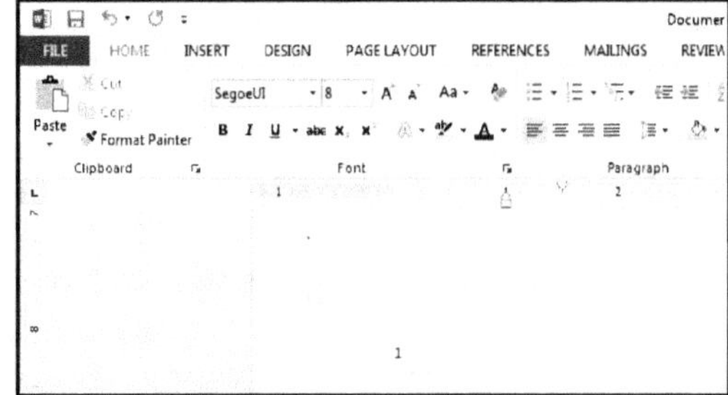

Fig. 2.98

Page Layout:

- The page layout affects how content appears and includes the page's orientation, margins, and size.

Page Orientation:

- Word offers two page orientation options i.e., landscape and portrait. Landscape means the page is oriented horizontally, while portrait means the page is oriented vertically.

To change page orientation:

Step 1 :　Select the PAGE LAYOUT tab.

Step 2 :　Click the Orientation command in the Page Setup group, (See Fig. 2.99).

Step 3 :　A drop-down menu will appear. Click either Portrait or Landscape to change the page orientation, (See Fig. 2.100).

Step 4 :　The page orientation of the document will be changed.

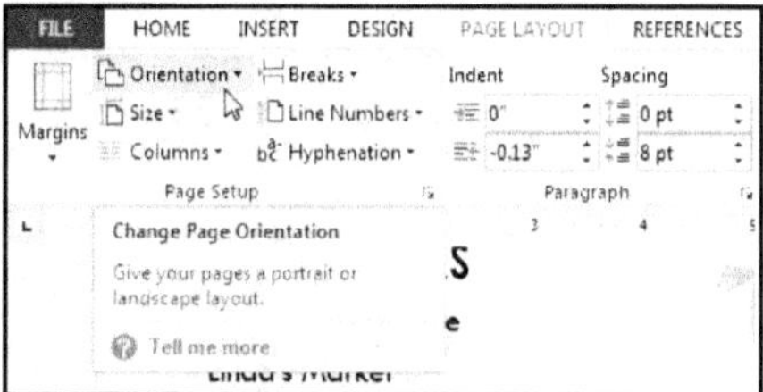

Fig. 2.99

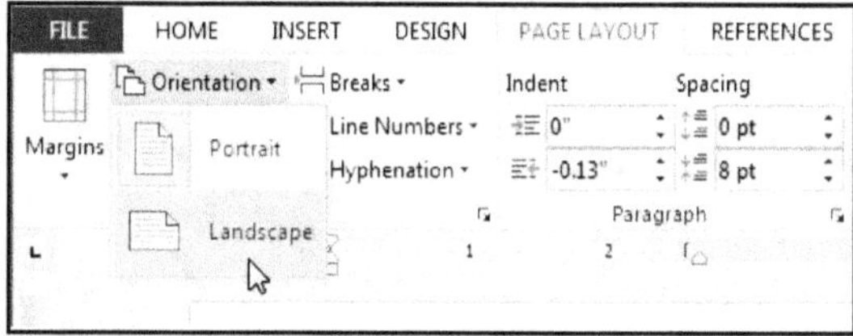

Fig. 2.100

Page Margins:

- A margin is the space between the text and the edge of the document.

- By default, a new document's margins are set to Normal, which means it has a one-inch space between the text and each edge. Depending on the needs, Word allows us to change the document's margin size.

To format page margins:

- Word has a variety of pre-defined margin sizes to choose from. To format/change page margins follow the following steps:

Step 1 :　Select the PAGE LAYOUT tab, then click the Margins command, (See Fig. 2.101).

Step 2 :　A drop-down menu will appear (See Fig. 2.102). Click the pre-defined margin size we want.

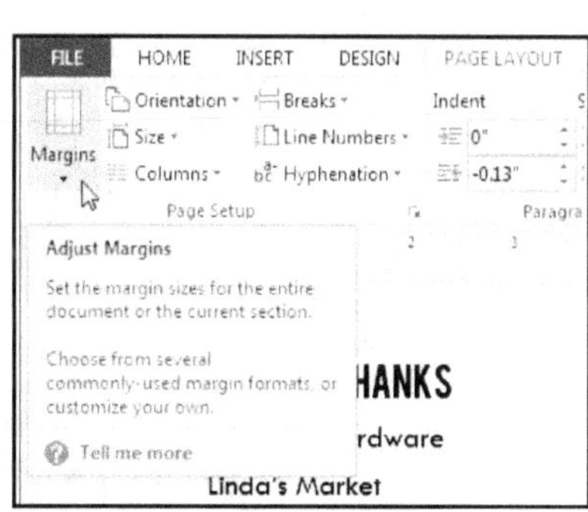

Fig 2.101

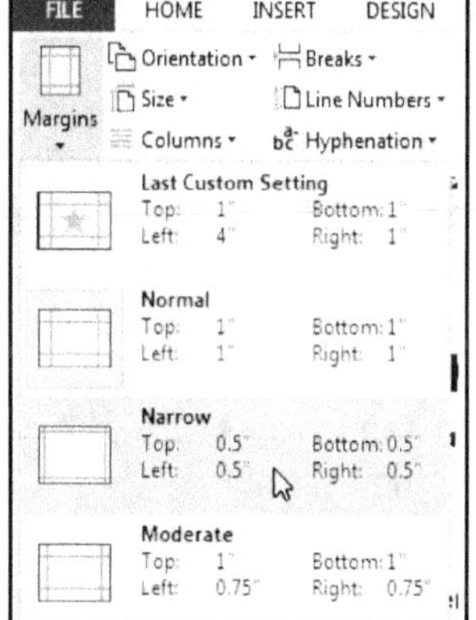

Fig. 2.102

Step 3 :　The margins of the document will be changed.

- Word also allows us to customize the size of the margins in the Page Setup dialog box.

    Step 1 : From the PAGE LAYOUT tab, click Margins. Select Custom Margins... from the drop-down menu.

    Step 2 : Them Page Setup dialog box will appear.

    Step 3 : Adjust the values for each margin, then click OK.

    Step 4 : The margins of the document will be changed.

Page Size:

- By default, the page size of a new document is 8.5 inches by 11 inches.

- Depending on the word document, we may need to adjust the document's page size.

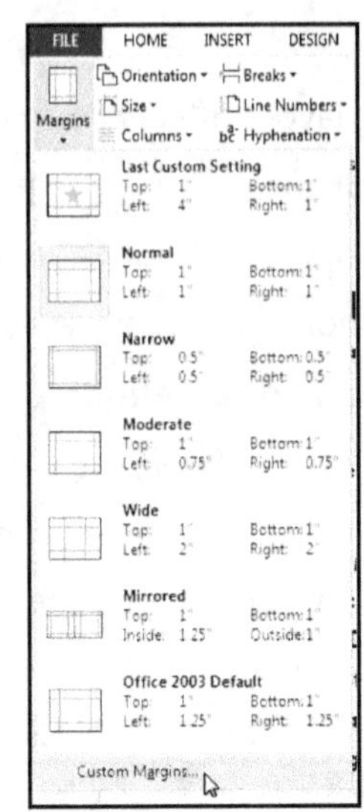

Fig. 2.103

To change the page size:

- Word has a variety of pre-defined page sizes to choose from.

    Step 1 : Select the PAGE LAYOUT tab, then click the Size command, (See Fig. 2.104).

    Step 2 : A drop-down menu will appear, (See Fig. 2.105). The current page size is highlighted. Click the desired pre-defined page size.

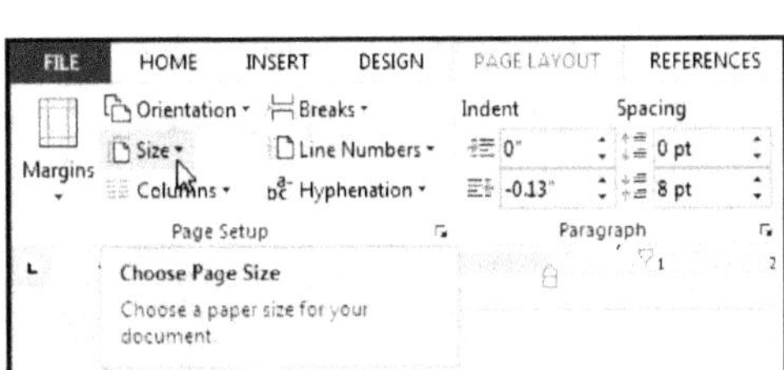

Fig. 2.104

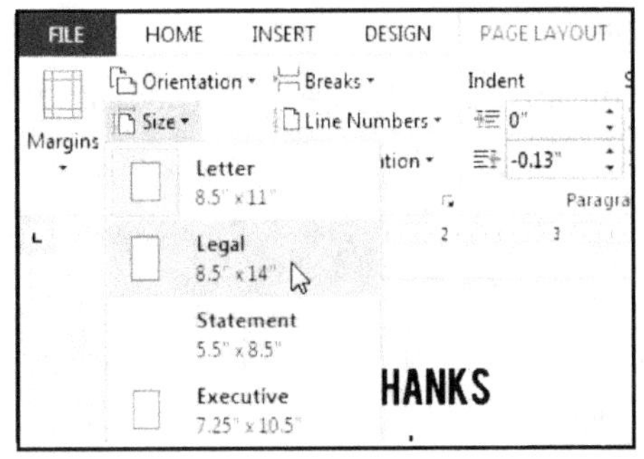

Fig. 2.105

    Step 3 : The page size of the document will be changed.

To use a custom page size:

- Word also allows us to customize the page size in the Page Setup dialog box.

    Step 1 : From the PAGE LAYOUT tab, click Size. Select More Paper Sizes... from the drop-down menu.

    Step 2 : The Page Setup dialog box will appear.

    Step 3 : Adjust the values for Width and Height, then click OK.

    Step 4 : The page size of the document will be changed.

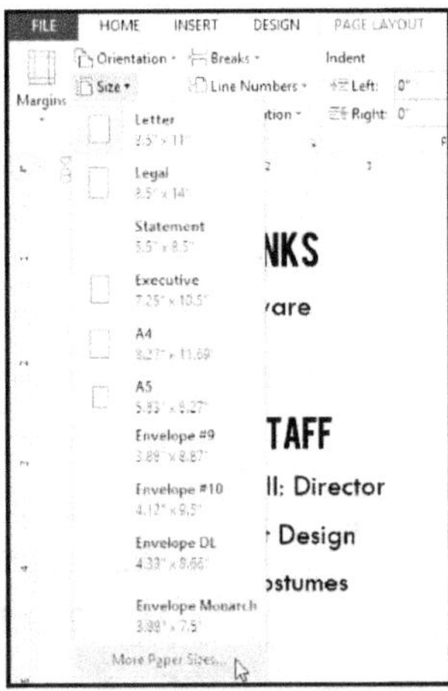

Fig. 2.106

Footnotes:

- Some types of academic writing utilize footnotes. To insert a footnote follow the following steps:

  Step 1 : Click the REFERENCES Tab on the Ribbon.

  Step 2 : Click Insert Footnote (or Insert Endnote depending on the needs).

  Step 3 : Begin typing the footnote.

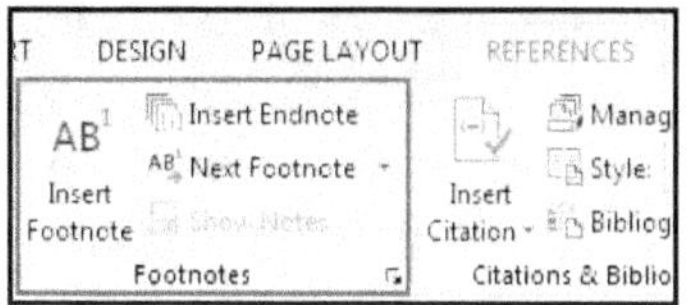

Fig. 2.107

Tables:

- A table is a grid of cells arranged in rows and columns. Tables are useful for various tasks such as presenting text information and numerical data.

- In Word, we can create a blank table, convert text to a table, and apply a variety of styles and formats to existing tables.

To insert a blank table:

Step 1 : Place the insertion point where we want the table to appear, then select the INSERT tab.

Step 2 : Click the Table command, (See Fig. 2.108).

Step 3 : A drop-down menu containing a grid of squares will appear. Hover the mouse over the grid to select the number of columns and rows in the table, (See Fig. 2.109).

Fig. 2.108

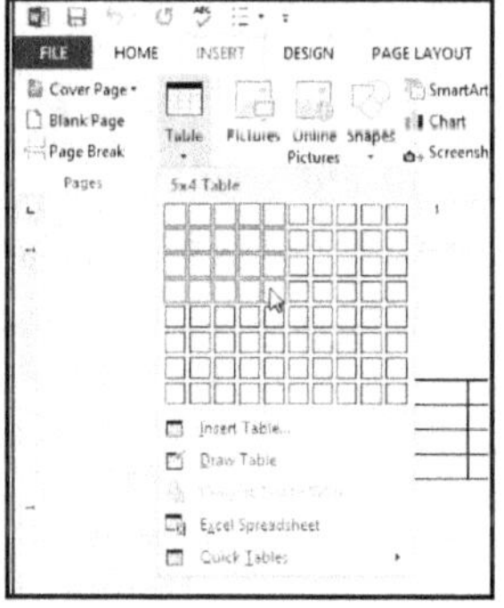

Fig. 2.109

Step 4 : Click the mouse, and the table will appear in the document.

Step 5 : We can now place the insertion point anywhere in the table to add text.

Monday	Tuesday	Wed		

To add a row or column:

Step 1 : Hover the mouse near the location where we want to add a row or column, then click the plus sign that appears, (See Fig. 2.110).

Step 2 : A new row or column will appear (See Fig. 2.111) in the table.

Menu Items	$828.45
Paper Items (Plates, silverware, cups)	$135.15
Rental Equipment (Tables, chairs, linens)	$227.75
Service Fee (18% of menu items ordered)	$122.33

Fig. 2.110

Menu Items	$828.45
Paper Items (Plates, silverware, cups)	$135.15
Rental Equipment (Tables, chairs, linens)	$227.75
Service Fee (18% of menu items ordered)	$122.33

Fig. 2.111

To delete a row or column:

Step 1 : Place the insertion point in the row or column we want to delete.

Step 2 : Right-click the mouse, then select Delete Cells... from the menu that appears, (See Fig. 2.112).

Step 3 : A dialog box will appear. Select Delete entire row or Delete entire column, then click OK. The column or row will be deleted.

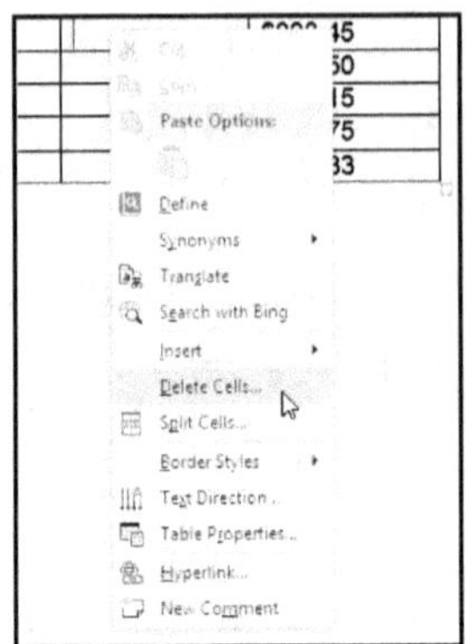

Fig. 2.112

Formulas

- We can perform calculations and logical comparisons in a table by using formulas. The Formula command is found on the Table Tools, LAYOUT tab, in the Data group.

- A formula in Word automatically updates when we open the document that contains the formula. We can also update a formula result manually.

Use of Formula:

- First we need to create a table with numeric data in the cells. To create a table, click the Table icon from the Insert tab and drag the mouse cursor to select an area from the appeared menu with 4 columns and 7 rows i.e., 4×7 Table.

- Click the left mouse button after we finish with the dragging and the table will be inserted in the Word document.

- After we create the table and enter the numeric data in the cells, select the cell where we wish to insert the formula.

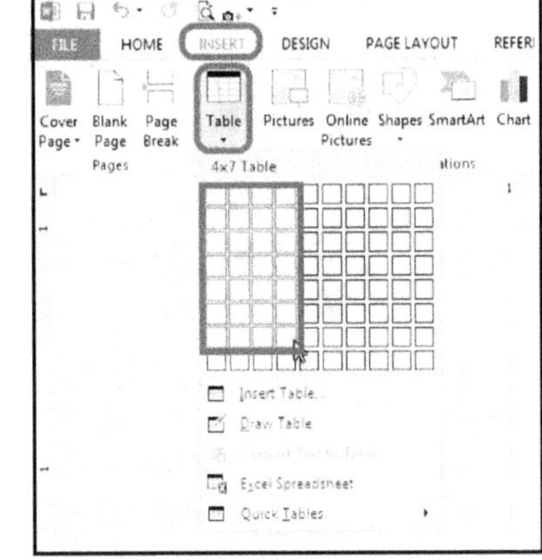

Fig. 2.113

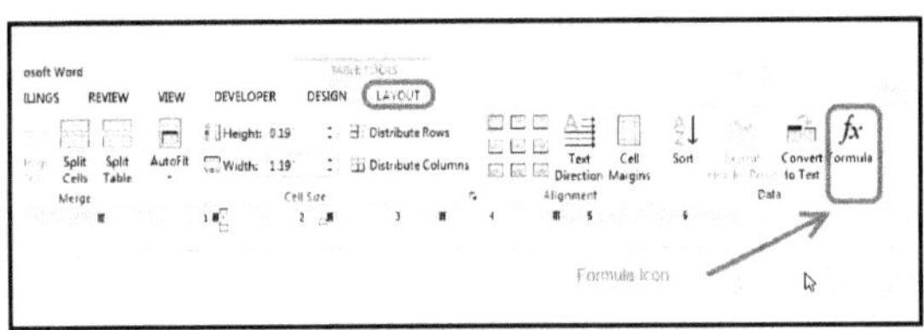

Fig. 2.114

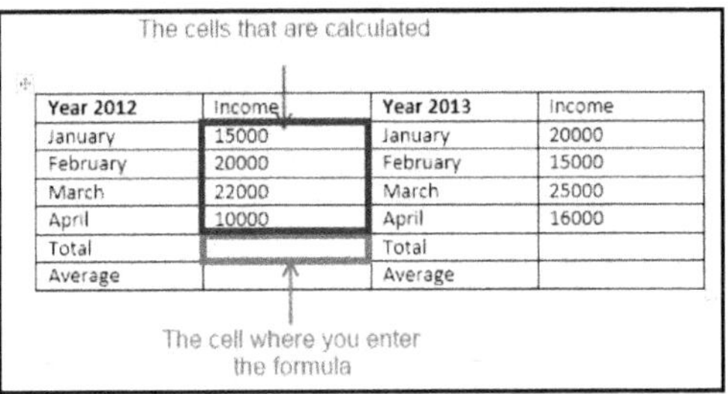

The cells that are calculated

Year 2012	Income	Year 2013	Income
January	15000	January	20000
February	20000	February	15000
March	22000	March	25000
April	10000	April	16000
Total		Total	
Average		Average	

The cell where you enter the formula

Fig. 2.115

- This formula will calculate the sum of all numeric cells that are located above the formula cell. As an example, select the cell that is located in the income column, on the right side of the Total cell. Next, select the LAYOUT tab from the Table Tools section on the ribbon and in the Data group click the Formula button.

- New Formula dialogue box with few options will appear (See Fig. 2.117). A default formula "=SUM(ABOVE)" is written in the Formula label textbox. This formula calculates the sum of every numeric cell from the table that is located above the formula cell.

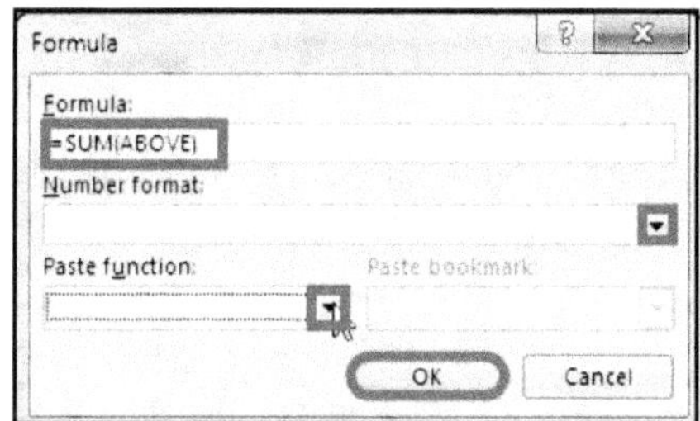

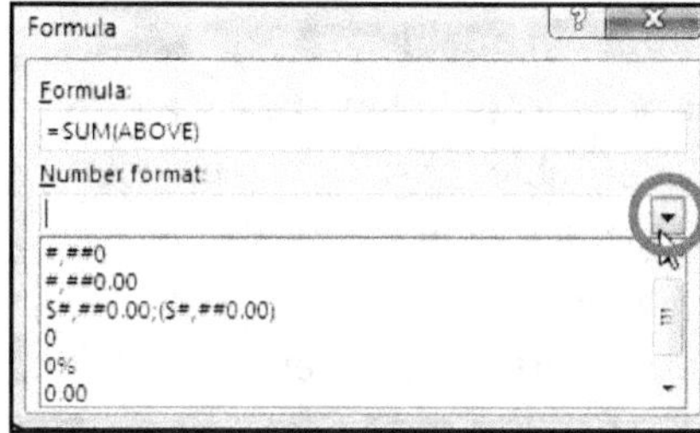

Fig. 2.116   **Fig. 2.117**

- After clicking on the drop down arrow from the Number format label, we can choose the numeric format of the formula cell. We can switch to percentage, double decimal, whole number format or some other format, (See Fig. 2.118).

- After selecting the drop down arrow from the Paste function label, we can choose from variety of formulas which can be inserted into the cell. Some of the most popular formulas are AVERAGE, MAX, MIN, COUNT and so on.

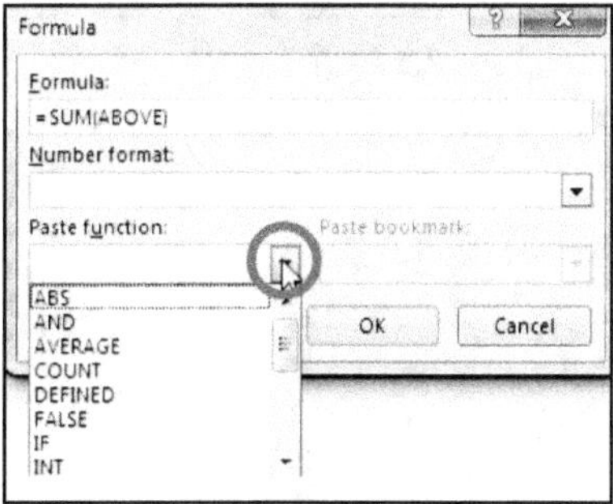

Year 2012	Income	Year 2013	Income
January	15000	January	20000
February	20000	February	15000
March	22000	March	25000
April	10000	April	16000
Total	67000	Total	
Average		Average	

Fig. 2.118   **Fig. 2.119**

- When we click on the OK button from the Formula dialogue box, the formula cell will show the number which is the result calculated by the inserted formula.

- In the word the Formula dialog box contains some important functions to be used as a formula in a cell.

Formula	Description
SUM( )	The sum of a list of cells.
PRODUCT( )	The multiplication of a list of cells.
AVERAGE( )	The average of a list of cells.
COUNT( )	The number of items in a list of cells.
MIN( )	The smallest value in a list of cells.
MAX( )	The largest value in a list of cells.
IF()	Evaluates the first argument. Returns the second argument if the first argument is true; returns the third argument if the first argument is false.
INT()	Rounds the value inside the parentheses down to the nearest integer.
MOD()	Takes two arguments (must be numbers or evaluate to numbers). Returns the remainder after the second argument is divided by the first. If the remainder is 0 (zero), returns 0.0.
NOT()	Takes one argument. Evaluates whether the argument is true. Returns 0 if the argument is true, 1 if the argument is false. Mostly used inside an IF formula.
OR()	Takes two arguments. If either is true, returns 1. If both are false, returns 0.
SIGN()	Takes one argument that must either be a number or evaluate to a number. Evaluates whether the item identified inside the parentheses if greater than, equal to, or less than zero (0). Returns 1 if greater than zero, 0 if zero, -1 if less than zero.
TRUE()	Takes one argument. Evaluates whether the argument is true. Returns 1 if the argument is true, 0 if the argument is false. Mostly used inside an IF formula.
DEFINED()	Evaluates whether the argument inside the parentheses is defined. Returns 1 if the argument has been defined and evaluates without error, 0 if the argument has not been defined or returns an error.

Sort a single column in a table:

Step 1 : Select the column that we want to sort, (See Fig. 2.120 (a)).

Step 2 : Under Table tools, on the LAYOUT tab, in the Data group, click Sort, (See Fig. 2.120 (b)).

Step 3 : Under My list has, click Header row or No header row.

Step 4 : Click Options.

Step 5 : Under Sort options, select the Sort column only check box, (See Fig. 2.121).

Step 6 : Click OK. The sorted list displayed on the document as shown in Fig. 2.122.

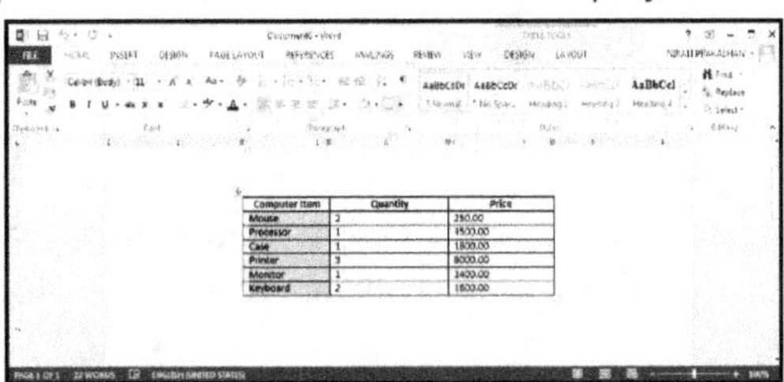

Fig. 2.120 (a)

Fig. 2.120 (b)

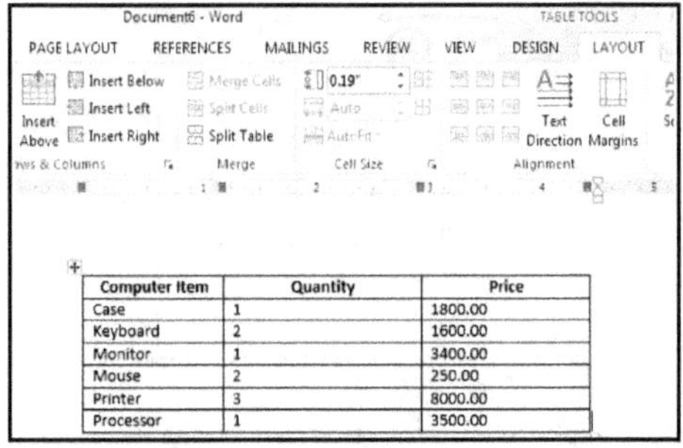

Fig. 2.121

Fig. 2.122

Watermarks/Washout:

- A watermark is a picture that shows up faintly behind the text on a Word document page.
- A watermark is a translucent image that appears behind the primary text in a document.
- To insert a watermark follow the following steps:

Step 1 : Click the DESIGN tab in the Ribbon, (See Fig. 2.123).

Step 2 : Click the Watermark Button in the Page Background Group

Step 3 : Click the Watermark we want for the document or click Custom Watermark and create the own watermark.

- The water displayed on the word document as shown in Fig. 2.124.

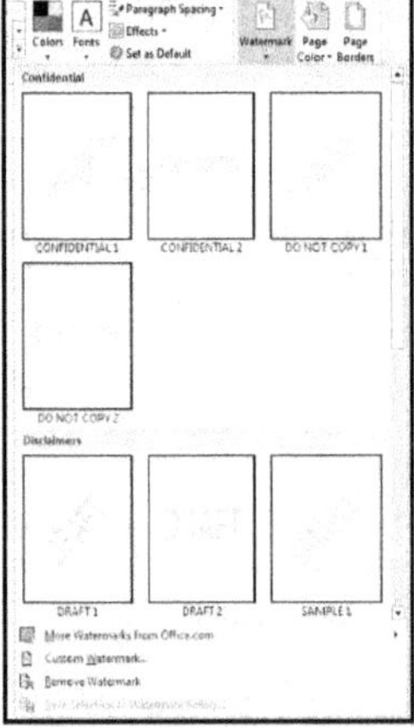

Fig. 2.123

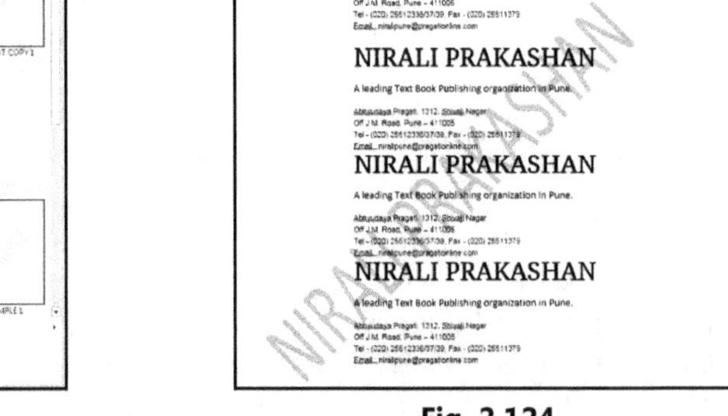

Fig. 2.124

- To remove a watermark, follow the steps above, but click Remove Watermark.

Macros:

- Macros are advanced features that can speed up editing or formatting we may perform often in a Word document.
- Macros are recorded sequences of menu selections that we choose so that a series of actions can be completed in one step.

Recording a Macro:

Step 1 : Click the VIEW Tab on the Ribbon.

Step 2 : Click Macros.

Step 3 : Click Record Macro, (See Fig. 2.125).

Step 4 : Enter a name (without spaces).

Step 5 : Click whether we want it assigned to a Button (on the Quick Access Toolbar) or the Keyboard (a sequence of keys).

Step 6 : To assign the macro a button on the Quick Access Toolbar:

- Click Button.
- Under the Customize Quick Access Toolbar, select the document for which we want the Macro available.

Step 7 : Under Choose Commands: Click the Macro that we are recording.

Step 8 : Click Add.

Step 9 : Click OK to begin Recording the Macro.

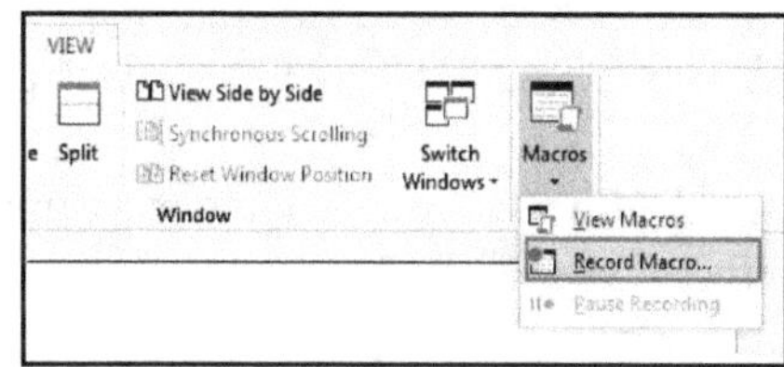

Fig. 2.125

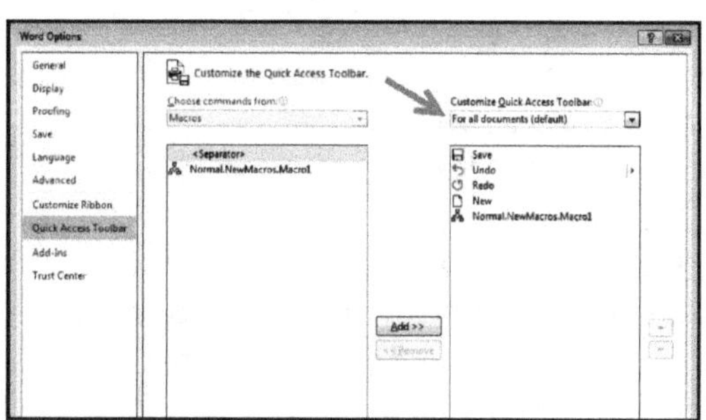

Fig. 2.126

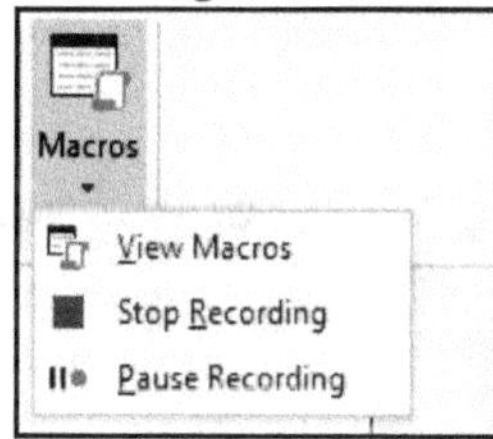

Fig. 2.127

Step 10 : Perform the actions we want recorded in the Macro.

Step 11 : Click on Macros.

Step 12 : Click on Stop Recording Macros, (See Fig. 2.127).

- To assign a macro button to a keyboard shortcut:

Step 1 : Click Keyboard.

Step 2 : In the Press New Shortcut Key box, type the key sequence that we want and click Assign.

Step 3 : Click Close to begin recording the Macro.

Step 4 : Perform the actions we want recorded in the Macro.

Step 5 : Click on Macros.

Step 6 : Click on Stop Recording Macros.

Running a Macro:

- Running a macro depends on whether it is been added to the Quick Access Toolbar or if it is been given a Keyboard Shortcut.

Step 1 : To run a Macro from the Quick Access Toolbar, simply click the Macro Icon, as shown in Fig. 2.129.

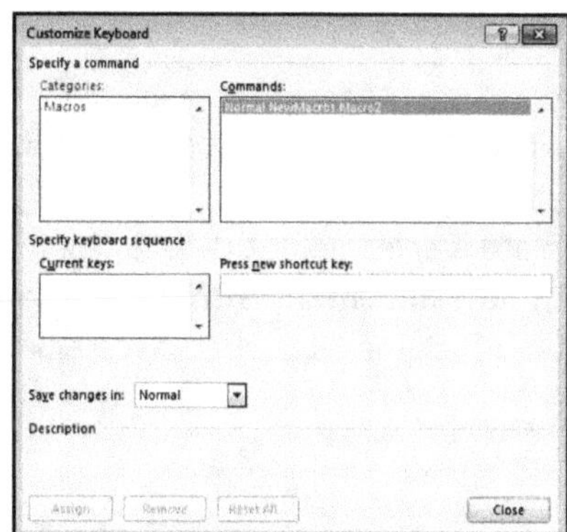

Fig. 2.128

Fig. 2.129

Step 2 : To run a Macro from the Keyboard shortcut, simply press the keys that we have programmed to run the Macro.

Themes:

- A theme is a set of colors, fonts, and effects that determines the overall look of the document. Themes are a great way to change the tone of the entire document quickly and easily.
- A theme consists of following three elements:
    1. **Colors:** A set of colors is chosen to format the text foreground and background, any graphics or design elements in the theme, plus hyperlinks.
    2. **Fonts:** Two fonts are chosen as part of the theme, one for the heading styles and a second for the body text.
    3. **Graphical Effects:** These effects are applied to any graphics or design elements in the document. The effects can include 3-D, shading, gradation, drop shadows and so on.

To insert the theme:

Step 1 : From the DESIGN tab, click the Themes command, (See Fig. 2.130).

Step 2 : Select the desired theme from the drop-down menu.

Step 3 : The selected theme will appear.

Mail Merge:

- Mail Merge is a useful tool that allows us to produce multiple letters, labels, envelopes, name tags, and more using information stored in a list, database or spreadsheet.

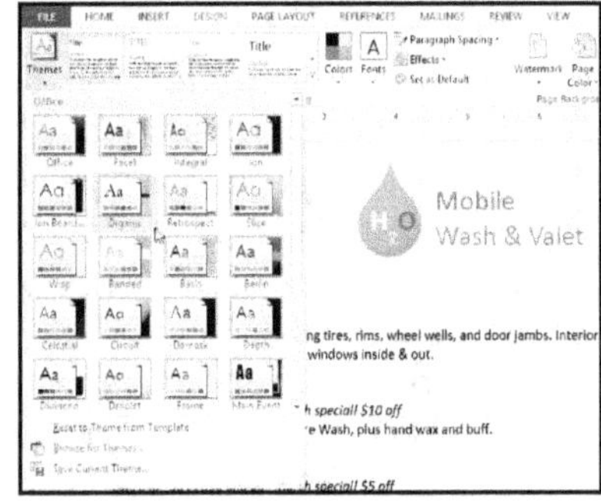

Fig. 2.130

- When performing a Mail Merge, we will need a Word document and a recipient list, which is typically an Excel workbook.

To use Mail Merge:

- Open an existing Word document, or create a new one.
- From the MAILINGS tab, click the Start Mail Merge command and select Step by Step Mail Merge Wizard from the drop-down menu.
- The Mail Merge pane appears and will guide we through the six main steps to complete a merge and follow the steps to complete mail merge.

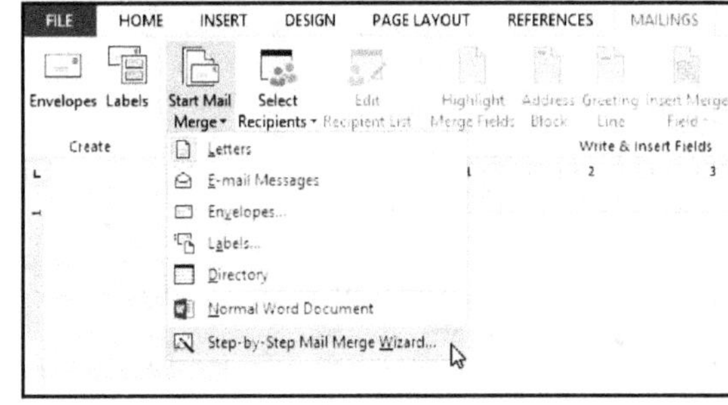

Fig. 2.131

Templates:

- A template is a predesigned document we can use to create a new document quickly.
- Templates often include custom formatting and designs, so they can save we a lot of time and effort when starting a new project.
- A template is selected at the time when we create a new blank document.

To create a new document from a template:

Step 1 : Click the FILE tab to access Backstage view, (See Fig. 2.132).

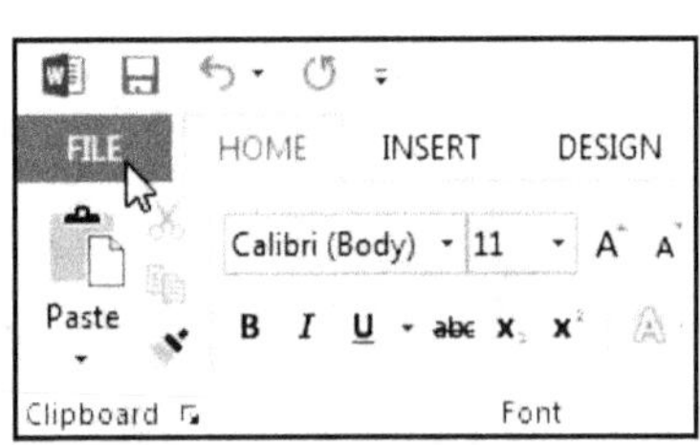

Fig. 2.132

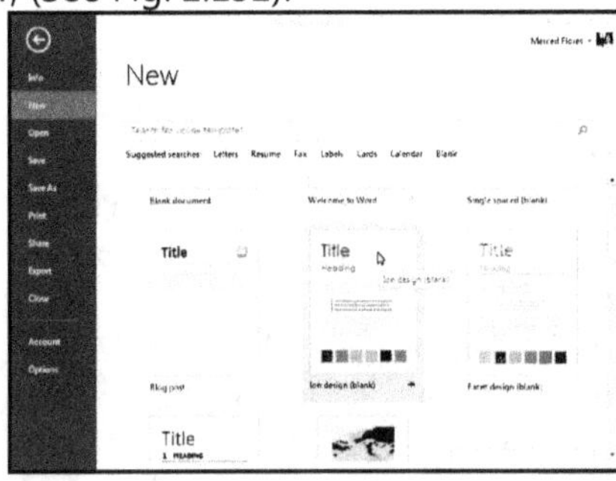

Fig. 2.133

Step 2 : Select New. Several templates will appear below the Blank document option.

Step 3 : Select a template to review it, (See Fig. 2.133).

Step 4 : A preview of the template will appear, along with additional information on how the template can be used.

Step 5 : Click Create to use the selected template, (See Fig. 2.134).

Step 6 : A new workbook will appear with the selected template, (See Fig. 2.135).

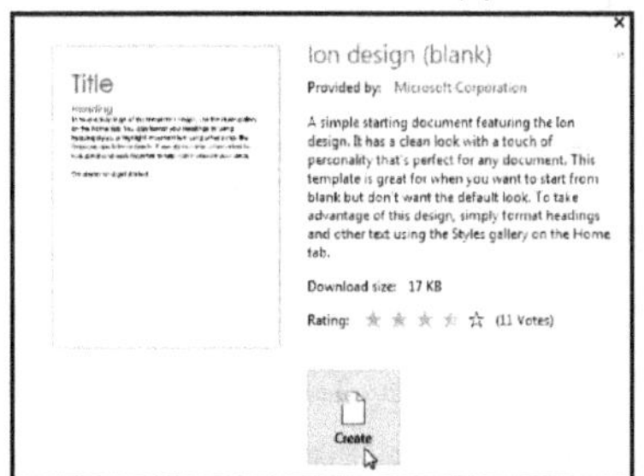

Fig. 2.134

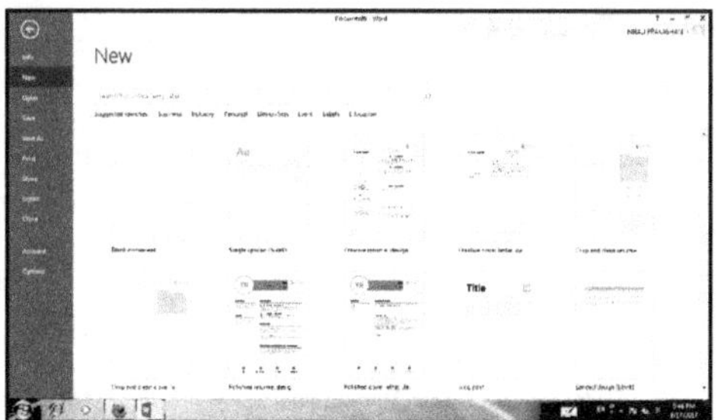

Fig. 2.135

- When we select Business template which display various categories as shown in Fig. 2.136.

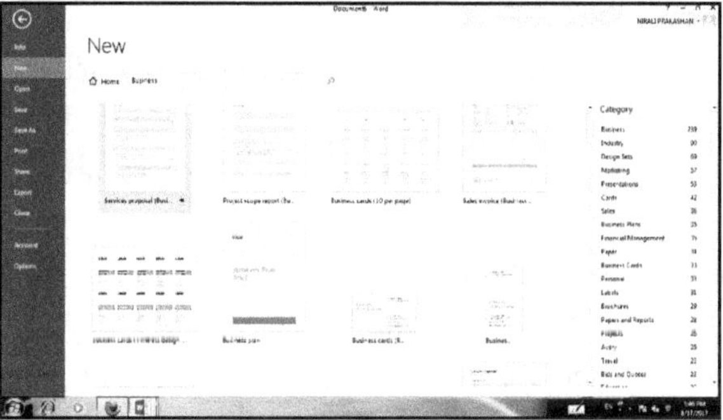

Fig. 2.136

Versions:

- In Word 2013, the only way we can access previous versions of a document in Word is to open previous automatically saved (autosaved) versions of the document.

- Word used to have a formal versioning feature that allowed us to save different versions of a document within the document itself. That feature has gone away and the only way we can retrieve previous versions of a document is through the Auto-Save feature or by accessing unsaved documents, if available.

- By default, Word automatically saves the document at certain intervals. This is the Auto-Save feature. To go back to a previously, automatically saved copy of the document, click the "FILE" tab and select an "(autosave)" item from the list under "Versions", (See Fig. 2.138).

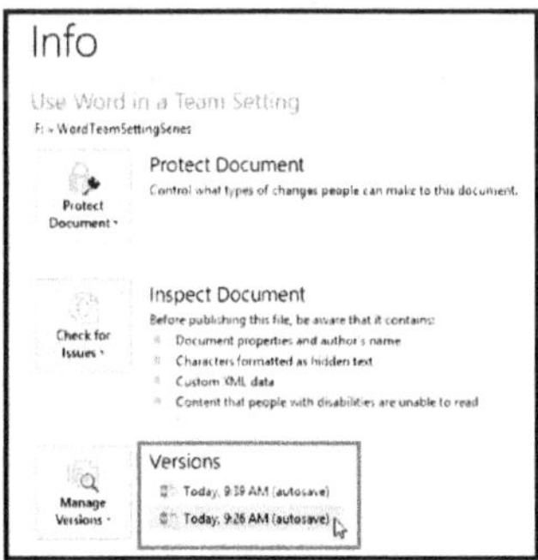

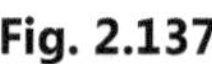

Fig. 2.137

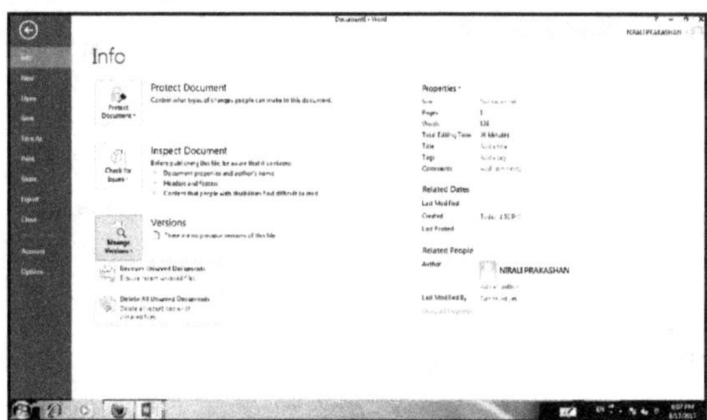

Fig. 2.138

- If any unsaved files were available, they would be listed on the "Open" dialog box that displays, (See Fig. 2.139).

- There are alternatives for adding a versioning system to Word. One is an add-in for Word, called Save Versions, and another is a plugin that works with an external version control program called Perforce.

- The Save Versions add-in allows us to easily save numbered versions of a Word document from within the file.

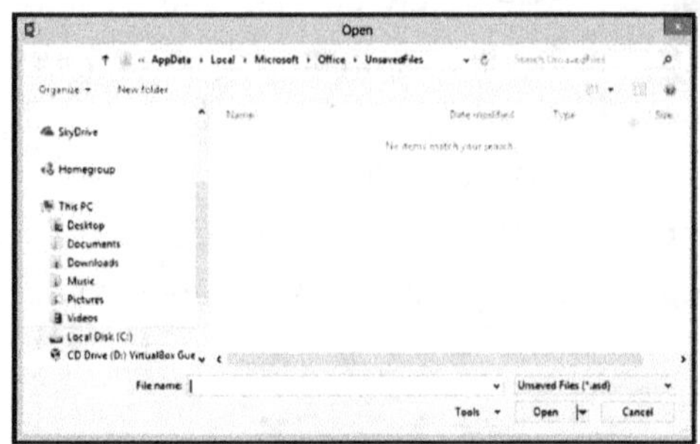

Fig. 2.139

- Perforce is a version control program that is available for free for up to 20 users. They provide a plugin for Microsoft Office that allows Microsoft Word, Excel, PowerPoint, and Project files to be easily stored and managed in Perforce from within the programs.

Compare Two Versions of the Same Document:

- The "Compare" and "Combine" features are similar, but each has its purpose. If we only have two documents we want to compare, and the neither one contains tracked changes, use the "Compare" feature.

- If we have two or more documents that contain tracked changes, and we need to keep track of who changed what and when, use the "Combine" feature.

- To compare the original and revised versions of a document, click "Compare" in the "Compare" section of the REVIEW tab and select "Compare" from the drop-down menu.

- On the "Compare Documents" dialog box, select the "Original document" from the drop-down list, (See Fig. 2.141).

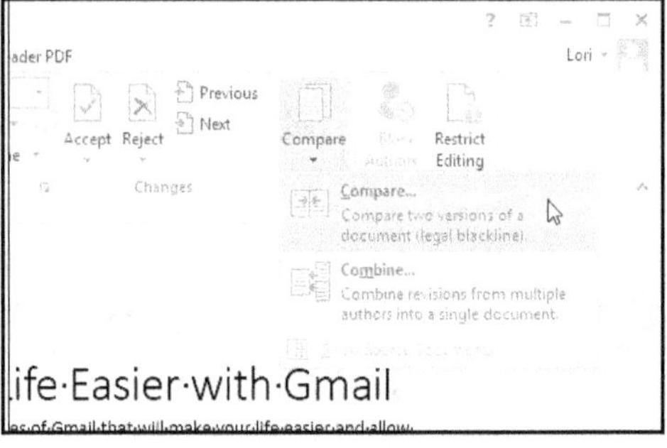
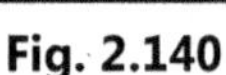

Fig. 2.140

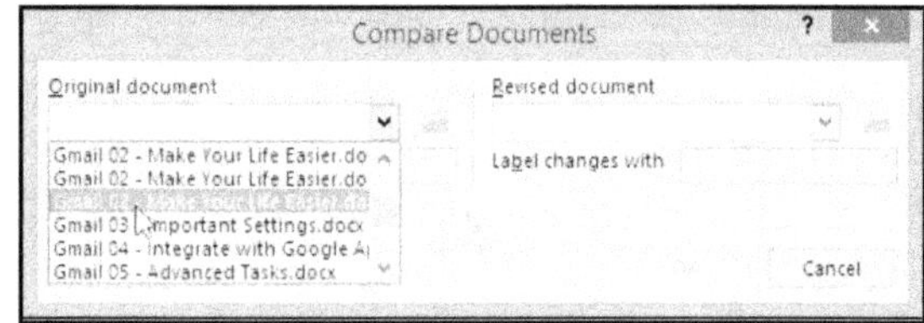

Fig. 2.141

- Select the "Revised document" from the drop-down list. Under the "Revised document," specify a label to apply to the changes so we know who made them in the "Label changes with" edit box.

- The comparison between the two documents is made and the changes are displayed in the specified document.

To combine Documents:

- When we have multiple documents containing tracked changes that we need to compare, it helps to be able to keep track of who made which changes and when. The "Combine" command allows us to merge the tracked changes from each document, two at a time, until all changes from all documents have been incorporated into one document.

- To start combining the first two documents, click "Compare" in the "Compare" section of the Review tab and select "Combine" from the drop-down menu, (See Fig. 2.142).

- The "Combine Documents" dialog box is essentially the same as the "Compare Documents" dialog box displayed.
- For the "Original document" select the earliest version of the original document and then select one of the reviewers' versions of the document as the "Revised document." Enter labels for the "Original document" and the "Revised document" using the corresponding "Label unmarked changes with" edit box. This allows us to see who made which changes.
- Just as we did when comparing documents, select the "Comparison settings," the level for "Show changes at," and which document to "Show changes in." When combining more than two documents, it might be a good idea to show the changes in the "Original document," making sure we make a copy of it first, and use the same original document when comparing to each reviewer's document. This allows us to "gather" all the changes into the original document.

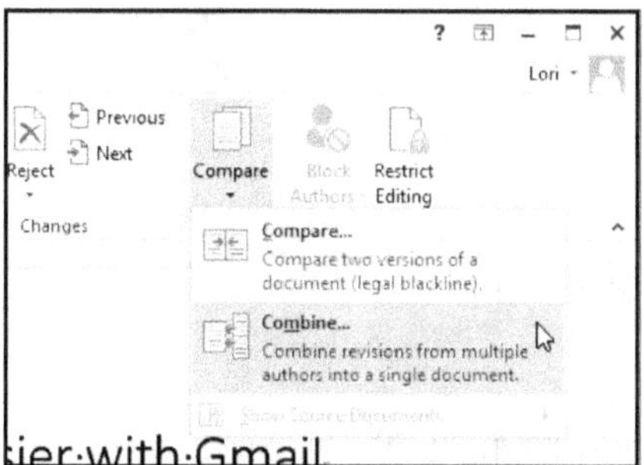

Fig. 2.142

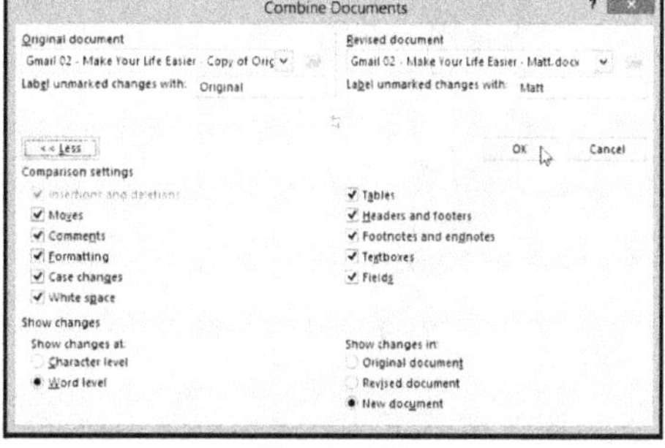

Fig. 2.143

- The "Combined Document" displays with all the changes tracked and marked with the specified labels and times.

Merge Two different Documents:

- There may be times when multiple users are working on different parts of a document and we need to combine those separate parts into one document. That is easily accomplished in Word.
- Open the main document into which we want to add the other documents and place the cursor at the point where we want to insert another file. Click the Insert tab, (See Fig. 2.144).
- In the "Text" section, click the down-arrow on the "Object" button and select "Text from File" from the drop-down list, (See Fig. 2.145).

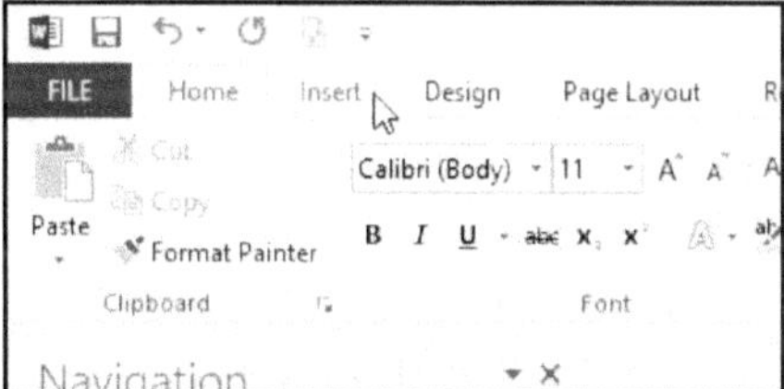

Fig. 2.144

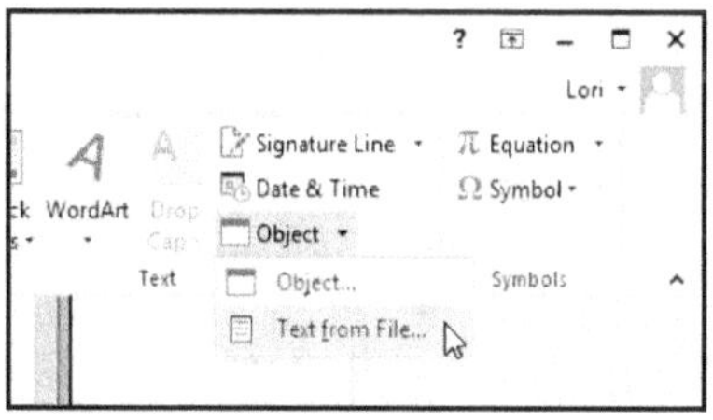

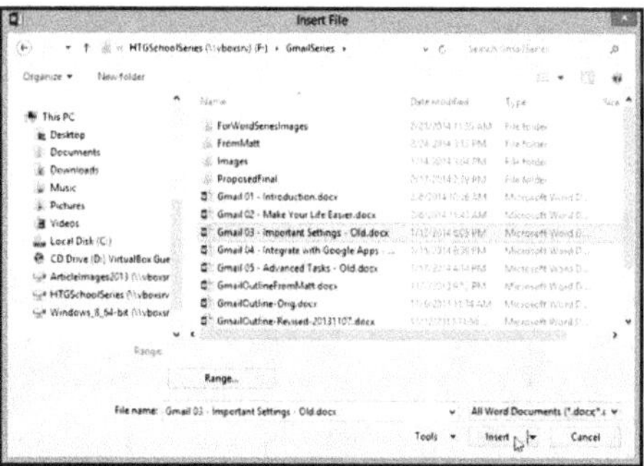

Fig. 2.145

- On the "Insert File" dialog box, navigate to the location of the file we want to insert, select the file, and click "Insert", (See Fig. 2.146).
- When we insert the contents of a file using the "Text from File" option, the styles of the current document are applied to the contents of the incoming file. If styles were used in the incoming file, the formatting of the inserted text will most likely change to the formatting of the current document. However, any information about paper size, orientation, margins, and other Page Layout settings will be discarded and replaced with the settings from the current document.

Fig. 2.146

Importing a File:

- We can import a file into Word; such as .txt files and PDFs.

To import a file:

    Step 1 : From the FILE menu, select Open, (See Fig. 2.147).

    Step 2 : Select where the file is located and from the Open dialog box, select the file and click Open.

Fig. 2.147

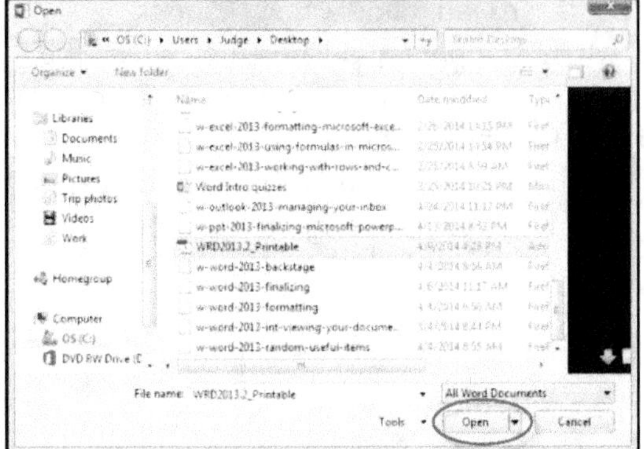

Fig. 2.148

    Step 3 : The file is imported into Word.

Charts:

- A chart is a tool we can use to communicate data graphically. Including a chart in the document can allow the reader to see the meaning behind the numbers, and it can make showing comparisons and trends easier.

Types of Charts:

1. **Column charts** use vertical bars to represent data. They can work with many different types of data, but they are most frequently used for comparing information.
2. **Line charts** are ideal for showing trends. The data points are connected with lines, making it easy to see whether values are increasing or decreasing over time.
3. **Pie charts** makes easy to compare proportions. Each value is shown as a slice of the pie, so it is easy to see which values make up the percentage of a whole.
4. **Bar charts** work just like column charts, but they use horizontal bars instead of vertical bars.
5. **Area charts** are similar to line charts, except the areas under the lines are filled in.
6. **Surface charts** allow us to display data across a 3D landscape. They work best with large data sets, allowing we to see a variety of information at the same time.

To insert a chart:

    Step 1 : Place the cursor in the location in the document in which we would like to insert the chart.

    Step 2 : From the INSERT tab, in the Illustrations group, click Chart, (See Fig. 2.149).

    Step 3 : In the Insert Chart dialog box, (See Fig. 2.150), select the type of chart we want by clicking on it in the left column.

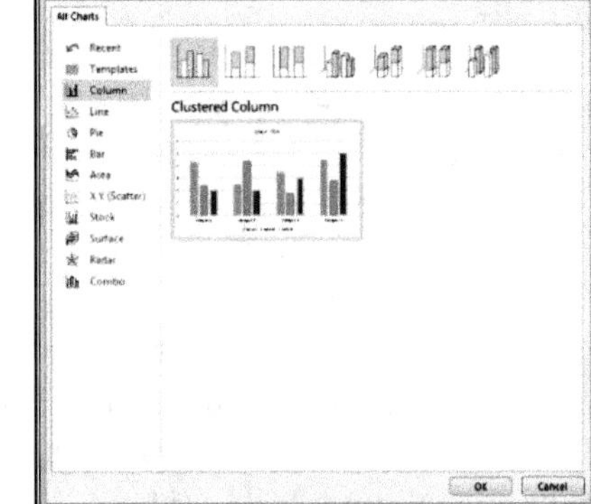

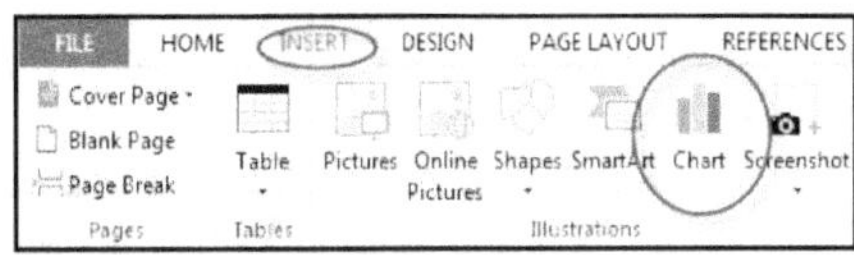

Fig. 2.149

Fig. 2.150

Step 4 :　Select a specific chart on the right and click OK, (See Fig. 2.151).

Step 5 :　Note that the chart now appears in the document and a Microsoft Excel spreadsheet opens up as shown in Fig. 2.152. For change chart type we use change chart type option as shown in Fig. 2.153.

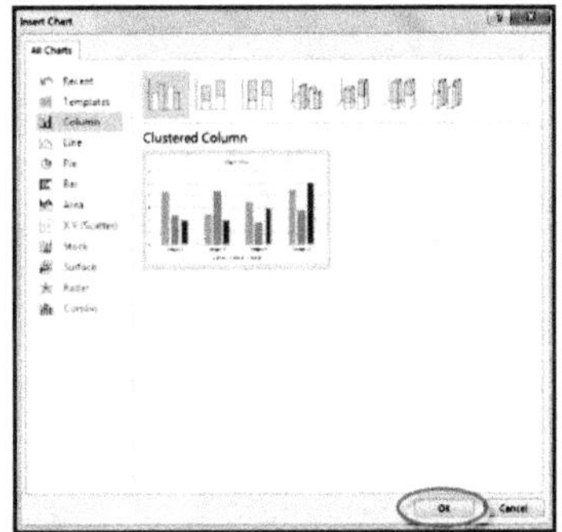

Fig. 2.151

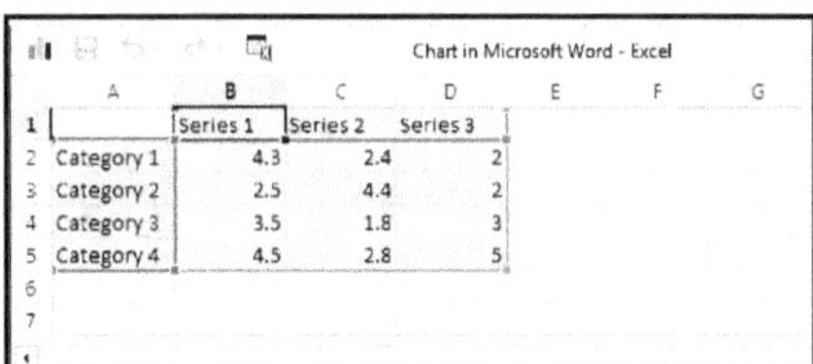

Fig. 2.152

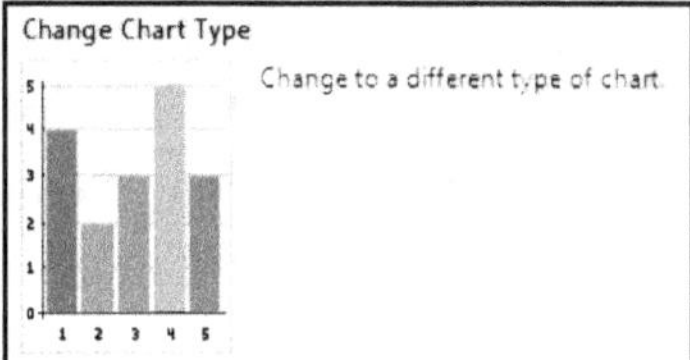

Fig. 2.153

2.3 MS EXCEL

- Excel is a commercial spreadsheet application written and distributed by Microsoft for Microsoft Windows OS and Mac OS.

- MS Excel is a spreadsheet program that enables us to store and manipulate data in a tabular (table) form.

- Microsoft Excel is a spreadsheet tool capable of performing calculations, analyzing data and integrating information from different programs.

- By default, documents saved in Excel 2013 are saved with the .xlsx (Excel 2007-2016) file extension whereas the file extension of the prior Excel versions (Excel 97-2003 Workbook) is .xls.

2.3.1 Starting MS Excel

- To start Microsoft Excel follow the following steps:

Step 1 :　Click the Start Button.

Step 2 :　Select All Programs.

Step 3 :　Select Microsoft Office 2013, (See Fig. 2.154).

Step 4 :　Select Excel 2013, (See Fig. 2.155).

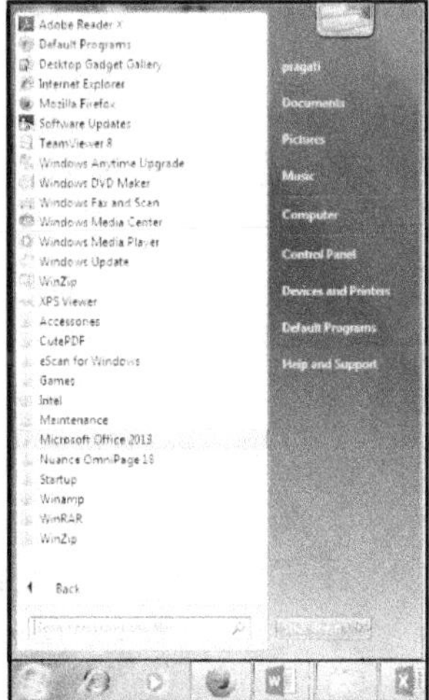

Fig. 2.154

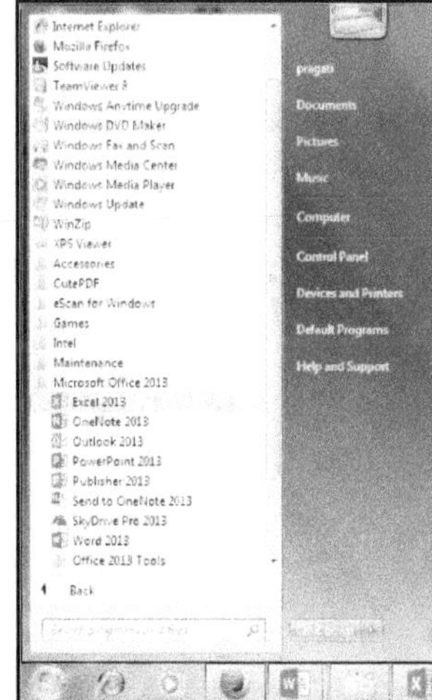

Fig. 2.155

- When to start Excel, the program displays the default startup screen, similar to the displayed in the Fig. 2.156. To create a new workbook, click the on Blank workbook. The first screen (window) we will see a new blank worksheet that contains grid of cells.

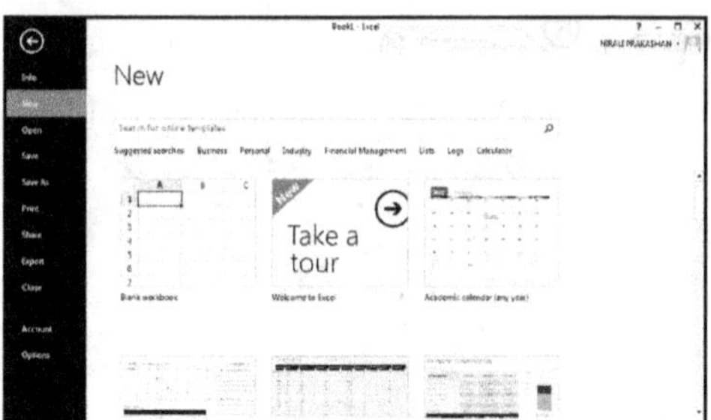

Fig. 2.156

2.3.2 Basics of Spreadsheets

- A spreadsheet is an interactive computer application for organization, analysis and storage of data in tabular form i.e., in table format.
- A spreadsheet is an application software consists of sheets having a matrix of rows and columns.
- The spreadsheet allows users to create tables and financial schedules by entering data and formulae into rows and columns arranged as a grid on a display screen.
- Spreadsheet software is also known as a spreadsheet program or spreadsheet application.
- Spreadsheet software is a powerful tool that helps in managing, manipulating and analyzing large set of numeric data.
- Spreadsheet module provides competency in creating and modifying workbooks; manipulating worksheet data by applying mathematical formulae and functions; restructuring and formatting worksheets; manipulating the display of worksheets; creating and using multiple worksheets; printing worksheets and workbooks; and creating comments, graphics, diagrams and charts.
- The spreadsheet software is used to maintain budget, financial statements, grade sheets, and sales records. Initially only used by accountants; now they are used by marketing professionals, students, teachers and financial analysts.
- Spreadsheets also called worksheets are made up of columns and rows. Rows are numbered numerically and columns are labeled alphabetically.
- The intersection of rows and columns is called a cell, which can store text, number data and mathematical formulas. MS Excel contains a group of worksheets, which is known as a workbook.

Uses of Spreadsheets:

- Spreadsheets are tools for anyone who needs to record, organize, or analyze numbers as rows and columns of data.
- Some typical uses of spreadsheets are given below:
  1. Spreadsheets are used for maintaining and analyzing inventory, payroll and other accounting records by accountants.
  2. Spreadsheets are used for recording grades of students and carrying out various types of analysis of grades by educators.
  3. Spreadsheets are used for tracking stocks and keeping records of investor accounts by stockbrokers.
  4. Spreadsheets are used for preparing budgets and bid comparisons by business analysts.
  5. Spreadsheets are used for creating and tracking personal budgets, loan payments, etc. by individuals.
  6. Spreadsheets are used for anlaysing experimental results by scientists and researchers.
- Most common spreadsheet programs are Lotus 1-2-3, Microsoft Excel, Corel Quattro Pro, Google Sheets and so on.
- Microsoft Excel 2013 is a most common and popular spreadsheet program that allows us to store, organize, and analyze information.

2.3.3 Screen Components of MS Excel

- From the Excel start screen, locate and select Blank workbook to access the Excel interface. Fig. 2.157 shows application Window (Screen) of Excel 2013.

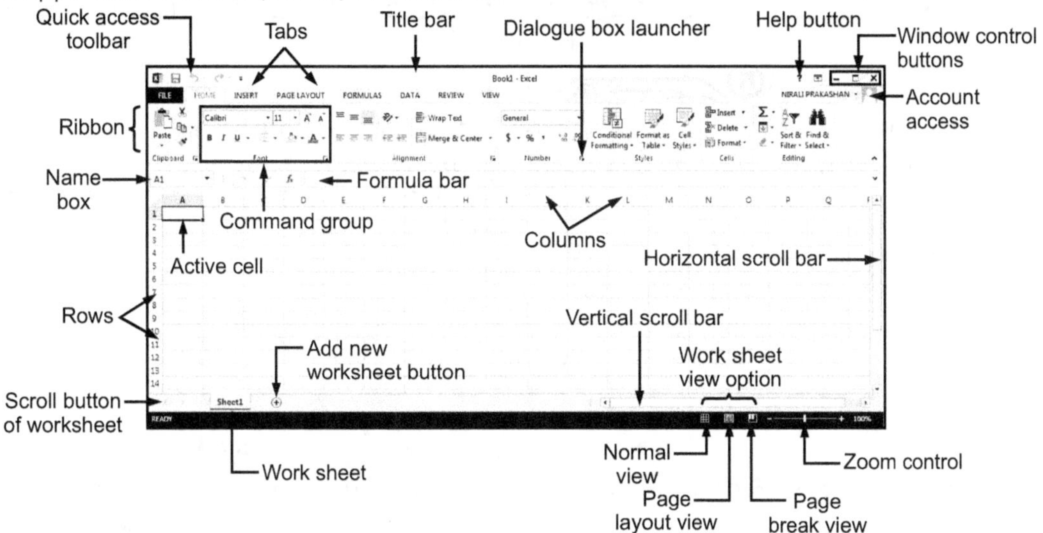

Fig. 2.157

- Fig. 2.157 shows following screen elements components of Excel 2013:

   1. **FILE Tab:** It replaces the Office button from Excel 2013. We can click it to check the Backstage view (See Fig. 2.156), where we come when we need to open or save files, create new sheets, print a sheet, and do other file-related operations.

   2. **Title Bar:** This lies in the middle and at the top of the window. Title bar shows the program and the sheet titles.

   3. **Help:** The Help icon can be used to get excel related help anytime we like. This provides nice explanation on various topics related to excel.

   4. **Quick Access Toolbar:** The Quick Access Toolbar lets us to access common commands no matter which tab is selected. By default, it includes the Save, Undo, and Redo commands. We can add other commands depending on our preference.

   5. **Ribbon:** The Ribbon contains all the commands are will need to perform common tasks in Excel. Microsoft Excel is a powerful program which is used to analyze and present data, perform calculations, and much more. To make it easier for users to find the specific commands they are looking for, commands are organized onto. The main tabs of Excel are explained below:

      (i) HOME: The HOME tab includes commands for formatting worksheets, cells and data and commands for inserting and deleting columns and rows.

      (ii) INSERT: Use the INSERT tab to insert tables, illustrations, charts, links, sparklines, headers and footers, custom text and symbols, and more.

      (iii) PAGE LAYOUT: Use the PAGE LAYOUT tab to change the margins, change the page background, change the page orientation, and more.

      (iv) FORMULAS: Use the FORMULAS tab to browse and select formulas and functions, to define names, to audit formulas, and more.

      (v) DATA: Use the DATA tab to access external data, to sort and filter, to access data tools, to group cells together, to add subtotals, and more.

      (vi) REVIEW: Use the REVIEW tab to check spelling, add comments, protect the worksheet or workbook, and more.

(vii) VIEW: Use the VIEW tab to change the workbook view, show or hide gridlines, headings, the formula bar and the ruler, arrange windows, freeze panes, zoom in or out, and more. course

Fig. 2.158

o In addition to the main tabs, there are numerous tool tabs which include less commonly used commands. Some of the most commonly used tool tabs are Smartart, Chart, Drawing, Picture, Pivottable, Pivotchart, Header and Footer.

o That they will appear when we select commands which have related tool tabs. For example, when we insert a PivotTable, two PivotTable-specific tool tabs (ANALYZE and DESIGN) will appear as shown in Fig. 2.159.

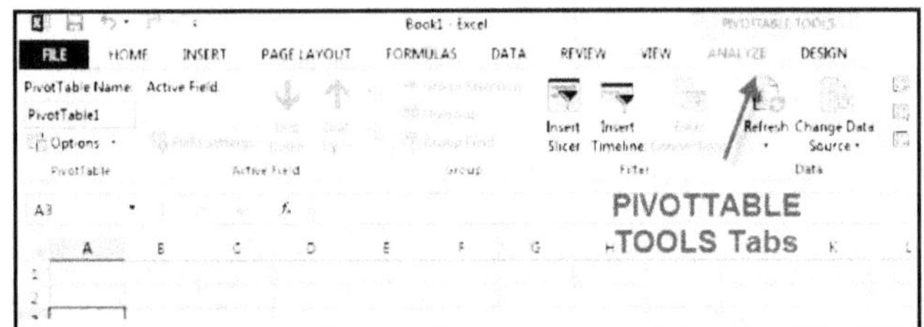

Fig. 2.159

o **Command Group:** Each group contains a series of different commands. Simply click any command to apply it. Some groups also have an arrow in the bottom-right corner, which you can click to see even more commands. To further organize the many commands available in Microsoft Excel, commands are organized in groups on each tab. Each group contains three or more related commands.

o **Commands** are controls that enable we to accomplish specific tasks, such as bolding a word, wrapping text, changing the format of a number to percent, or adding a column.

6. **Dialog Box Launcher:** This appears as a very small arrow in the lower-right corner of many groups on the Ribbon. Clicking this button opens a dialog box or task pane that provides more options about the group.

7. **Name Box:** The Name box displays the location or name of a selected cell.

8. **Row:** A row is a group of cells that runs from the left of the page to the right. In Excel, rows are identified by numbers. Row 1 is selected in the Fig. 2.157.

9. **Worksheets:** Excel files are called workbooks. Each workbook holds one or more worksheets (also known as spreadsheets).

10. **Worksheet View Options:** There are three ways to view a worksheet. Simply click to select the desired view:

    (i) Normal View is selected by default, and shows us, an unlimited number of cells and columns.

    (ii) Page Layout View divides the spreadsheet into pages.

    (iii) Page Break View lets us to see an overview of the worksheet, which is especially helpful when adding page breaks.

11. **Zoom Control:** Click and drag the slider to use the Zoom control. The number to the right of the slider reflects the zoom percentage.

12. **Vertical and Horizontal Scroll Bars:** The spreadsheet may frequently have more data than we can see on the screen at once. Click, hold and drag the vertical or horizontal scroll bar depending on what part of the page we want to see.

13. **Column:** A column is a group of cells that runs from the top of the page to the bottom. In Excel, columns are identified by letters. Column A is selected in the Fig. 2.157.

14. **Formula Bar:** In the formula bar, we can enter or edit data, a formula, or a function that will appear in a specific cell.

15. **Workspace:** The working area is divided into cells, rows and columns.

2.3.4 Elementary Working with Excel

Cell Basics:

- Cells are the basic building blocks of a worksheet. Every worksheet is made up of thousands of rectangles, which are called cells.

- A cell is the intersection of a row and a column. Columns are identified by letters (A, B, C), while rows are identified by numbers (1, 2, 3) as shown in Fig. 2.160.

- Each cell has its own name or cell address based on its column and row as shown in Fig. 2.161. In this example, the selected cell intersects column C and row 5, so the cell address is C5. The cell address will also appear in the Name box.

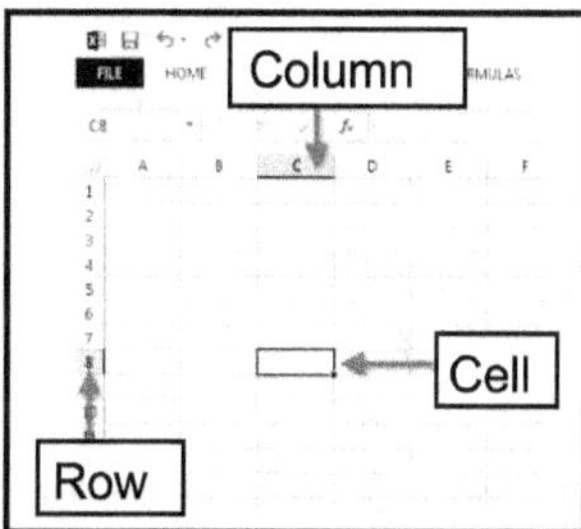

Fig. 2.160

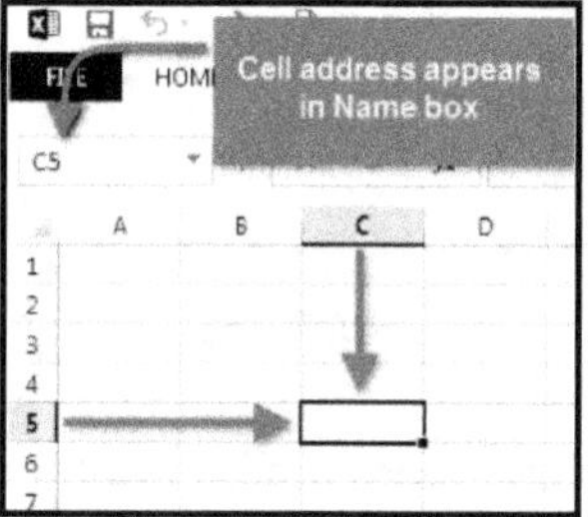

Fig. 2.161

- We can also select multiple cells at the same time. A group of cells is known as a cell range.

To Select a Cell:

- To input or edit cell content, we will first need to select the cell.

    Step 1 : Click a cell to select it.

    Step 2 : A border will appear around the selected cell, and the column heading and row heading will be highlighted. The cell will remain selected until we click another cell in the worksheet.

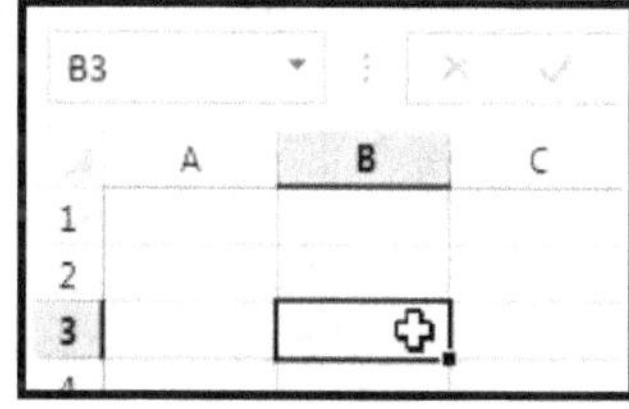

Fig. 2.162

- We can also select cells using the arrow keys on the keyboard.

Cell Content:

- Any information we enter into a spreadsheet will be stored in a cell. Each cell can contain different types of content, as given below:

    1. **Text:** Cells can contain text, such as letters, numbers, and dates.

    2. **Formatting Attributes:** Cells can contain formatting attributes that change the way letters, numbers, and dates are displayed.

    3. **Formulas and Functions**: Cells can contain formulas and functions that calculate cell values.

    To Insert Content:

    Step 1 : Click a cell to select it, (See Fig. 2.163).

    Step 2 : Type content into the selected cell, then press Enter on the keyboard. The content will appear in the cell and the formula bar, (See Fig. 2.164). We can also input and edit cell content in the formula bar.

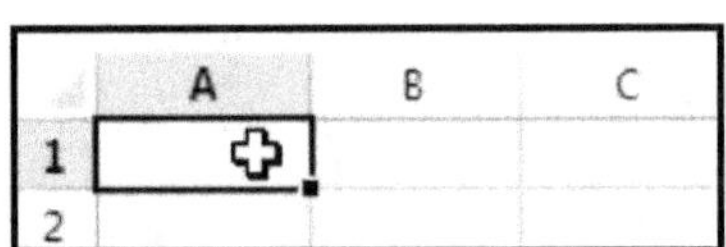

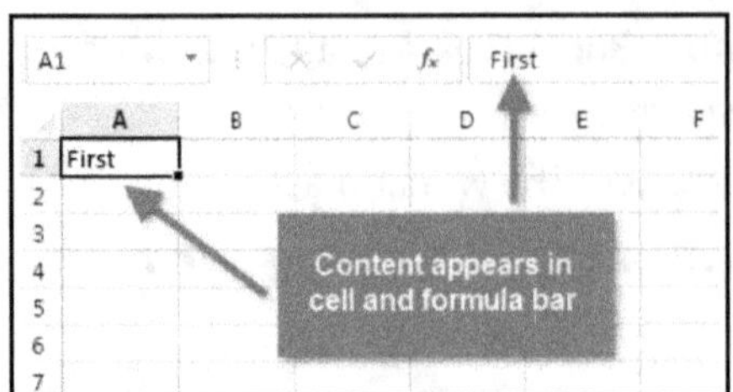

Fig. 2.163   **Fig. 2.164**

To Delete Cell Content:

Step 1 : Select the cell with content we want to delete, (See Fig. 2.165).

Step 2 : Press the Delete or Backspace key on the keyboard. The cell's contents will be deleted as shown in Fig. 2.166.

	A	B	C
1			
2	First Name	Middle Name	Last Name
3	Heidi	Lauren	Lee
4	Josie	Marie	Gates
5	Wendy	Anne	Crocker
6	Loretta	Susan	Johnson

	A	B	C
1			
2	First Name	Middle Name	Last Name
3	Heidi		Lee
4	Josie	Marie	Gates
5	Wendy	Anne	Crocker
6	Loretta	Susan	Johnson

Fig. 2.165   **Fig. 2.166**

Formatting Cells:

- All cell content uses the same formatting by default, which can make it difficult to read a workbook with a lot of information.
- Basic formatting can customize the look and feel of the workbook, allowing we to draw attention to specific sections and making the content simpler and easier to view and understand.

To Change the Font:

- By default, the font of each new workbook is set to Calibri. However, Excel provides many other fonts that can use to customize the cell text.
- To change font in Excel follow the following steps:

Step 1 : Select the cell(s) we want to modify, (See Fig. 2.167).

Step 2 : Click the drop-down arrow next to the Font command on the HOME tab. The Font drop-down menu will appear, (See Fig. 2.168).

Step 3 : Select the desired font. A live preview of the new font will appear as we hover the mouse over different options. In our example, we will choose Georgia.

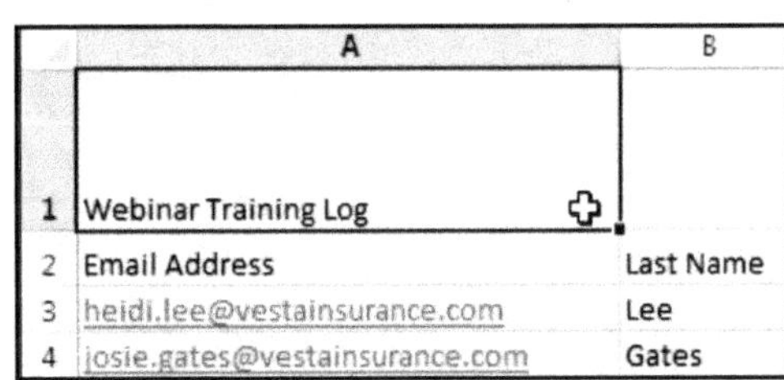

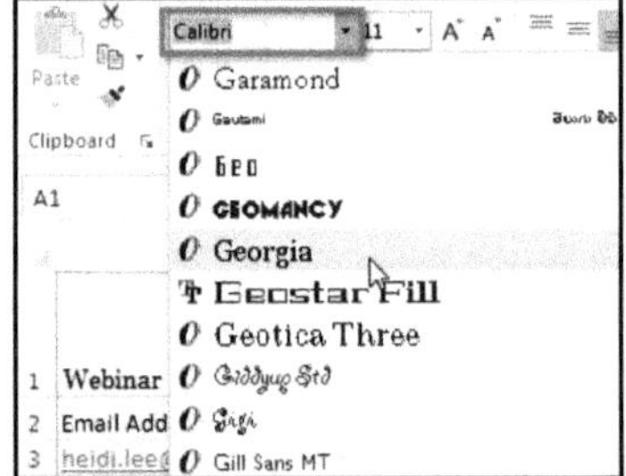

Fig. 2.167   **Fig. 2.168**

Step 4 : The text will change to the selected font.

To Change the Font Size:

Step 1 : Select the cell(s) we want to modify, (See Fig. 2.169).

Step 2 : Click the drop-down arrow next to the Font Size command on the HOME tab. The Font Size drop-down menu will appear, (See Fig. 2.170).

Step 3 : Select the desired font size. A live preview of the new font size will appear as we hover the mouse over different options. In this example, we will choose 16 to make the text larger.

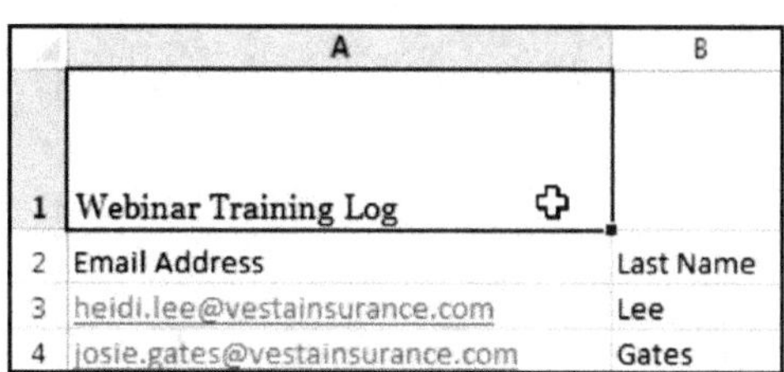

Fig. 2.169

Fig. 2.170

Step 4 :   The text will change to the selected font size.

To Change the Font Color:

- Follow the following steps for change the font color.

    Step 1 :   Select the cell(s) we want to modify, (See Fig. 2.171).

    Step 2 :   Click the drop-down arrow next to the Font Color command on the HOME tab. The Color menu will appear, (See Fig. 2.172).

    Step 3 :   Select the desired font color. A live preview of the new font color will appear as we hover the mouse over different options. In our example, we will choose Green.

Fig. 2.171

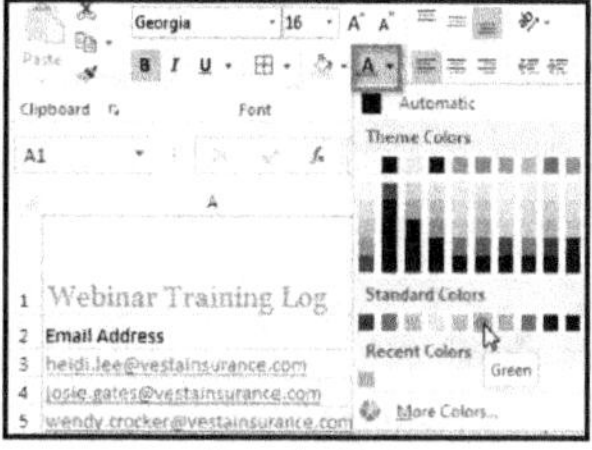

Fig. 2.172

Step 4 :   The text will change to the selected font color.

To Use the Bold, Italic, and Underline Commands:

Step 1 :   Select the cell(s) we want to modify, (See Fig. 2.171).

Step 2 :   Click the Bold (B), Italic (I), or Underline (U) command on the Home tab. In this example, we will make the selected cells bold, (See Fig. 2.173).

Step 3 :   The selected style will be applied to the text, (See Fig. 2.174).

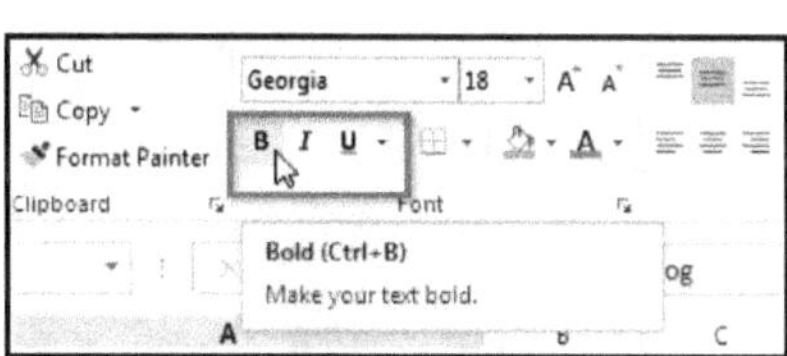

Fig. 2.173

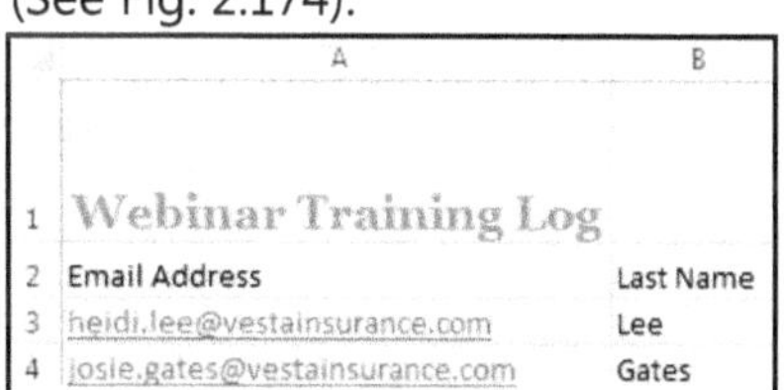

Fig. 2.174

Format Cells using Dialog Box:

- In Excel, we can also apply specific formatting to a cell. To apply formatting to a cell or group of cells as shown in Fig. 2.175.

    Step 1 :   Select the cell or cells that will have the formatting.

    Step 2 :   Click the Dialog Box arrow on the Alignment group of the HOME tab.

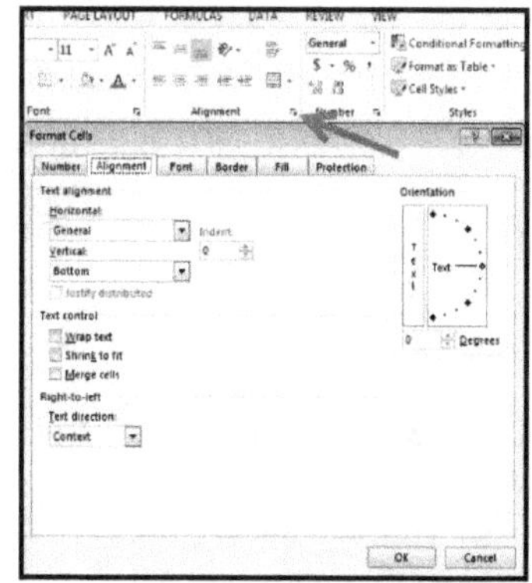

Fig. 2.175

- There are several tabs on this dialog box that allow us to modify properties of the cell or cells. These are:

  1. **Number**: Allows for the display of different number types and decimal places.

  2. **Alignment**: Allows for the horizontal and vertical alignment of text, wrap text, shrink text, merge cells and the direction of the text.

  3. **Font**: Allows for control of font, font style, size, color, and additional features.

  4. **Border**: Border styles and colors.

  5. **Fill**: Cell fill colors and styles.

To Add Borders and Colors to Cells:

- Borders and colors can be added to cells manually or through the use of styles.

- To add borders manually use following steps:

  Step 1 : Click the Borders drop down menu on the Font group of the HOME tab.

  Step 2 : Choose the appropriate border and then click.

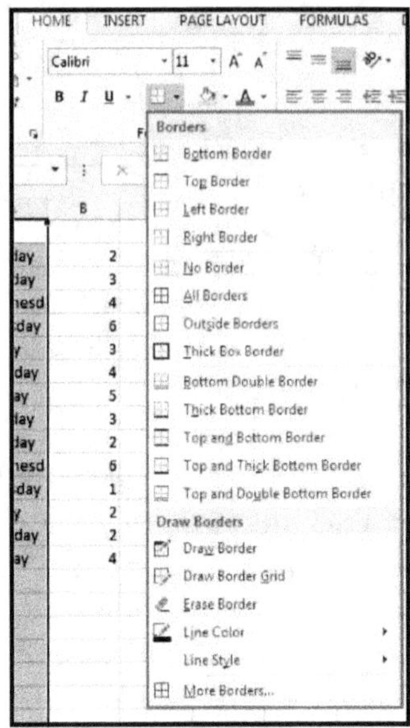

Fig. 2.176

Copy and Paste:

To copy and paste cell content:

- Excel allows us to copy content that is already entered into the spreadsheet and paste that content to other cells, which can save we time and effort.

  Step 1 : Select the cell(s) we want to copy, (See Fig. 2.177).

  Step 2 : Click the Copy command on the HOME tab, (See Fig. 2.178) or press Ctrl+C on the keyboard.

Fig. 2.177

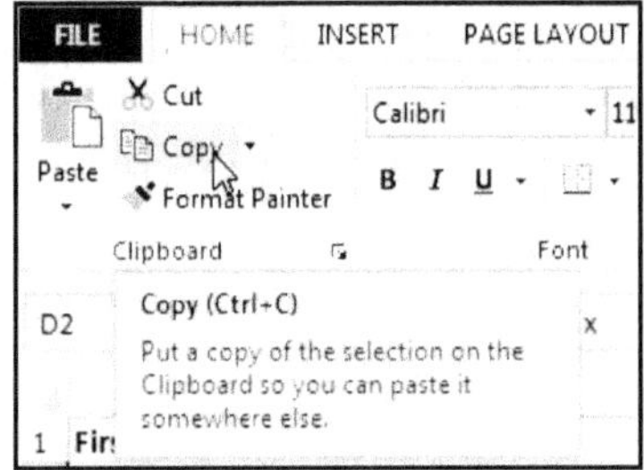

Fig. 2.178

  Step 3 : Select the cell(s) where we want to paste the content. The copied cells will now have a dashed box around them as shown in Fig. 2.179.

  Step 4 : Click the Paste command on the HOME tab (See Fig. 2.180) or press Ctrl+V on the keyboard.

  Step 5 : The content will be pasted into the selected cells.

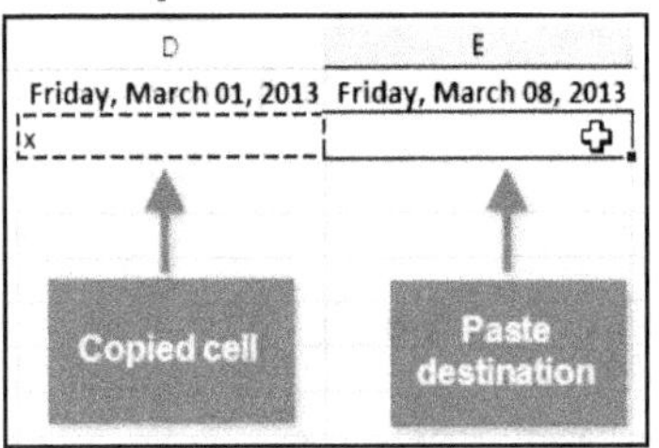

Fig. 2.179

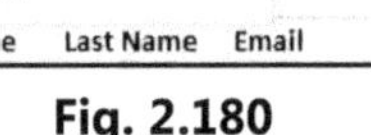

Fig. 2.180

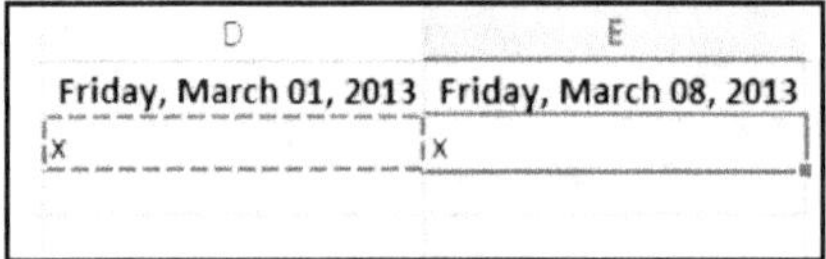

Fig. 2.181

To cut and paste the cell content:

- Unlike copying and pasting, which duplicates cell content, cutting allows us to move content between cells.

　　Step 1 : Select the cell(s) we want to cut, (See Fig. 2.182).

　　Step 2 : Click the Cut command on the HOME tab, (See Fig. 2.183) or press Ctrl+X on the keyboard.

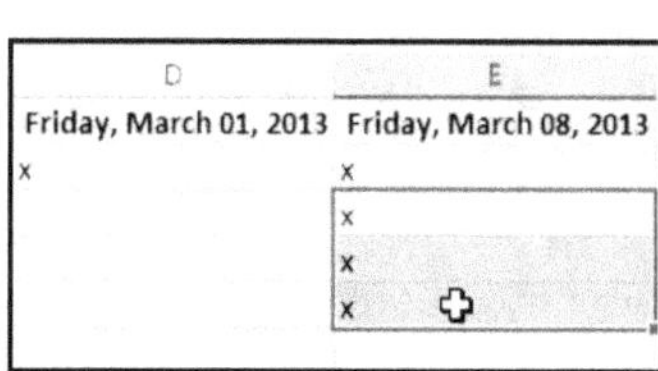

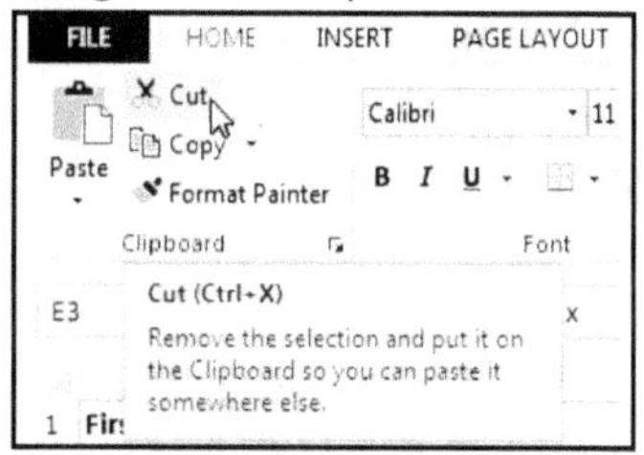

Fig. 2.182　　　　　　　　**Fig. 2.183**

　　Step 3 : Select the cells where we want to paste the content. The cut cells will now have a dashed box around them, (See Fig. 2.184).

　　Step 4 : Click the Paste command on the Home tab, (See Fig. 2.185). or press Ctrl+V on the keyboard.

　　Step 5 : The cut content will be removed from the original cells and pasted into the selected cells.

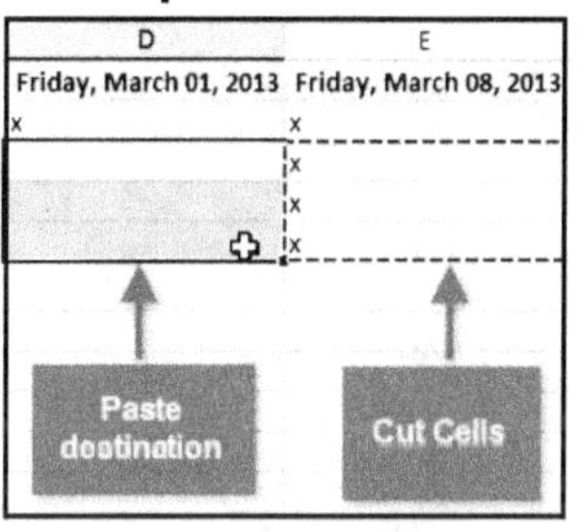

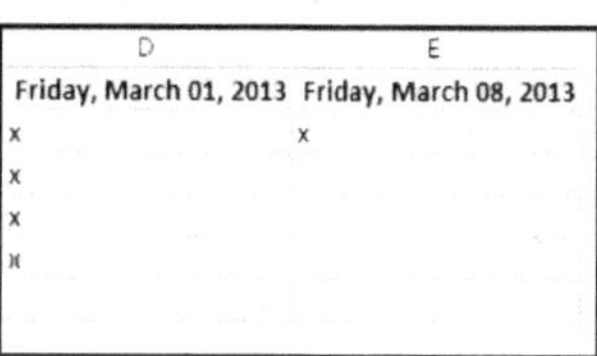

Fig. 2.184　　　　　　　**Fig. 2.185**　　　　　　　**Fig. 2.186**

Modifying Columns and Rows:

To modify column width:

- In the example below, some of the content in column A cannot be displayed. We can make all of this content visible by changing the width of column A.

　　Step 1 : Position the mouse over the column line in the column heading so the white cross becomes a double arrow, (See Fig. 2.187).

　　Step 2 : Click, hold, and drag the mouse to increase or decrease the column width, (See Fig. 2.188).

　　Step 3 : Release the mouse. The column width will be changed, (See Fig. 2.189).

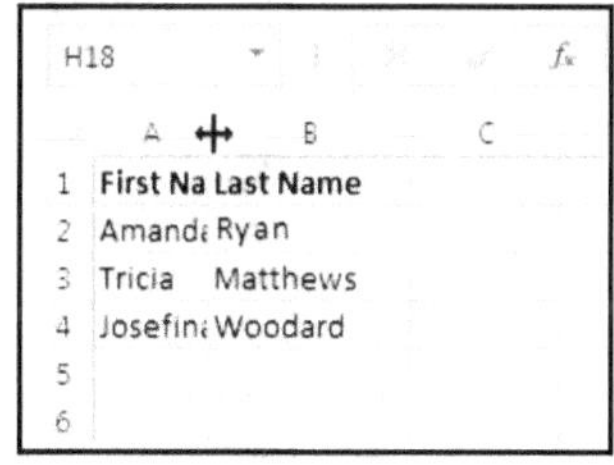

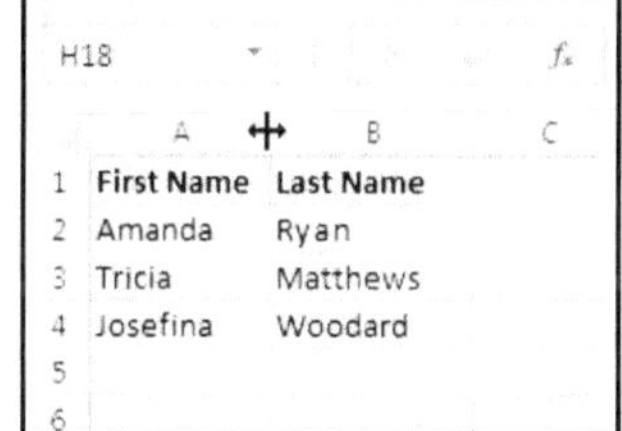

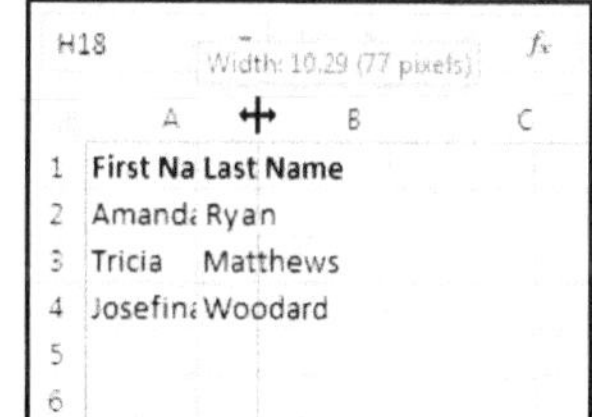

Fig. 2.187　　　　　　　**Fig. 2.188**　　　　　　　**Fig. 2.189**

To modify row height:

　　Step 1 : Position the cursor over the row line so the white cross becomes a double arrow, (See Fig. 2.190).

　　Step 2 : Click, hold, and drag the mouse to increase or decrease the row height, (See Fig. 2.191).

　　Step 3 : Release the mouse. The height of the selected row will be changed, (See Fig. 2.192).

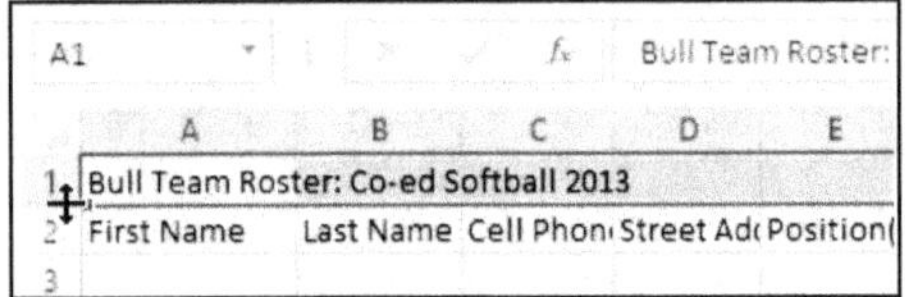

Fig. 2.190

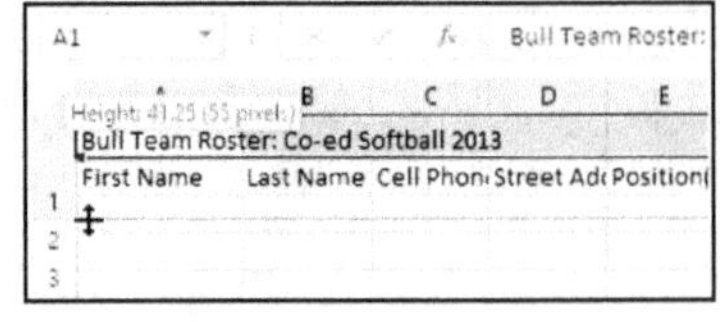

Fig. 2.191

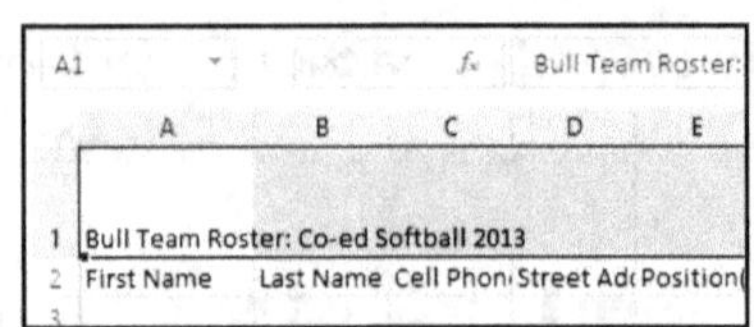

Fig. 2.192

Inserting Rows and Columns:

- After we have been working with a workbook for a while, we may find that we want to insert new columns or rows, delete certain rows or columns, move them to a different location in the worksheet, or even hide them.

To insert rows:

Step 1 :   Select the row heading below where we want the new row to appear. For example, if we want to insert a row between rows 7 and 8, select row 8, (See Fig. 2.193).

Step 2 :   Click the Insert command on the Home tab, (See Fig. 2.194).

Step 3 :   The new row will appear above the selected row, (See Fig. 2.195).

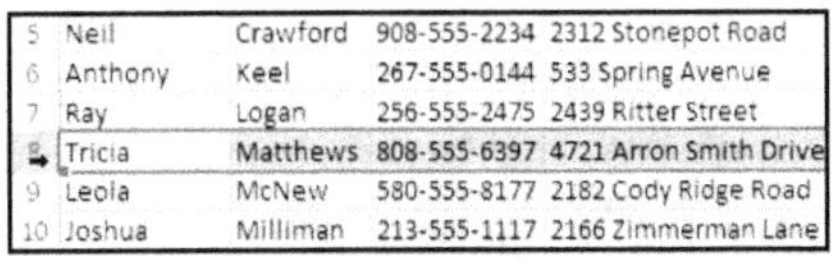

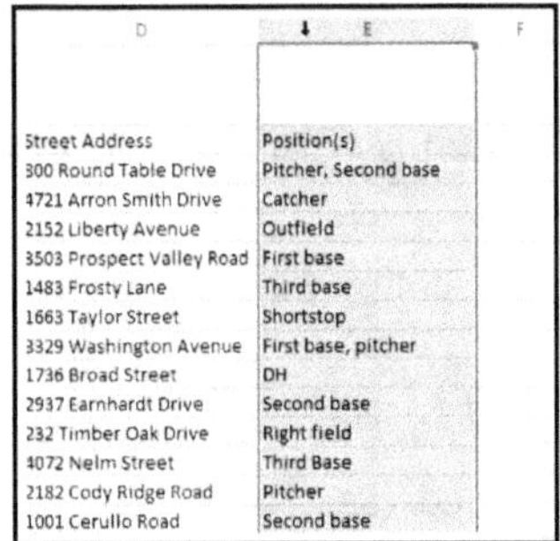

Fig. 2.193

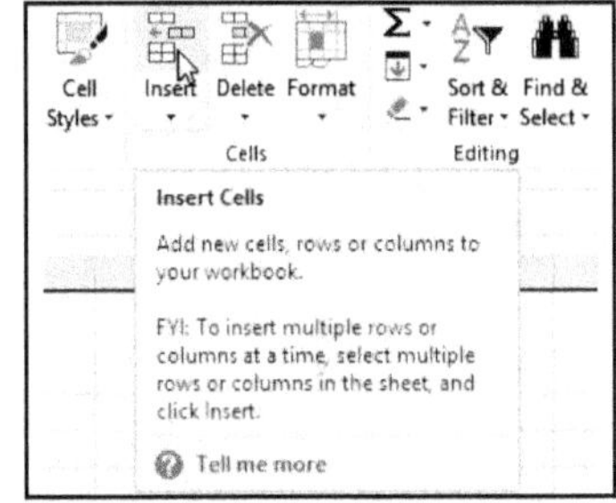

Fig. 2.194

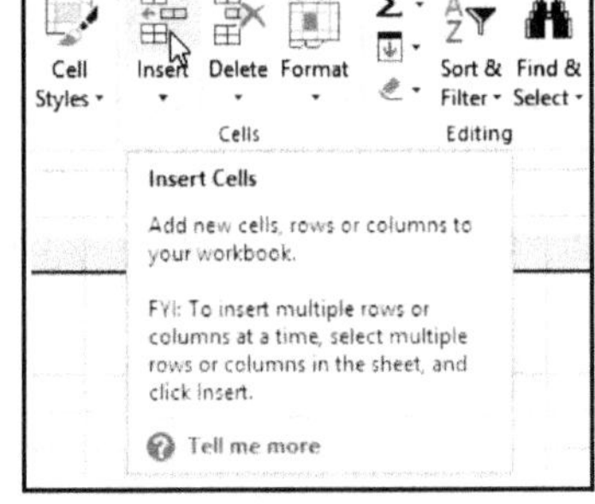

Fig. 2.195

To insert columns:

Step 1 :   Select the column heading to the right of where we want the new column to appear. For example, if we want to insert a column between columns D and E, select column E.

Step 2 :   Click the Insert command on the HOME tab, (See Fig. 2.197).

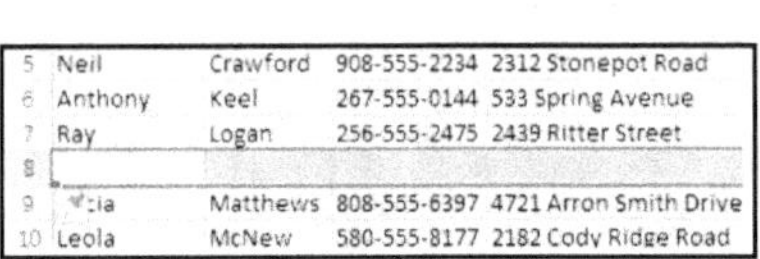

Fig. 2.196

Fig. 2.197

Step 3 :   The new column will appear to the left of the selected column.

Fig. 2.198

To delete rows:

- It is easy to delete any row that we no longer need in the workbook. For this select HOME tab and select cells groups (See Fig. 2.199).

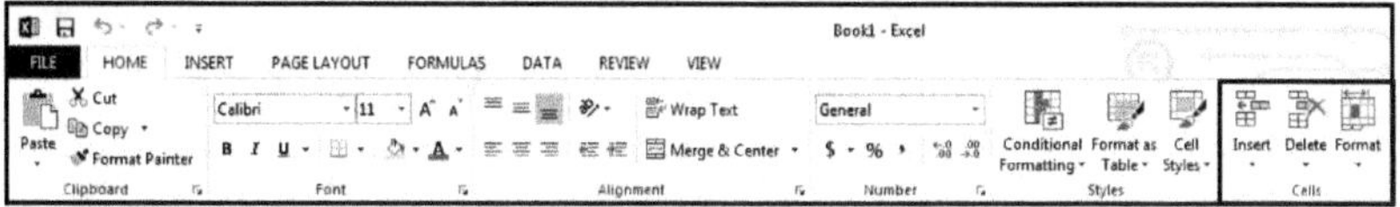

Fig. 2.199

Step 1 :   Select the row(s) we want to delete. In this example, we will select rows 6-8, (See Fig. 2.200 (a)).

Step 2 :   Click the Delete command on the HOME tab. (See Fig. 2.200 (b)).

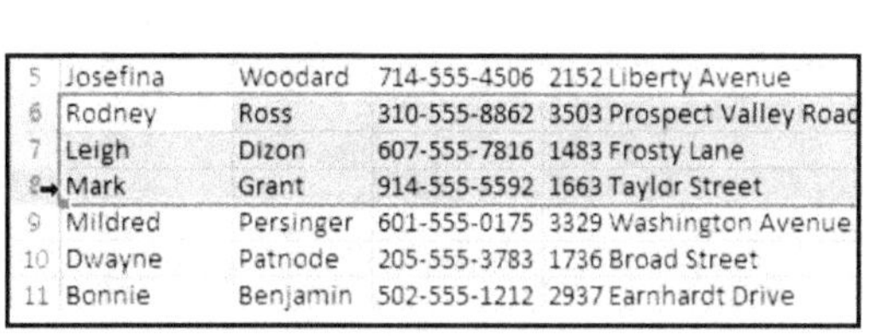

(a)

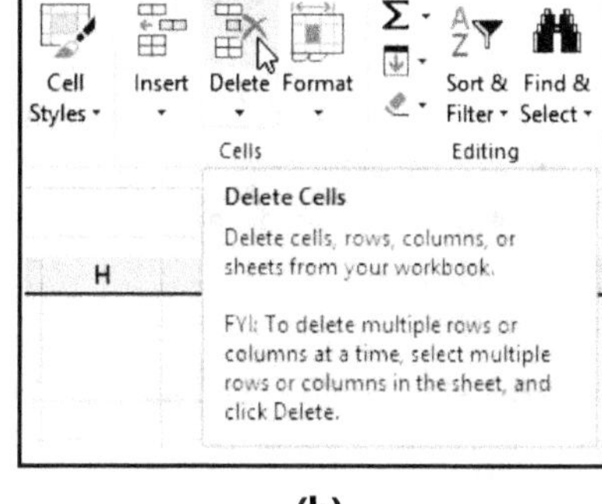

(b)

Fig. 2.200

Step 3 :   The selected row(s) will be deleted, and the rows below will shift up. In this example, rows 9-11 are now rows 6-8.

5	Josefina	Woodard	714-555-4506	2152 Liberty Avenue
6	Mildred	Persinger	601-555-0175	3329 Washington Avenue
7	Dwayne	Patnode	205-555-3783	1736 Broad Street
8	Bonnie	Benjamin	502-555-1212	2937 Earnhardt Drive
9	Eva	Ramer	805-555-8514	232 Timber Oak Drive
10	Carol	Pena	571-555-0704	4072 Nelm Street
11	Leola	McNew	580-555-8177	2182 Cody Ridge Road

Fig. 2.201

To delete columns:

Step 1 :   Select the columns(s) we want to delete. In this example, we will select column E, (See Fig. 2.202).

Step 2 :   Click the Delete command on the HOME tab, (See Fig. 2.203).

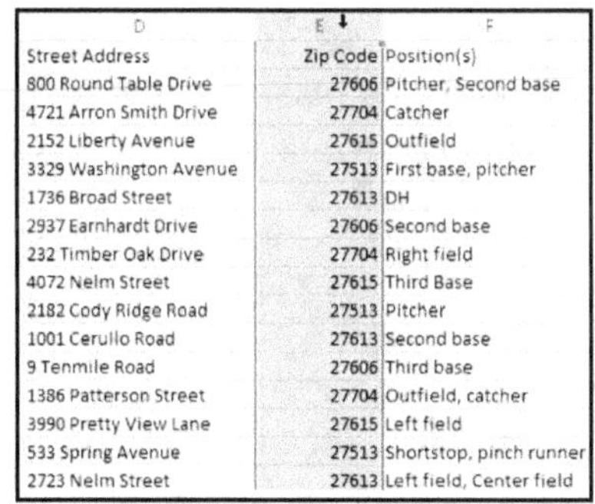

Fig. 2.202

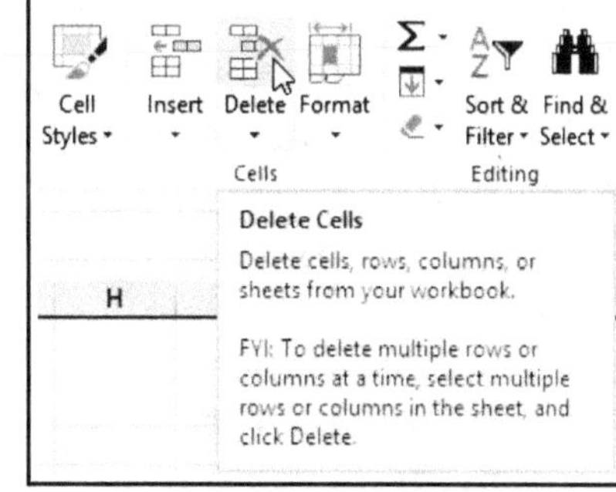

Fig. 2.203

Step 3 :   The selected columns(s) will be deleted, and the columns to the right will shift left. In this example, Column F is now Column E.

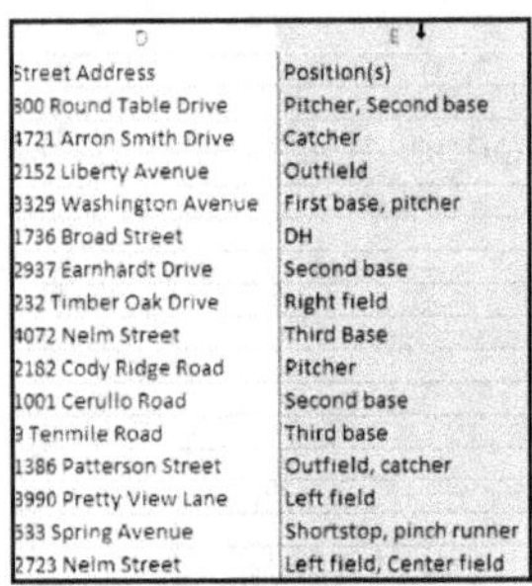

Fig. 2.204

To Merge Cells:

- To merge cells select the cells we want to merge and click the Merge & Center button on the Alignment group of the Home tab, (See Fig. 2.205).
- The four choices for merging cells are:

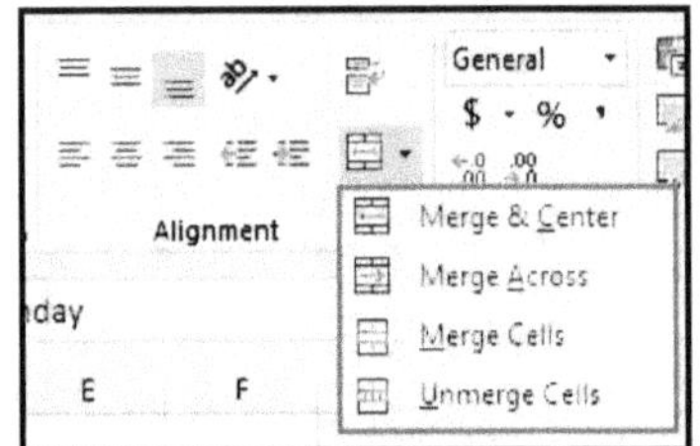

  1. **Merge and Center**: Combines the cells and centers the contents in the new, larger cell.
  2. **Merge Across**: Combines the cells across columns without centering data.
  3. **Merge Cells**: Combines the cells in a range without centering.
  4. **Unmerge Cells**: Splits the cell that has been merged.

Fig. 2.205

Align Cell Contents:

- To align cell contents, click the cell or cells we want to align and click on the options within the Alignment group on the HOME tab, (See Fig. 2.206).

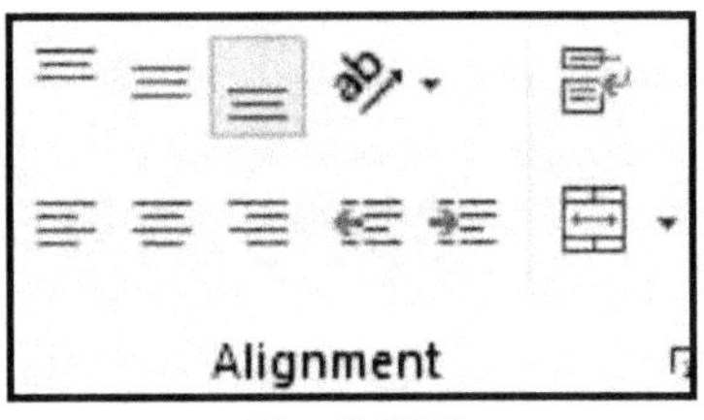

- There are several options for alignment of cell contents:
  1. **Top Align**: Aligns text to the top of the cell.
  2. **Middle Align**: Aligns text between the top and bottom of the cell.

Fig. 2.206

  3. **Bottom Align**: Aligns text to the bottom of the cell.
  4. **Align Text Left**: Aligns text to the left of the cell.
  5. **Center**: Centers the text from left to right in the cell.
  6. **Align Text Right**: Aligns text to the right of the cell.
  7. **Decrease Indent**: Decreases the indent between the left border and the text.
  8. **Increase Indent**: Increase the indent between the left border and the text.
  9. **Orientation**: Rotate the text diagonally or vertically.

Working with Worksheet:

- Every workbook contains at least one worksheet by default. When working with a large amount of data, we can create multiple worksheets to help organize the workbook and make it easier to find content.
- We can also group worksheets to quickly add information to multiple worksheets at the same time.

To Rename a Worksheet:

- Whenever, we create a new Excel workbook, it will contain one worksheet named Sheet1. We can rename a worksheet to better reflect its content.
- In this example, we will create a training log organized by month.

 Step 1 : Right-click the worksheet we want to rename, then select Rename from the worksheet menu, (See Fig. 2.207).

 Step 2 : Type the desired name for the worksheet, (See Fig. 2.208).

 Step 3 : Click anywhere outside of the worksheet, or press Enter on the keyboard. The worksheet will be renamed, (See Fig. 2.209).

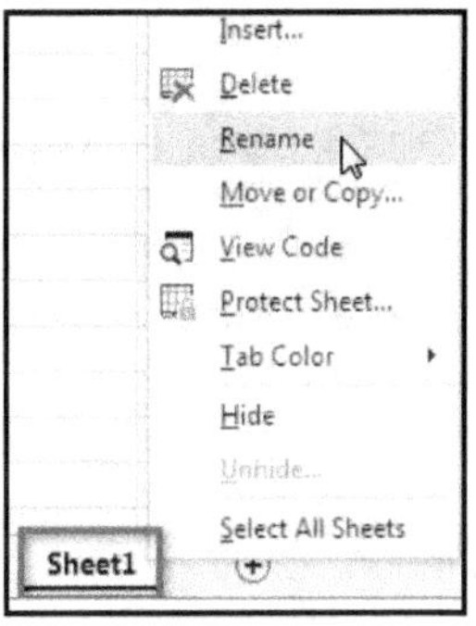

Fig. 2.207

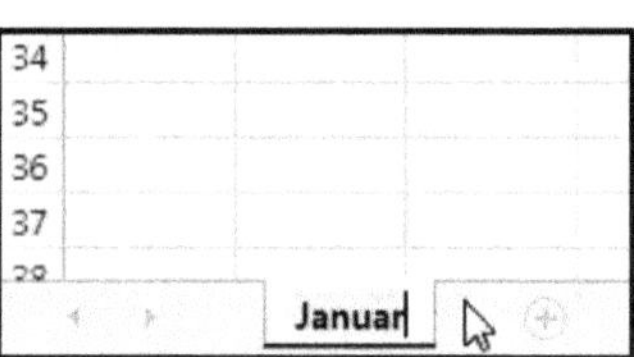

Fig. 2.208

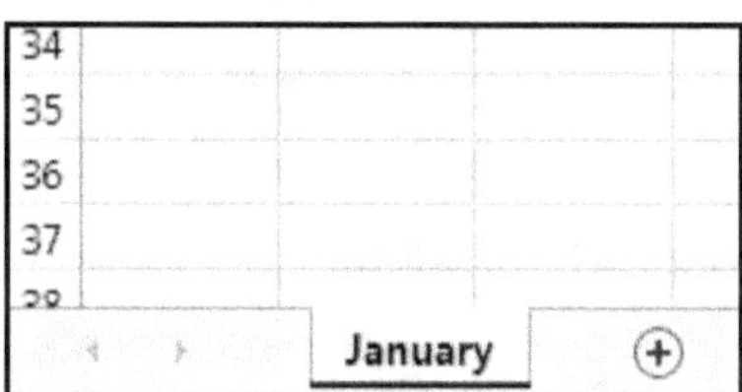

Fig. 2.209

To Insert a New Worksheet:

- Follow the following steps for inserting a new worksheet:

	Step 1 : Locate and select the New sheet button, (See Fig. 2.210).

	Step 2 : A new blank worksheet will appear, (See Fig. 2.211).

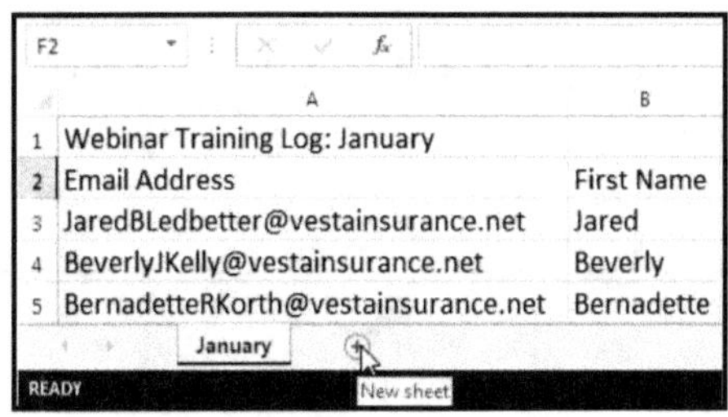

Fig. 2.210

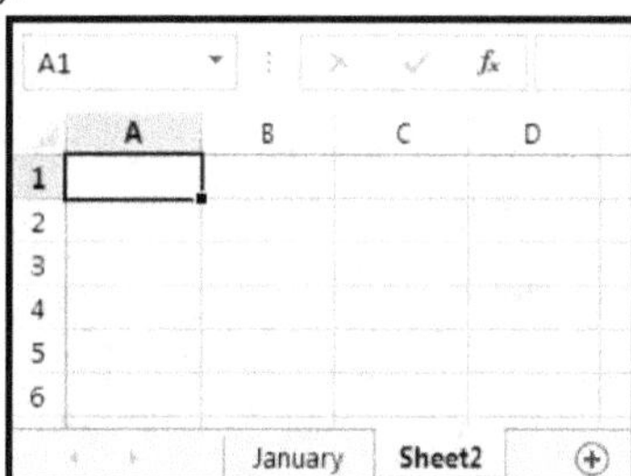

Fig. 2.211

To Delete a Worksheet:

	Step 1 : Right-click the worksheet we want to delete, then select Delete from the worksheet menu, (See Fig. 2.212).

	Step 2 : The worksheet will be deleted from the workbook, (See Fig. 2.213).

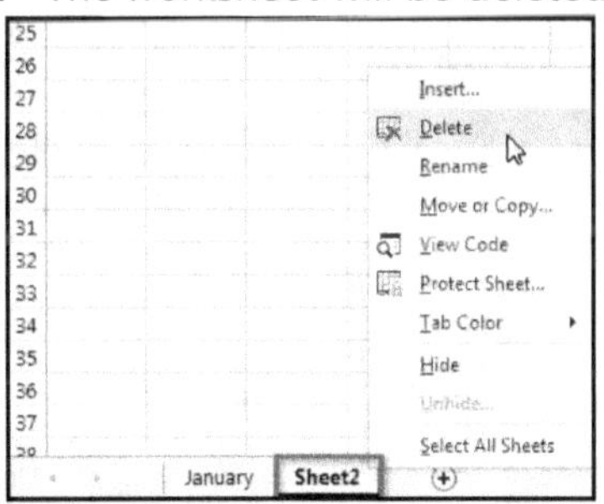

Fig. 2.212

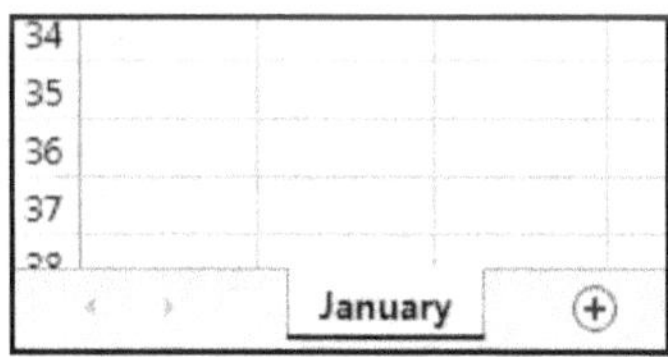

Fig. 2.213

Insert or Delete Rows and Columns in Excel:

- To insert a single row, select either the whole row or a cell in the row above which we want to insert the new row, (See Fig. 2.214).
- To insert a single column, select the column or a cell in the column immediately to the right of where we want to insert the new column, (See Fig. 2.215).

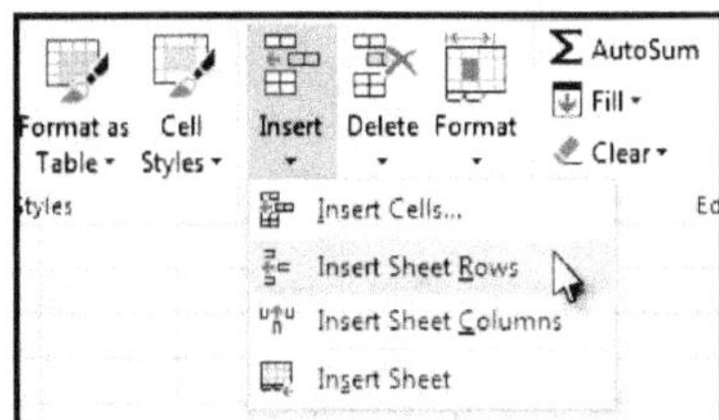

Fig. 2.214

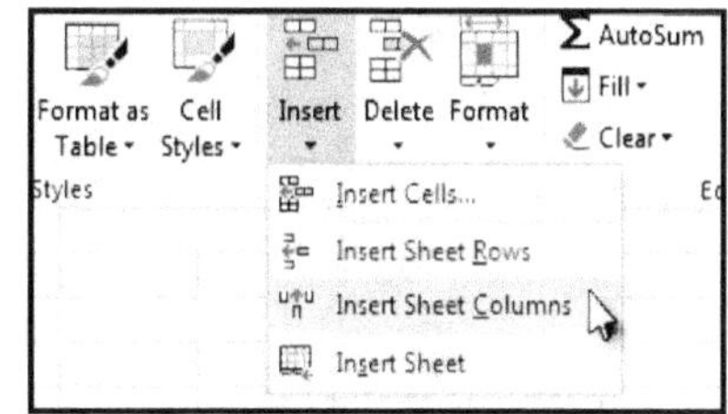

Fig. 2.215

Page Layout in Excel:

- Many of the commands we will use to prepare the workbook for printing and PDF export can be found on the PAGE LAYOUT tab.

- These commands let us to control the way the content will appear on a printed page, including the page orientation and margin size.

Page Orientation:

- Excel offers two page orientation options i.e., Landscape and Portrait as shown in Fig. 2.216.

- Landscape orients the page horizontally, while portrait orients the page vertically. Portrait is especially helpful for worksheets with a lot of rows, while landscape is best for worksheets with a lot of columns.

To Change Page Orientation:

Step 1 : Click the PAGE LAYOUT tab on the Ribbon.

Step 2 : Select the Orientation command, then choose either Portrait or Landscape from the drop-down menu.

Step 3 : The page orientation of the workbook will be changed.

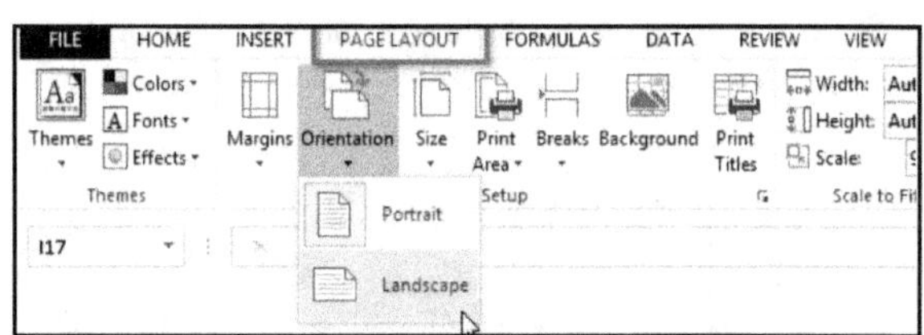

Fig. 2.216

To Format Page Margins:

- A margin is the space between the content and the edge of the page. By default, every workbook's margins are set to Normal, which is a one-inch space between the content and each edge of the page.

- Sometimes we may need to adjust the margins to make the data fit more comfortably on the page.

- Excel includes a variety of predefined margin sizes. To change page margins follow the following steps:

Step 1 : Click the PAGE LAYOUT tab on the Ribbon, then select the Margins command.

Step 2 : Select the desired margin size from the drop-down menu. In this example, we will select Narrow to fit more of this content on the page.

Step 3 : The margins will be changed to the selected size.

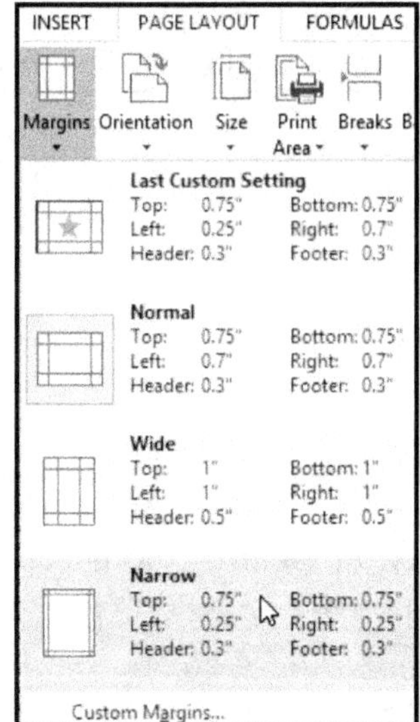

Fig. 2.217

To Use Custom Margins:

- Excel also allows us to customize the size of the margins in the Page Setup dialog box.

Step 1 : From the PAGE LAYOUT tab, click Margins. Select Custom Margins... from the drop-down menu, (See Fig. 2.218).

Step 2 : The Page Setup dialog box will appear, (See Fig. 2.219).

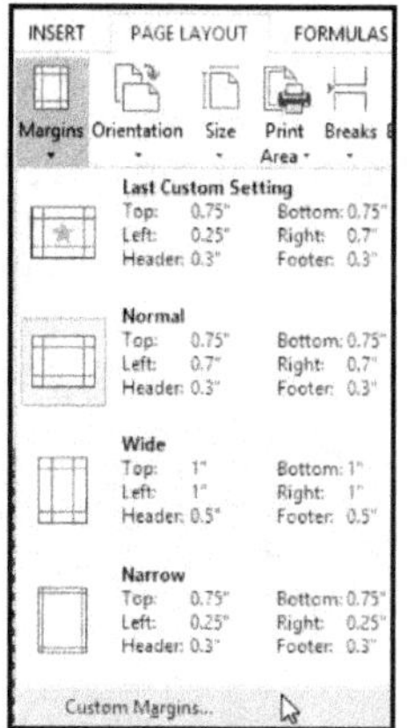

Fig. 2.218

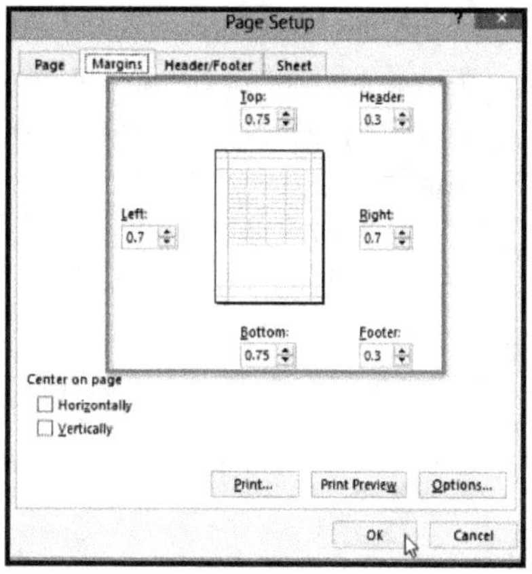

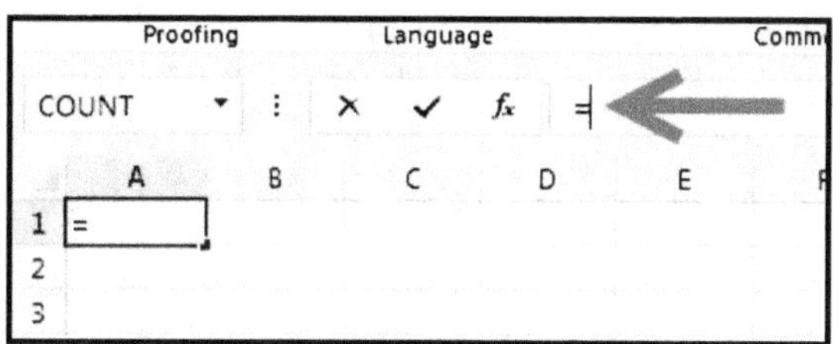

Fig. 2.219

Step 3 : Adjust the values for each margin, then click OK.

Step 4 : The margins of the workbook will be changed.

Formulas in Excel:

- One of the most powerful features in Excel is the ability to calculate numerical information using formulas. Just like a calculator, Excel can add, subtract, multiply, and divide.

- A formula is a set of mathematical instructions that can be used in Excel to perform calculations. Formals are started in the formula box with an = (equals) sign.

Fig. 2.220

- There are many elements to and excel formula.

    1. **References:** The cell or range of cells that we want to use in the calculation.

    2. **Operators:** Symbols (+, -, *, /, etc.) that specify the calculation to be performed.

    3. **Constants:** Numbers or text values that do not change.

    4. **Functions:** Predefined formulas in Excel.

To Create a Formula:

- In example below, we will use a simple formula and cell references to calculate a budget.

    Step 1 : Select the **cell** that will contain the formula. In this example, we will select cell B3, (See Fig. 2.221).

    Step 2 : Type the equals sign (=), as shown in Fig. 2.222. Notice how it appears in both the cell and the formula bar.

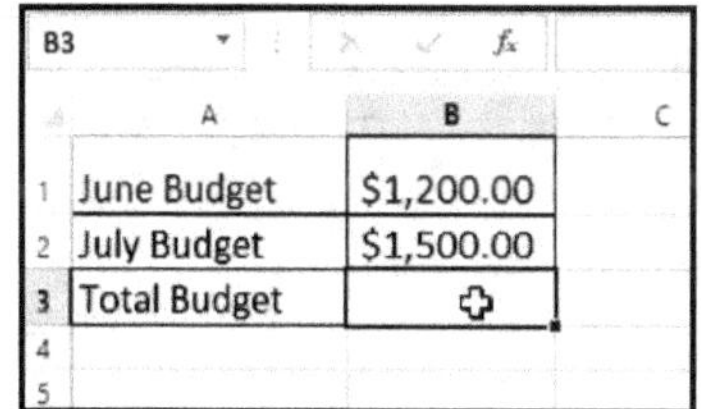

Fig. 2.221

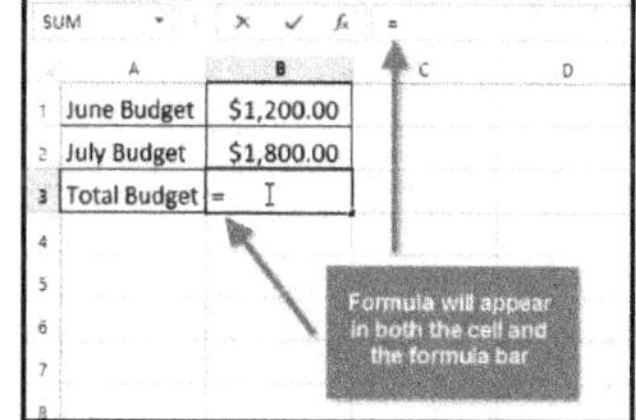

Fig. 2.222

    Step 3 : Type the cell address of the cell we want to reference first in the formula: cell B1 in this example. A blue border will appear around the referenced cell, (See Fig. 2.223).

    Step 4 : Type the mathematical operator we want to use. In this example, we'll type the addition sign (+).

Step 5 : Type the cell address of the cell we want to reference second in the formula: cell B2 in this example. A red border will appear around the referenced cell, (See Fig. 2.224).

Step 6 : Press Enter on the keyboard. The formula will be calculated, and the value will be displayed in the cell, (See Fig. 2.225).

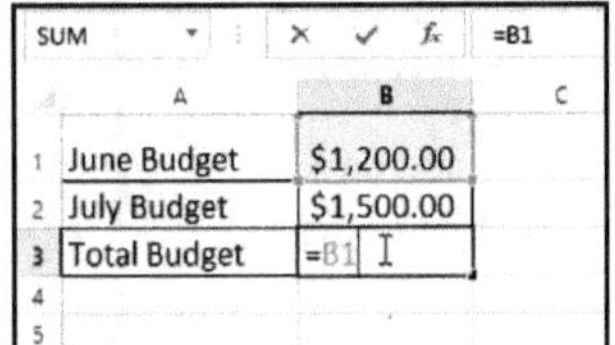

Fig. 2.223

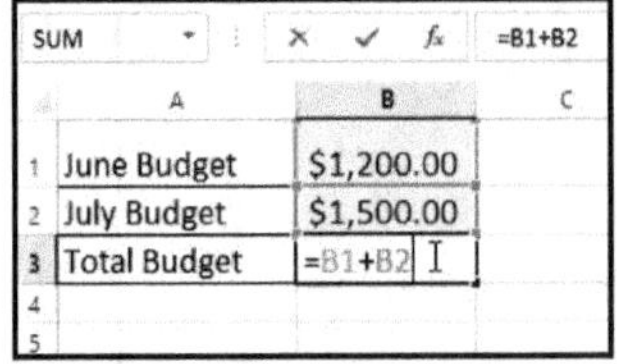

Fig. 2.224

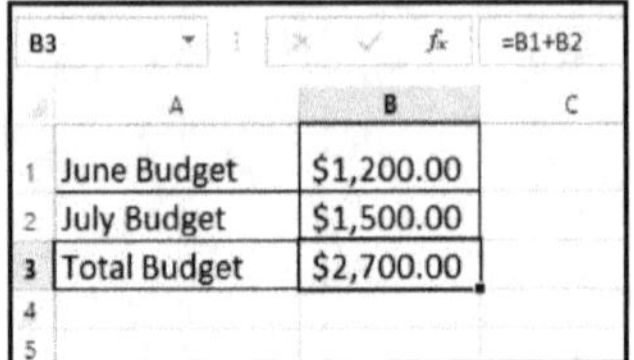

Fig. 2.225

Functions in Excel:

- A function is a built in formula in Excel. A function has a name and arguments (the mathematical function) in parentheses.

- A function is a pre-defined formula that performs calculations using specific values in a particular order. In order to work correctly, a function must be written a specific way, which is called the syntax.

- The basic syntax for a function is the equals sign (=), the function name (SUM, for example), and one or more arguments.

- Arguments contain the information we want to calculate. The function in the Fig. 2.226 would add the values of the cell range A1:A20.

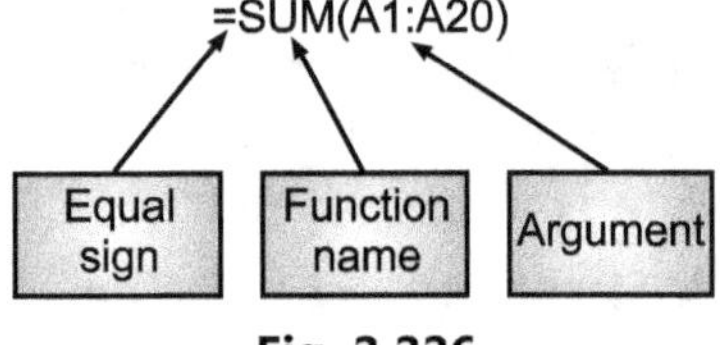

Fig. 2.226

Common Functions in Excel:

1. **SUM:** This function adds all of the values of the cells in the argument.

2. **AVERAGE:** This function determines the average of the values included in the argument. It calculates the sum of the cells and then divides that value by the number of cells in the argument.

3. **COUNT:** This function counts the number of cells with numerical data in the argument. This function is useful for quickly counting items in a cell range.

4. **MAX:** This function determines the highest cell value included in the argument.

5. **MIN:** This function determines the lowest cell value included in the argument.

Function Library:

- The function library is a large group of functions on the FORMULAS Tab of the Ribbon. These functions include:

1. **AutoSum**: Easily calculates the sum of a range.

2. **Recently Used**: All recently used functions.

3. **Financial**: Accrued interest, cash flow return rates and additional financial functions.

4. **Logical**: And, If, True, False, etc.

5. **Text**: Text based functions.

6. **Date & Time**: Functions calculated on date and time.

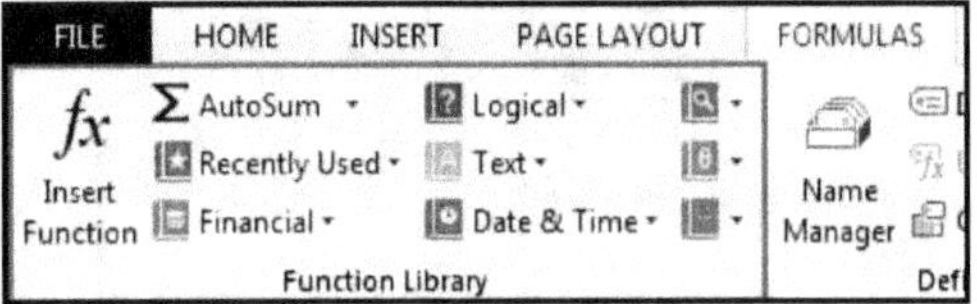

Fig. 2.227

Charts in Excel:

- Charts allow us to present information contained in the worksheet in a graphic format. Charts allows workbook data graphically, which makes it easy to visualize comparisons and trends.

- Excel has a variety of chart types, some of them are:
  1. **Column Charts** use vertical bars to represent data and can work with many different types of data.
  2. In **Line Charts** the data points are connected with lines, making it easy to see whether values are increasing or decreasing over time.
  3. In **Pie Charts** each value is shown as a slice of the pie, so it is easy to see which values make up the percentage of a whole.
  4. **Bar Charts** use horizontal bars instead of vertical bars.
  5. **Area Charts** are similar to line charts, except the areas under the lines are filled in.
  6. **Surface Charts** allow us to display data across a 3D landscape. They work best with large data sets, allowing we to see a variety of information at the same time.

To Insert a Chart: Follow the following steps for inserting a chart.

Step 1 : Select the cells we want to chart, including the column titles and row labels. These cells will be the source data for the chart. In this example, we will select cells A1:F6, (See Fig. 2.228).

	A	B	C	D	E	F	G
1	Genre	2008	2009	2010	2011	2012	
2	Classics	$18,580	$49,225	$16,326	$10,017	$26,134	
3	Mystery	$78,970	$82,262	$48,640	$49,985	$73,428	
4	Romance	$24,236	$131,390	$79,022	$71,009	$81,474	
5	Sci-Fi & Fantasy	$16,730	$19,730	$12,109	$11,355	$17,686	
6	Young Adult	$35,358	$42,685	$20,893	$16,065	$21,388	
7							
8							

Fig. 2.228

Step 2 : From the INSERT tab, click the desired Chart command. In this example, we will select Column chart as shown in Fig. 2.229.

Step 3 : Choose the desired chart type from the drop-down menu, (See Fig. 2.230).

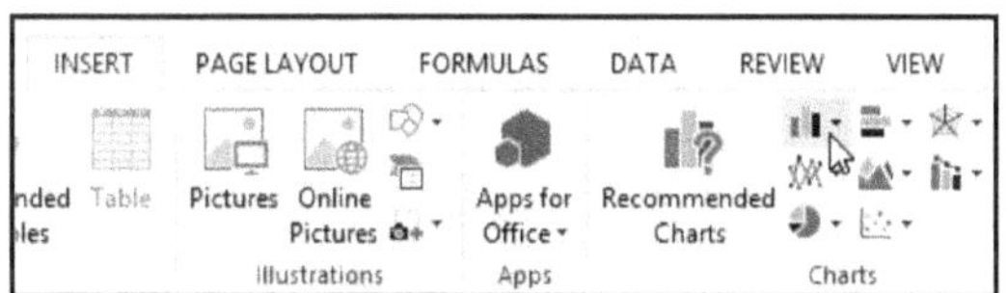

Fig. 2.229

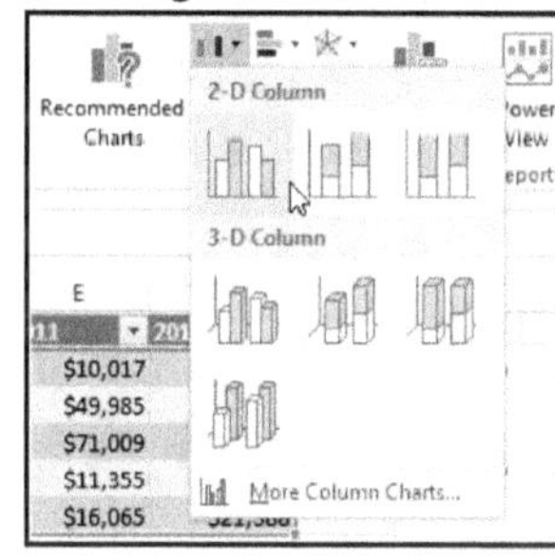

Fig. 2.230

Step 4 : The selected chart will be inserted in the worksheet, (See Fig. 2.231).

To Change the Chart Type: From the DESIGN tab, click the Change Chart Type command.

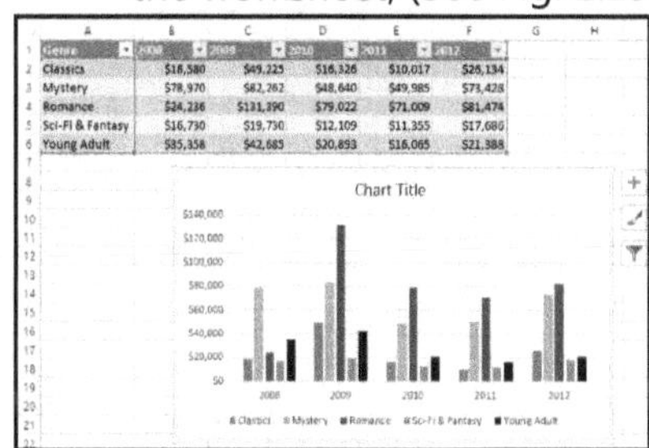

Fig. 2.231

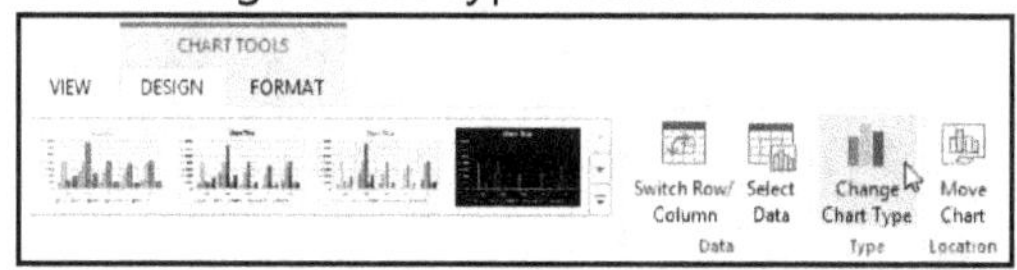

Fig. 2.232

Connect to (Import) External Data from Databases:

1. From Access Database to Workbook:

- Access databases are great for storing information, but sometimes we want to use or analyze its data in Excel.
- We can use the Data Connection Wizard to create a dynamic connection between an Access database and the Excel workbook.

- We get to the Data Connection Wizard through the DATA tab.

Step 1 : On the DATA tab, in the Get External Data group, click From Access.

Step 2 : In the Select Data Source dialog box, browse to the Access database.

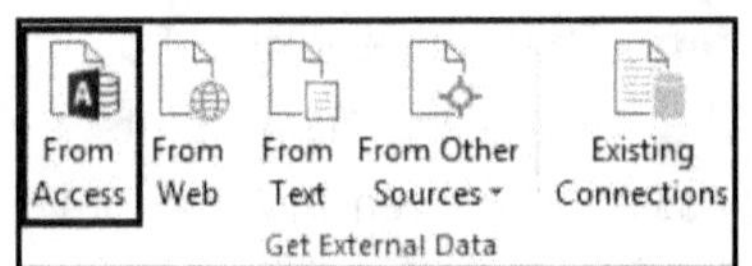

Fig. 2.233

Step 3 : In the Select Table dialog box, select the tables or queries we want to use, and click OK.

Step 4 : We can click Finish, or click Next to change details for the connection.

Step 5 : In the Import Data dialog box, choose where to put the data in the workbook and whether to view the data as a table, PivotTable report, or PivotChart.

Step 6 : Click the Properties button to set advanced properties for the connection, such as options for refreshing the connected data.

Step 7 : Optionally, we can add the data to the Data Model so that we can combine the data with other tables or data from other sources, create relationships between tables, and do much more than we can with a basic PivotTable report.

Step 8 : Click OK to finish.

2. From SQL Server:

- Databases are great for storing information, but sometimes we want to use or analyze its data in Excel.
- We can use the Data Connection Wizard to create a dynamic connection between a SQL Server database and the Excel workbook.
- We get to the Data Connection Wizard through the DATA tab.

Step 1 : On the Data tab, in the Get External Data group, click From Other Sources, (See Fig. 2.234).

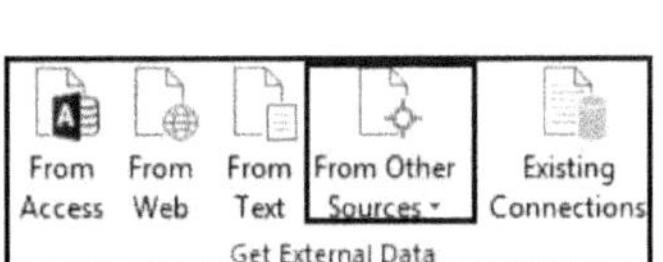

Fig. 2.234

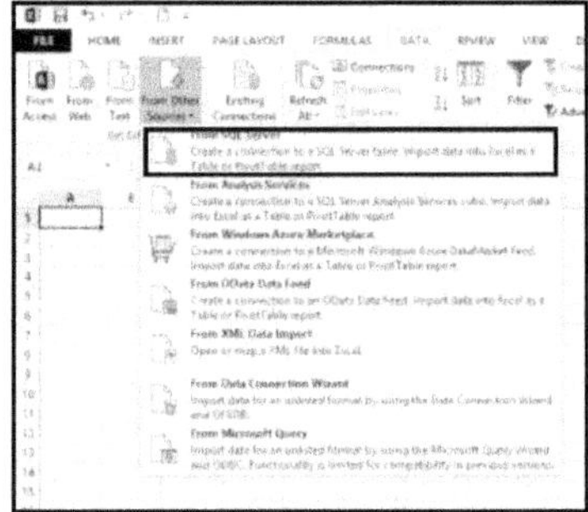

Fig. 2.235

Step 2 : Click From SQL Server, (See Fig. 2.235).

Step 3 : In the wizard, enter the server name and logon credentials, and click Next.

Step 4 : Choose the database and tables we want to work with, and click Next.

Step 5 : We can click Finish, or click Next to change details for the connection.

Step 6 : In the Import Data dialog box that appears, choose where to put the data in the workbook and whether to view the data as a table, PivotTable report, or PivotChart.

Step 7 : Click the Properties button to set advanced properties for the connection, such as OLAP settings or options for refreshing the connected data.

Step 8 : Optionally, we can add the data to the Data Model so that we can combine the data with other tables or data from other sources, create relationships between tables, and do much more than we can with a basic PivotTable report.

Step 9 : Click OK to finish.

3. From Analysis Services:

- Create a connection to a SQL Server Analysis Services cube to import data into Excel as a table, PivotTable report, or PivotChart.

  Step 1 : On the DATA tab, click From Other Sources then select. From Analysis Services.

  Step 2 : In the Data Connection Wizard, enter the server name and logon information.

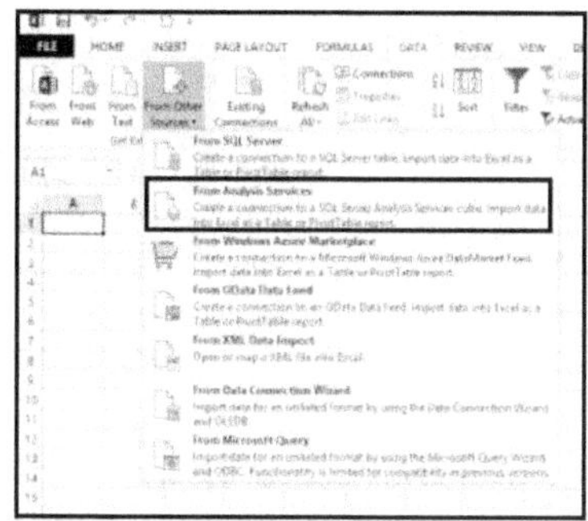

Fig. 2.236

  Step 3 : Choose the database we want to work with, and then choose a cube, perspective, or table.

  Step 4 : Click Finish, and then choose where we want to import the data.

2.4 MS POWERPOINT

- Presentation is the process of presenting a topic to an audience, for example, presenting a lesson to students about global warming.

- Many people make use of presentation software to support them when they have to give a presentation to others.

- Microsoft PowerPoint is a presentation software that allows users to create presentations and slide shows using a variety of media, including images, video and music.

2.4.1　Starting MS PowerPoint

- To start MS PowerPoint follow the following steps:

  Step 1 : Select the Start menu.

  Step 2 : Select All Programs.

  Step 3 : Select Microsoft Office, (See Fig. 2.237).

  Step 4 : Select Microsoft PowerPoint 2013, (See Fig. 2.238). Then Start Screen will appear (See Fig. 2.239). Select Blank Presentation. A new presentation will appear (See Fig. 2.240).

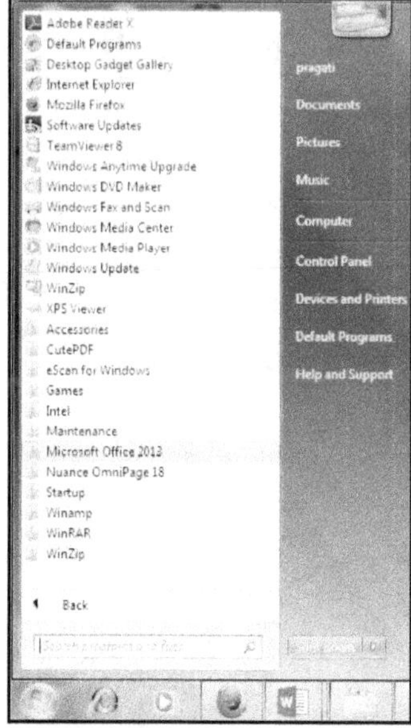

Fig. 2.237

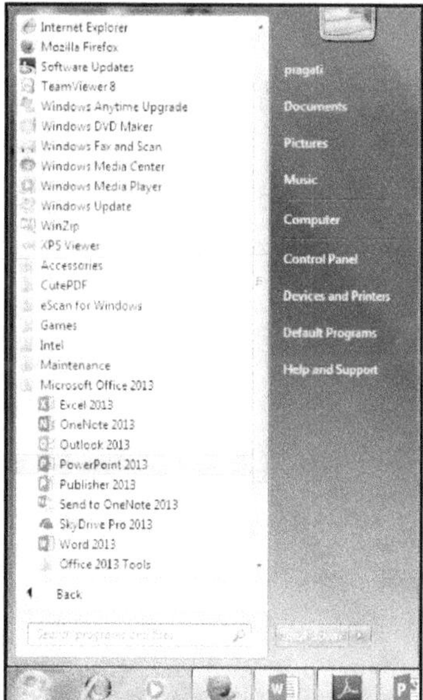

Fig. 2.238

Fig. 2.239

2.4.2 Basics of PowerPoint

- A presentation is a collection of data and information that is to be delivered to a specific audience.
- Presenting ideas in a clear and concise manner is important when sharing information with others. Presentation software enables visual organization and communication of concepts. One can customize the presentations with sound, animation, charts, graphics and video.
- The presentation software is used to display the information in the form of slide show. The three main functions of presentation software is editing that allows insertion and formatting of text, including graphics in the text and executing the slide shows.
- A PowerPoint presentation is a collection of electronic slides that can have text, pictures, graphics, tables, sound and video.
- By default, documents saved in PowerPoint 2013 are saved with the .pptx (PowerPoint 2007-2016) file extension whereas, the file extension of the prior PowerPoint versions is .ppt (PowerPoint 97-2003 Presentation).

2.4.3 Screen Elements of PowerPoint

- PowerPoint 2013 is a presentation software that allows us to create dynamic slide presentations. Slideshows can include animation, audios, images, videos, and much more.
- Fig. 2.240 shows screen components of PowerPoint 2013.

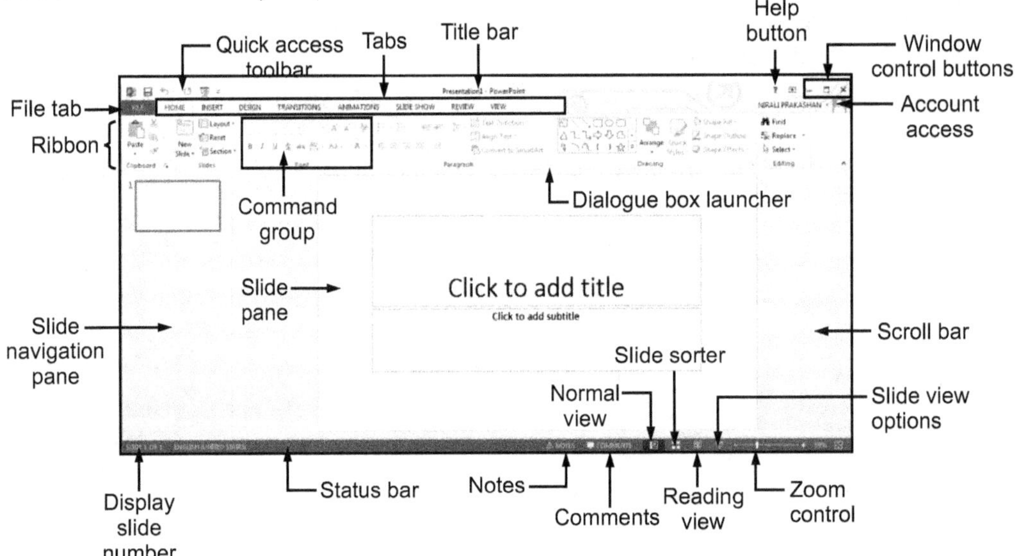

Fig. 2.240

- Fig. 2.240 shows following screen elements of PowerPoint:
1. **FILE tab:** This tab opens the Backstage view which basically allows us to manage the file and settings in PowerPoint. We can save presentations, open existing ones and create new presentations based on blank or predefined templates. The other file related operations can also be executed from this view.

2. **Quick Access Toolbar:** It lets us to access common commands no matter which tab is selected on the Ribbon.

3. **Microsoft Account:** We can access our Microsoft account information, view profile, and switch accounts.

4. **Ribbon:** It contains the commands we will need to perform common tasks in PowerPoint. It has multiple tabs, each with several groups of commands. PowerPoint has hundreds of commands for working with different presentations. To make it easier for users to find the specific commands they are looking for, commands are organized onto following main tabs as shown in Fig. 2.241 and explained below:

   (i) **HOME:** The HOME tab includes commands for formatting presentations.

   (ii) **INSERT:** Use the INSERT tab to insert tables, clip art, pictures, links, headers and footers, and more.

   (iii) **DESIGN:** Use the DESIGN tab to change the page setup, slide orientation, fonts, background styles, and more.

   (iv) **TRANSITIONS:** Use the TRANSITIONS tab to add transitions to a slide and to customize transition effects.

   (v) **ANIMATIONS:** Use the ANIMATIONS tab to add animation to text in a slide and to manage the animation order.

   (vi) **SLIDE SHOW:** Use the SLIDE SHOW tab to finalize the slide show details, timings, and more.

   (vii) **REVIEW:** Use the REVIEW tab to check spelling, to collaborate by adding comments, and to access editor tools.

   (viii) **VIEW:** Use the VIEW tab to access the various presentation views, to show and hide the ruler and grid lines, and more.

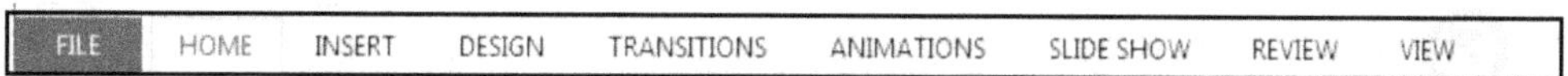

Fig. 2.241

- In addition to the main tabs, there are numerous tool tabs which include less commonly used commands. Some of the most commonly used tool tabs are Smartart, Chart, Drawing, Picture, Table.

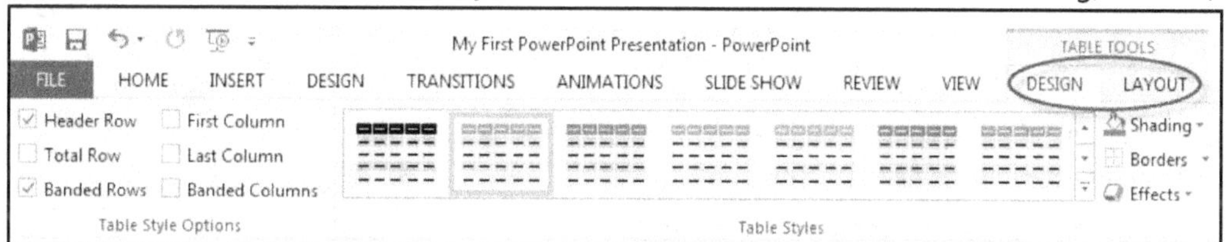

Fig. 2.242

5. **Title Bar:** This is the top section of the PowerPoint window. It shows the name of the current working file.

6. **Slide Area:** It is the area where the actual slide is created and edited. We can add, edit and delete text, images, shapes and multimedia in this section.

7. **Help:** This icon can be used to get PowerPoint related help anytime we need. Clicking on the "?" opens the PowerPoint Help window where we have a list of common topics to browse from.

8. **Slide View Options:** Adjust the slide view by choosing one of the following options:

   (i) **Normal View** is selected by default. This view shows the slide navigation pane and the selected slide.

   (ii) **Slide Sorter** displays smaller versions of all of the slides in the presentation.

   (iii) **Reading View** hides all editing tools to make your slides easier to review.

   (iv) **Play Slide Show** will play the slides as an actual presentation.

9. **Notes:** Click Notes to add notes to the current slide. Often called speaker notes.

10. **Comments:** Reviewers can leave comments on any slide. Click Comments to view comments for the current slide.

11. **Status Bar:** A horizontal strip that provides information about the opened presentation like slide number, applied Theme, etc. It also includes the zoom options.

12. **Slide Pane:** Here, we can view and edit the selected slide.

13. **Vertical and Horizontal Scroll Bars:** Click, hold and drag the vertical or horizontal scroll bar depending on what part of the slide we want to see.

14. **Zoom Control:** Click and drag the slider to use the zoom control. The number to the right of the slider reflects the zoom percentage. We can also click the Fit slide to current window button to the right of the zoom control.

2.4.4 Elementary Working with PowerPoint

- PowerPoint files are called presentations. Whenever we start a new presentation in PowerPoint, we will need to create a new presentation, which can either be blank or from a template. We will also need to know how to open an existing presentation.

To Create a New Presentation:

Step 1 : Select the FILE tab to go to Backstage view.

Step 2 : Select New on the left side of the window, then click Blank Presentation or choose a theme. A new presentation will appear.

To Open an Existing Presentation:

Step 1 : Select the FILE tab to go to Backstage view.

Step 2 : Select Open.

Step 3 : Select Computer, then click Browse.

Step 4 : The Open dialog box will appear. Locate and select the presentation, then click Open.

Using Templates:

- A template is a pre-designed presentation we can use to create a new slide show quickly.

- Templates often include custom formatting and designs, so they can save we a lot of time and effort when starting a new project.

To Create a New Presentation From a Template:

Step 1 : Click the FILE tab to access Backstage view.

Step 2 : Select New. We can click a suggested search to find templates or use the search bar to find something more specific. In this example, we will search for Business presentations as shown in Fig. 2.243.

Step 3 : Select a template to review it, (See Fig. 2.244).

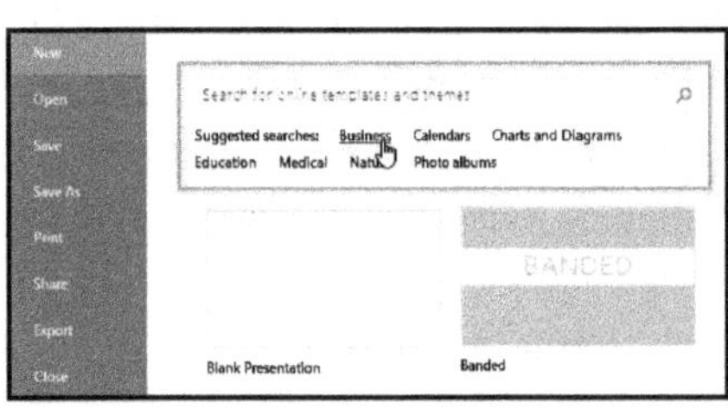

Fig. 2.243

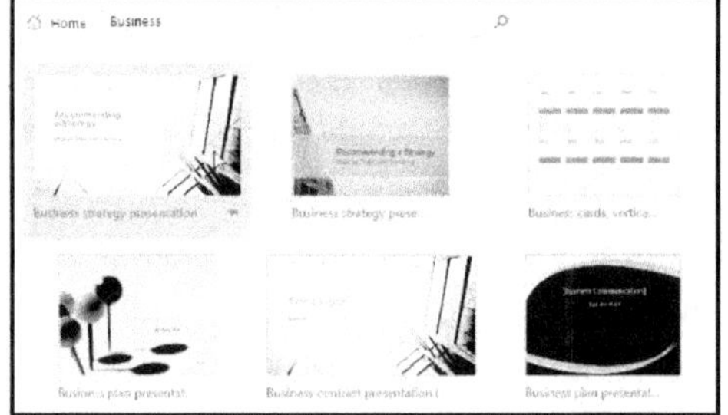

Fig. 2.244

Step 4 : A preview of the template will appear, along with additional information on how the template can be used.

Step 5 : Click Create to use the selected template. A new presentation will appear with the selected template.

Slide Basics in PowerPoint 2013:

- PowerPoint presentations are made up of a series of slides. Slides contain the information that will present to the audience, which might include text, pictures, and charts.

- Before we start creating presentations, we will need to know the basics of working with slides and slide layouts.

Understanding Slides and Slide Layouts:

- When we insert a new slide, it will usually have placeholders. Placeholders can contain different types of content, including text and images.

- Some placeholders have placeholder text, which we can replace with the own text. Others have thumbnail icons that allow we to insert pictures, charts, and videos, (See Fig. 2.245).

- Slides have different layouts for placeholders, depending on the type of information we want to include. Whenever we create a new slide, we will need to choose a slide layout that fits the content, (See Fig. 2.246).

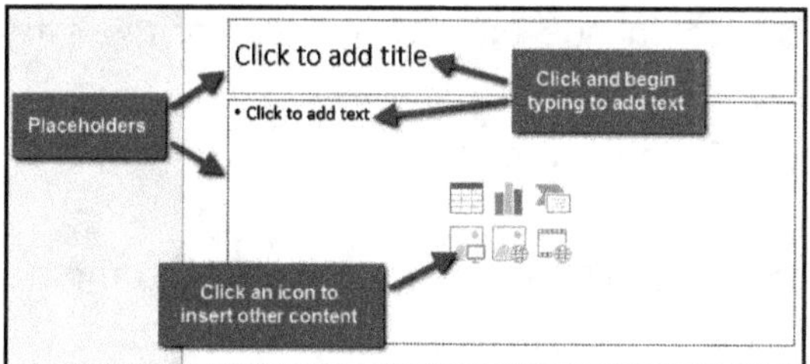

Fig. 2.245

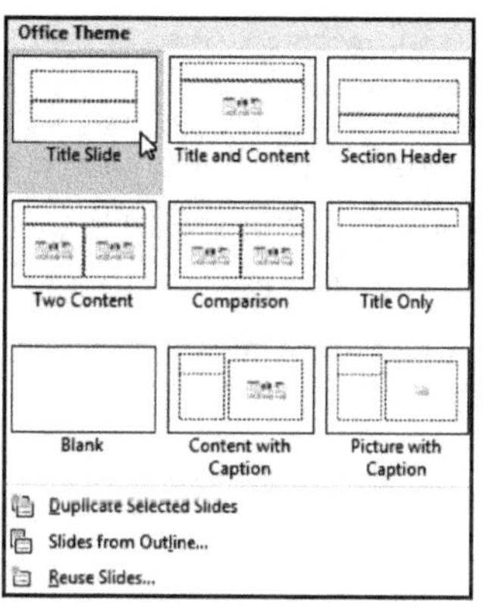

Fig. 2.246

To Insert a New Slide:

- Whenever, we start a new presentation, it will contain one slide with the Title Slide layout. We can insert as many slides as we need from a variety of layouts.

  Step 1 : From the HOME tab, click the bottom half of the New Slide command, (See Fig. 2.247).

  Step 2 : Choose the desired slide layout from the menu that appears, (See Fig. 2.248).

  Step 3 : The new slide will appear. Click any placeholder and begin typing to add text. We can also click an icon to add other types of content, such as a picture or a chart, (See Fig. 2.249).

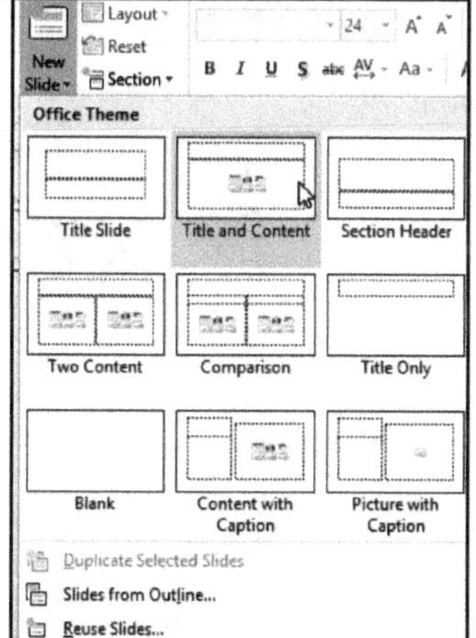

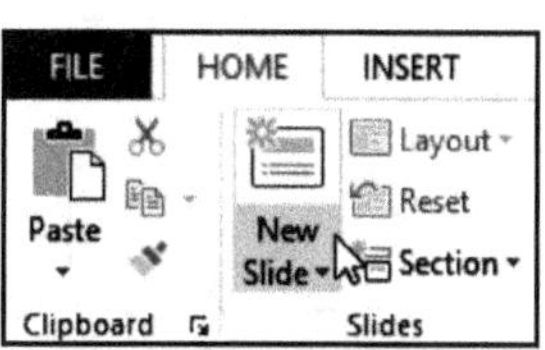

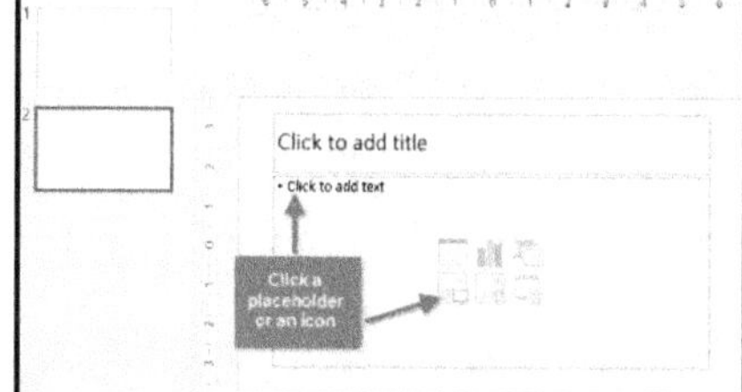

Fig. 2.247   **Fig. 2.248**   **Fig. 2.249**

Themes:

- Themes are design templates that can be applied to an entire presentation that allows for consistency throughout the presentation.
- By default, all slides in the presentation use a white background. It is easy to change the background style for some or all of the slides. Backgrounds can have a solid, gradient, pattern, or picture fill.

   Step 1 : Select the DESIGN tab, then click the Format Background command.

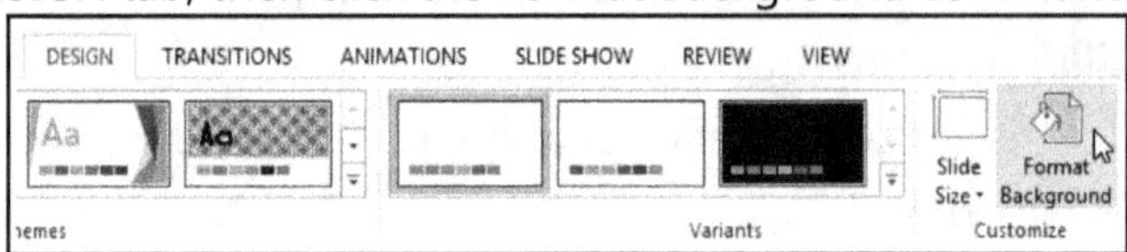

Fig. 2.250

   Step 2 : The Format Background pane will appear on the right. Select the desired fill options.

To Apply a Theme:

- A theme is a pre-defined combination of colors, fonts, and effects that can quickly change the look and feel of the entire slide show.
- Different themes also use different slide layouts, which can change the arrangement of the existing placeholders.

   Step 1 : Select the DESIGN tab on the Ribbon (See Fig. 2.251), then click the More drop-down arrow to see all of the available themes.

   Step 2 : Select the desired theme, (See Fig. 2.252). Then click the theme will be applied to the entire presentation.

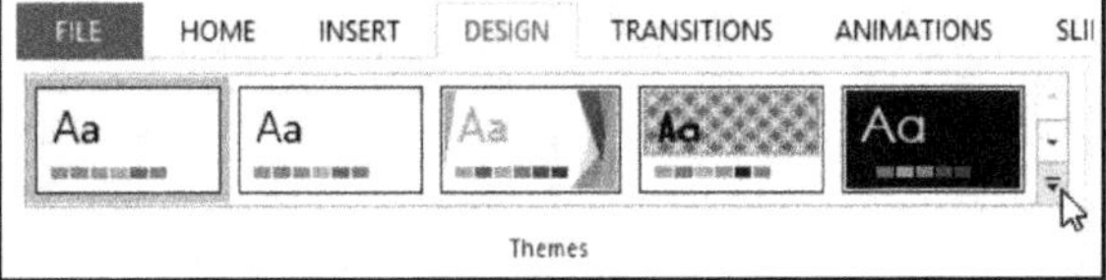

Fig. 2.251

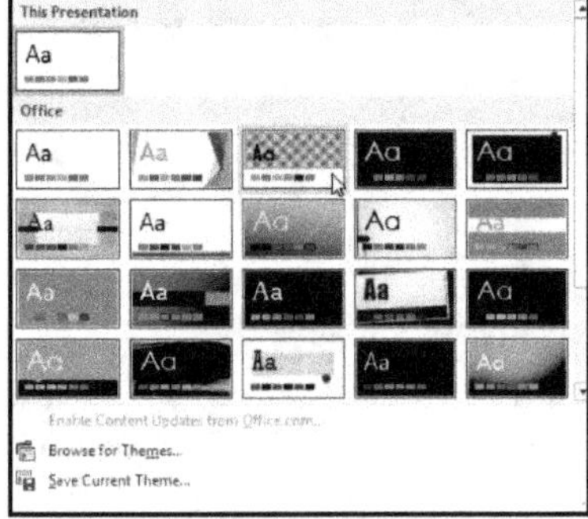

Fig. 2.252

To Enter Text:

- Follow the following steps for entering text in PowerPoint:

   Step 1 : Select the slide where we want to place the text box.

   Step 2 : On the INSERT tab, click Text Box, (See Fig. 2.253).

   Step 3 : Type in the text.

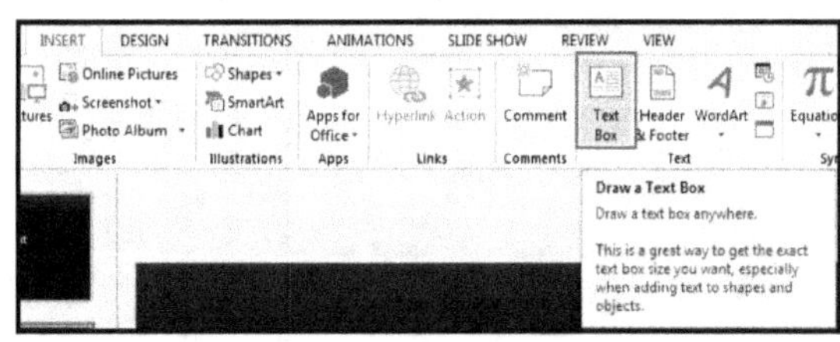

Fig. 2.253

To Cut, Copy and Paste Data:

   Step 1 : Select the item(s) that we wish to copy or cut.

   Step 2 : On the Clipboard group of the HOME Tab, click Cut or Copy.

   Step 3 : Select the items(s) where we would like to copy the data.

   Step 4 : On the Clipboard group of the HOME Tab, click Paste.

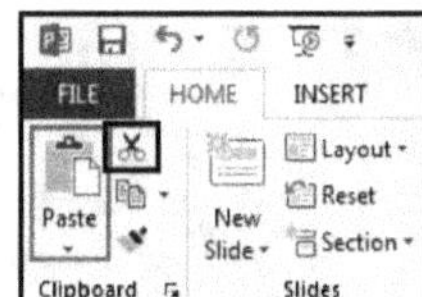

Fig. 2.254

Using Blank Slides:

- If we want even more control over the content, we may prefer to use a blank slide, which contains no placeholders. Blank slides can be customized by adding the own text boxes, pictures, charts, and more.

  Step 1 : To insert a blank slide, click the bottom half of the New Slide command.

  Step 2 : Then choose Blank from the menu that appears, (See Fig. 2.255).

- While blank slides offer more flexibility, keep in mind that we won't be able to take advantage of the predesigned layouts included in each theme.

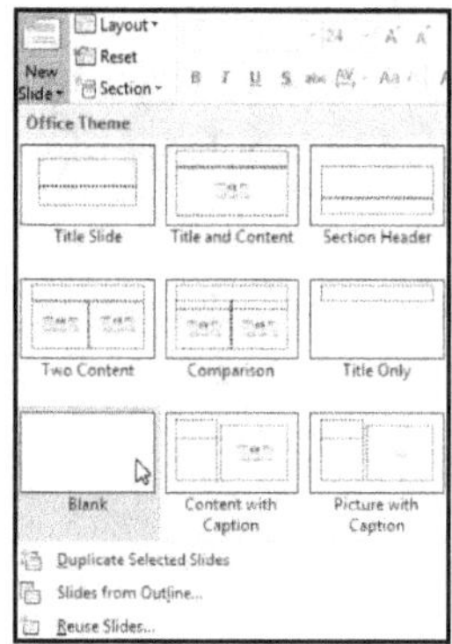

Fig. 2.255

To Play the Presentation or Slide Show:

- Once, we have arranged the slides, we may want to play the presentation. This is how we will present the slide show to an audience.

  Step 1 : Click the Start From Beginning command on the Quick Access toolbar to see the presentation.

  Step 2 : The presentation will appear in full-screen mode.

  Step 3 : We can advance to the next slide by clicking the mouse or pressing the spacebar on the keyboard. Alternatively, we can use the arrow keys on the keyboard to move forward or backward through the presentation.

  Step 4 : Press the Esc key to exit presentation mode.

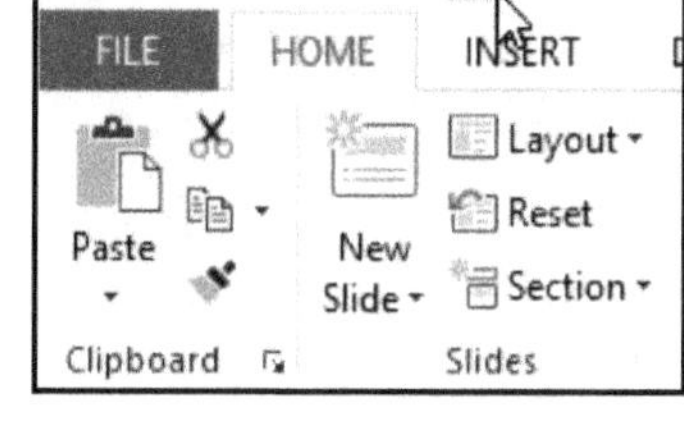

Fig. 2.256

Formatting Text in PowerPoint 2013:

- To change the font typeface follow the following steps:

  Step 1 : Click the arrow next to the font name and choose a font.

  Step 2 : Remember that we can preview how the new font will look by highlighting the text, and hovering over the new font typeface.

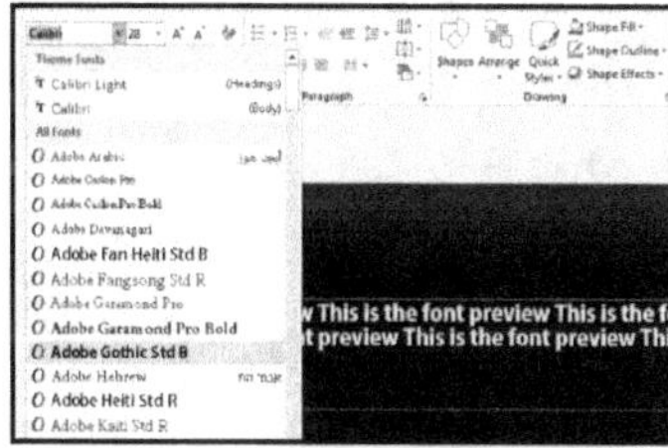

Fig. 2.257

To Change the Font Size:

  Step 1 : Click the arrow next to the font size and choose the appropriate size.　　　　　　　　　**OR**

  Step 2 : Click the increase or decrease font size buttons.

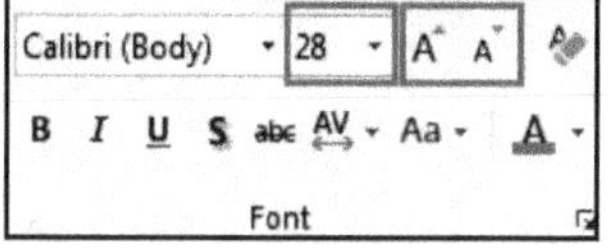

Fig. 2.258

To Change the Text Color:

  Step 1 : Select the text and click the Colors button included on the Font Group of the Ribbon.　　　　　　**OR**

  Step 2 : Highlight the text and right-click and choose the Colors tool.

  Step 3 : Select the color by clicking the down arrow next to the font color button.

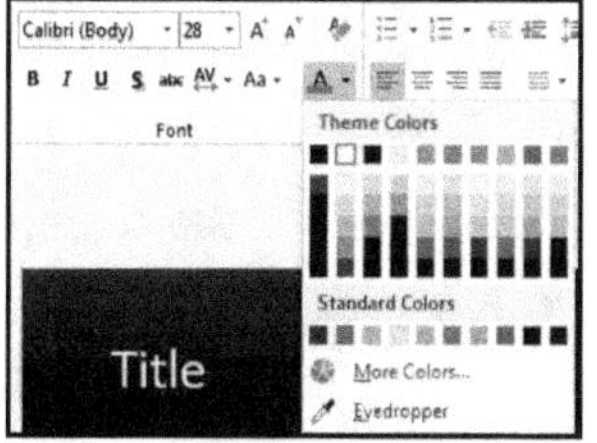

Fig. 2.259

WordArt:

- WordArt are styles that can be applied to text to create a visual effect. To apply WordArt follow following steps:

  Step 1 : Select the text.

  Step 2 : Click the INSERT tab.

  Step 3 : Click the WordArt button.

  Step 4 : Choose the WordArt, then click it.

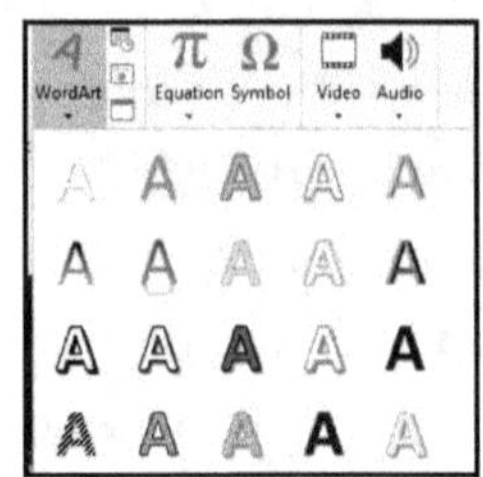

Fig. 2.260

Change Paragraph Alignment:

- The paragraph alignment allows us to set how we want text to flow relative to other content. To change the alignment follow the following steps:

  Step 1 : Click the HOME tab.

  Step 2 : Choose the appropriate button for alignment on the Paragraph Group.

  - **Align Left:** the text is aligned with left margin.

  - **Center:** The text is centered within margins.

  - **Align Right:** Aligns text with the right margin.

  - **Justify:** Aligns text to both the left and right margins.

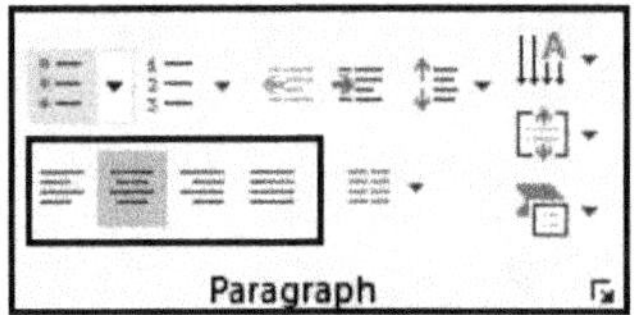

Fig. 2.261

To Add a Picture:

- Follow the following step for adding picture in PowerPoint.

  Step 1 : Click the INSERT tab.

  Step 2 : Click the Picture Button, (See Fig. 2.262).

  Step 3 : Browse to the desired picture within the files.

  Step 4 : Click the name of the picture.

  Step 5 : Click the Insert button.

  Step 6 : To move the graphic, click it and drag it to where we want it.

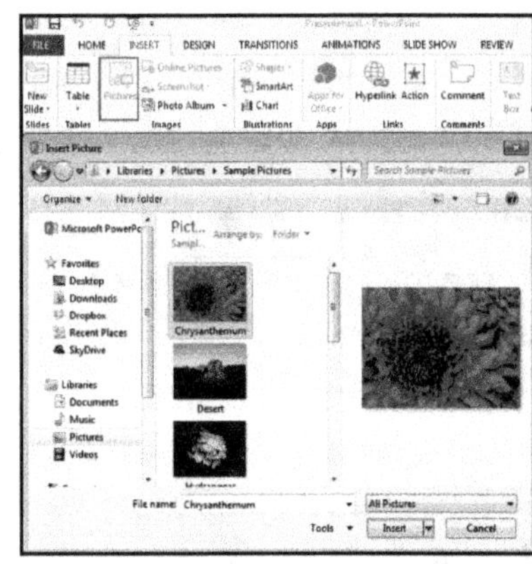

Fig. 2.262

To Add Clip Art:

- Follow the following steps to add ClipArt or Online pictures.

  Step 1 : Click the INSERT tab.

  Step 2 : Click the Online Pictures Button, (See Fig. 2.269).

  Step 3 : Search for the clip art using the search Clip Art dialog box.

  Step 4 : Click the clip art.

  Step 5 : Click the Insert button.

  Step 6 : To move the graphic, click it and drag it to where we want it.

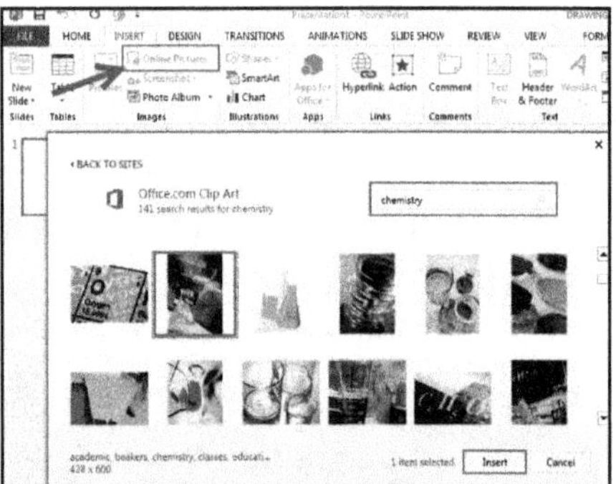

Fig. 2.263

To Add Shapes:

- Follow the following steps to add shapes.

    Step 1 : Click the INSERT tab.

    Step 2 : Click the Shapes Button, (See Fig. 2.264).

    Step 3 : Click the desired shape which display on presentation.

Tables:

- Tables are used to display data in a table format.

To Create a Table:

    Step 1 : Place the cursor on the page where we want the new table.

    Step 2 : Click the INSERT Tab of the Ribbon.

    Step 3 : Click the Table Button on the Tables group. We can create a table one of four ways:

    - Highlight the number of row and columns as shown in Fig. 2.265.

    - Click Insert Table and enter the number of rows and columns.

    - Click the Draw Table, create the table by clicking and entering the rows and columns.

    - Click Excel Spreadsheet and enter data.

- Place the cursor in the cell where we wish to enter the information and Begin typing.

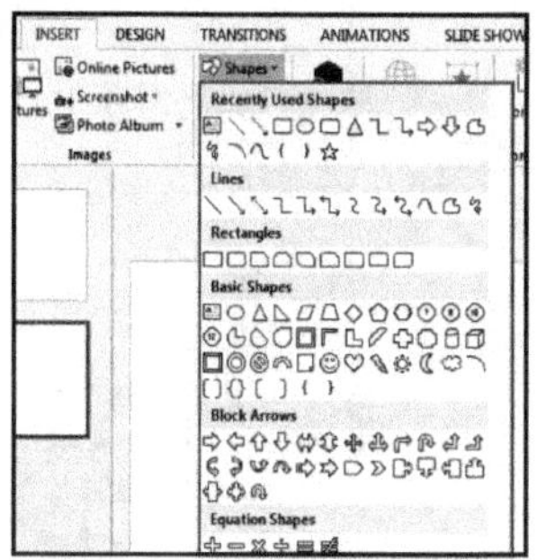

Fig. 2.264

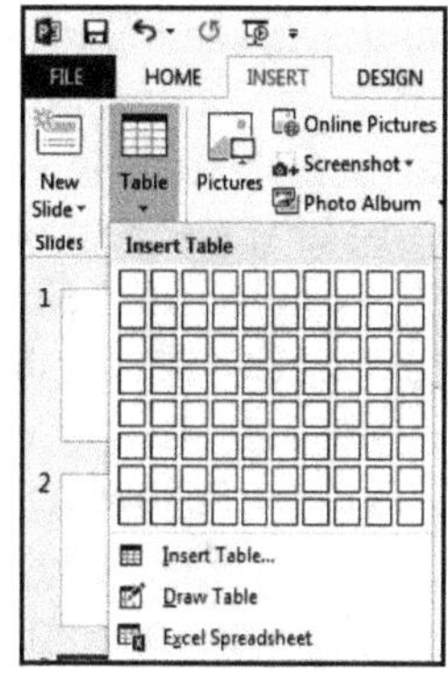

Fig. 2.265

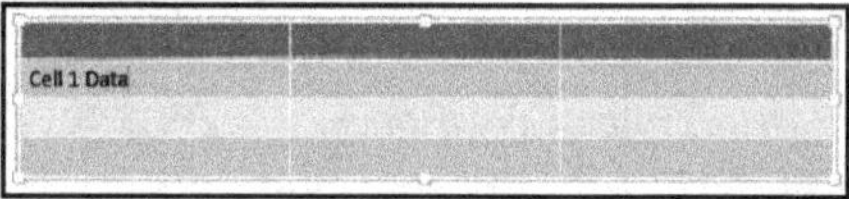

Lists:

- To create effective PowerPoint presentations, it is important to make the slides easy for the audience to read. One of the most common ways of doing this is to format the text as a bulleted or numbered list.

- By default, when we type text into a placeholder, a bullet is placed at the beginning of each paragraphautomatically creating a bulleted list.

- If we want, we can modify a list by choosing a different bullet style or by switching to a numbered list.

To Modify the Bullet Style:

    Step 1 : Select the list we want to format.

    Step 2 : On the HOME tab, click the Bullets drop-down arrow.

    Step 3 : Select the desired bullet style from the menu that appears, (See Fig. 2.266 (a)) The bullet style will appear in the list.

To Modify a Numbered List:

    Step 1 : Select an existing list we want to format.

    Step 2 : On the HOME tab, click the Numbering drop-down arrow.

    Step 3 : Select the desired numbering option from the menu that appears, (See Fig. 2.266 (b)). The numbering option will appear in the list.

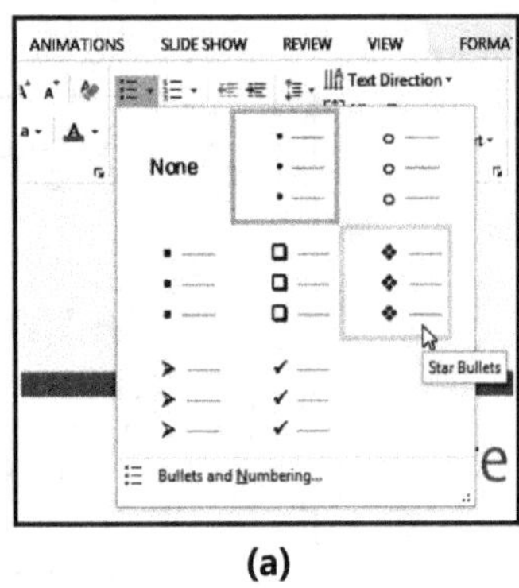

(a)

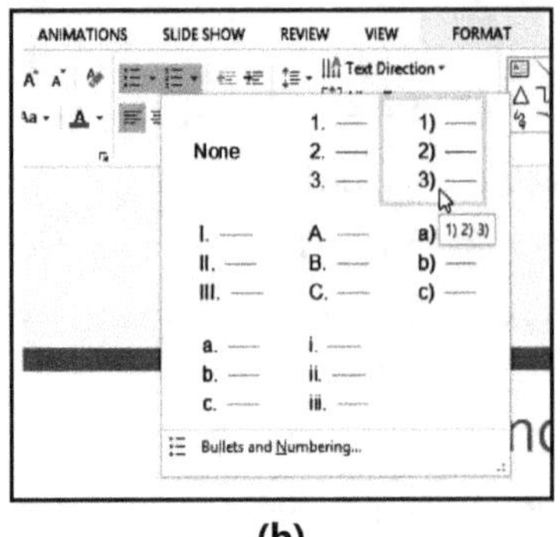

(b)

Fig. 2.266

Charts:

- A chart is a tool we can use to communicate data graphically. Including a chart in a presentation allows the audience to see the meaning behind the numbers, which makes it easy to visualize comparisons and trends.
- Various types of charts in PowerPoint 2013 are Line charts, Bar charts, Column charts, Surface charts, Pie charts, Area charts and so on.

Inserting Charts in PowerPoint 2013:

- PowerPoint uses a spreadsheet as a placeholder for entering chart data, much like Excel. The process of entering data is fairly simple.
- To insert a chart: Follow the following steps:

    Step 1 : Select the INSERT tab, then click the Chart command in the Illustrations group.

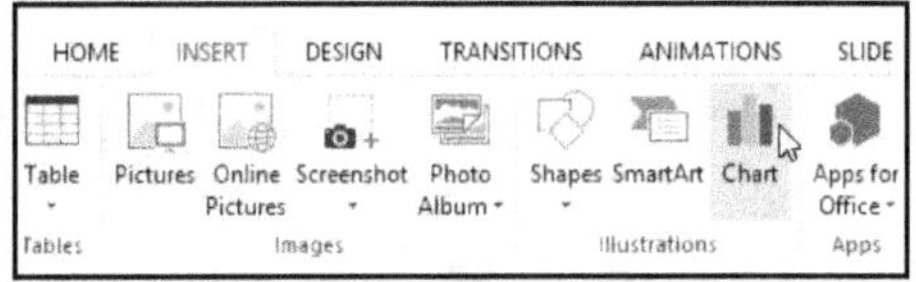

Fig. 2.267

Step 2 : A dialog box will appear, (See Fig. 2.268), select a category from the left pane, and review the charts that appear in the right pane.

Step 3 : Select the desired chart, then click OK, (See Fig. 2.268).

Step 4 : A chart and a spreadsheet will appear, (See Fig. 2.269). The data that appears in the spreadsheet is placeholder source data we will replace with the own information. The source data is used to create the chart.

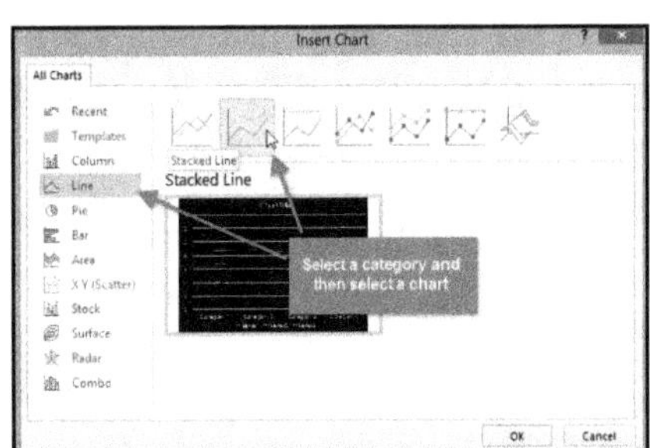

Fig. 2.268

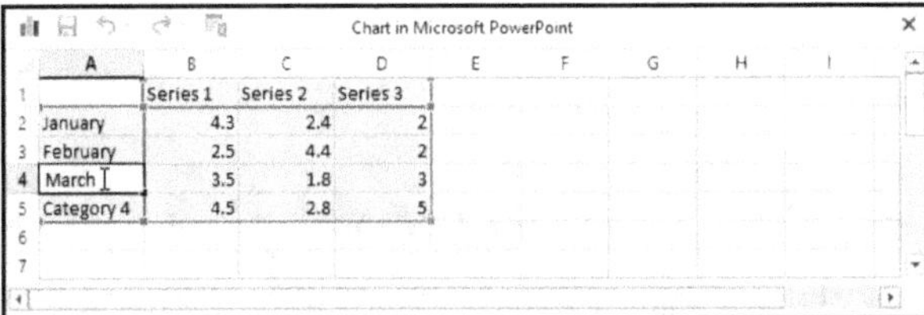

Fig. 2.269

Step 5 : Enter data into the worksheet.

Fig. 2.270

Step 6 : If necessary, click and drag the lower-right corner of the blue line to increase or decrease the data range for rows and columns, as shown in Fig. 2.270. Only the data enclosed by the blue lines will appear in the chart.

Step 7 : When we are done, click the ✕ to close the spreadsheet.

Step 8 : The chart will be completed, (See Fig. 2.271 (a) and (b)).

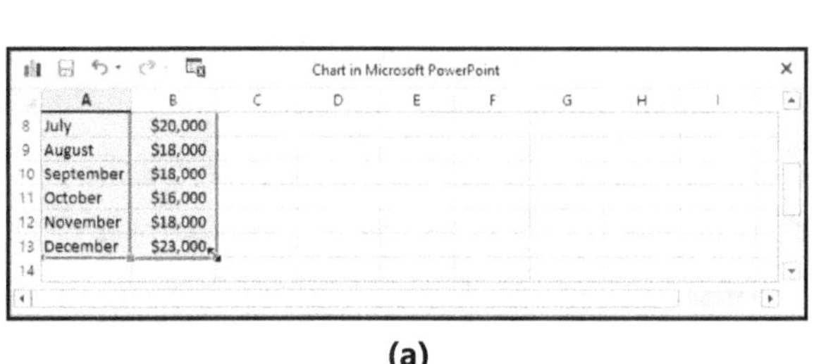

(a)

(b)

Fig. 2.271

To Change the Chart Type:

- If we find that the data is not well-suited to a certain chart, it is easy to switch to a new chart type.
- In this example, we will change this chart from a line chart to a column chart.

    Step 1 : Select the chart we want to change. The Design tab will appear on the right side of the Ribbon, (See Fig. 2.272).

    Step 2 : From the DESIGN tab click the Change Chart Type command, (See Fig. 2.273).

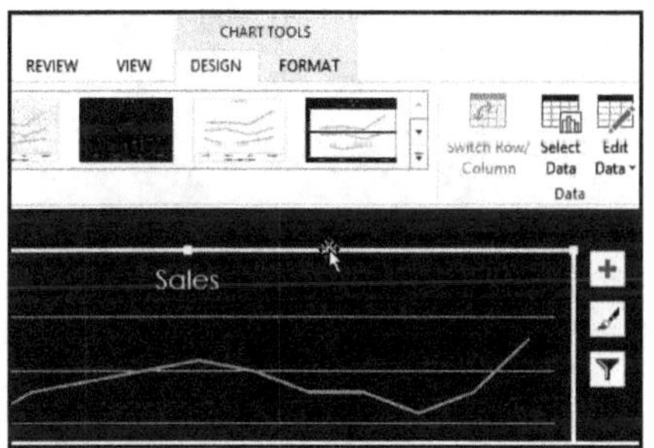

Fig. 2.272

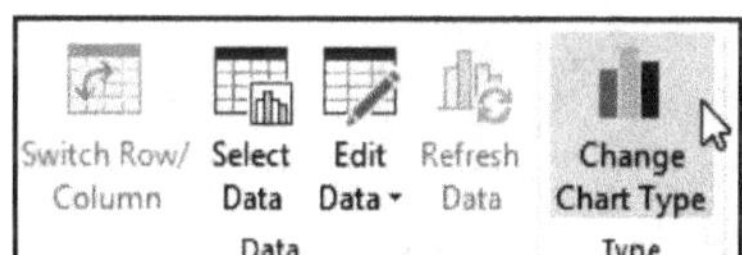

Fig. 2.273

Step 3 : A dialog box will appear, (See Fig. 2.274) then select the desired chart type, then click OK.

Step 4 : The new chart type will appear, (See Fig. 2.275).

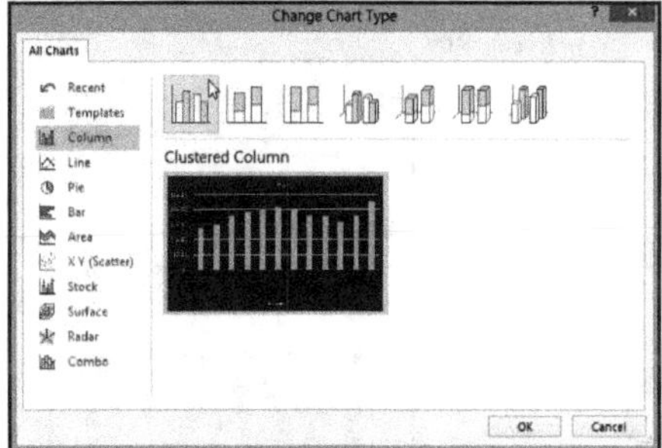

Fig. 2.274

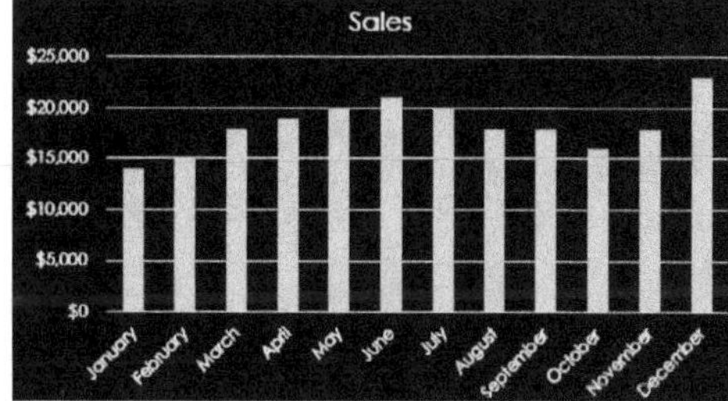

Fig. 2.275

Slide Transitions:

- A transition can be as simple as fading to the next slide or as flashy as an eye-catching effect. PowerPoint makes it easy to apply transitions to some or all of the slides, giving the presentation a polished, professional look.
- Transitions are effects that appear when we switch from one slide to the next during the presentation of the slide show.

- There are three categories of unique transitions to choose from, all of which can be found on the TRANSITIONS tab as shown in Fig. 2.278 and explained below:

  1. **Subtle:** These are the most basic types of transitions. They use simple animations to move between slides.

  2. **Exciting:** These use more complex animations to transition between slides. While they are more visually interesting than Subtle transitions, adding too many can make the presentation look less professional. However, when used in moderation they can add an nice touch between important slides.

  3. **Dynamic Content:** If we are transitioning between two slides that use similar slide layouts, dynamic transitions will move only the placeholders, not the slides themselves. When used correctly, dynamic transitions can help unify the slides and add a further level of polish to the presentation.

To Apply a Transition:

Step 1 : Select the desired slide from the Slide Navigation pane. This is the slide that will appear after the transition, (See Fig. 2.276).

Step 2 : Click the TRANSITIONS tab, then locate the Transition to This Slide group. By default, None is applied to each slide.

Step 3 : Click the More drop-down arrow to display all transitions, (See Fig. 2.277).

Step 4 : Click a transition to apply it to the selected slide. This will automatically preview the transition.

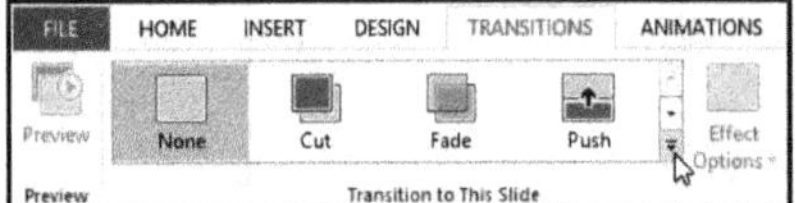
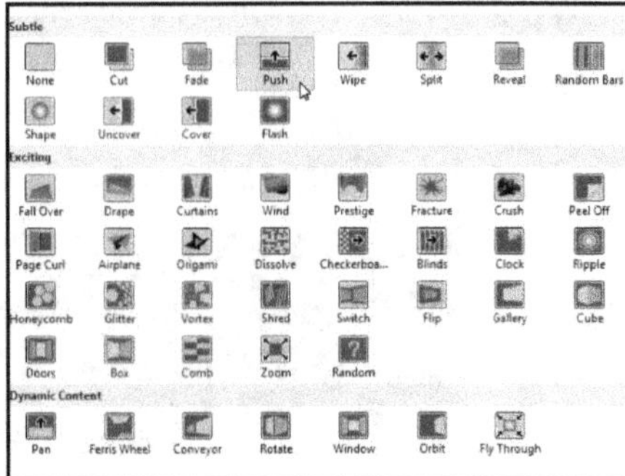

Fig. 2.276　　　　**Fig. 2.277**　　　　**Fig. 2.278**

To Preview a Transition:

- We can preview the transition for a selected slide at any time using either of these two methods:

  Step 1 : Click the Preview command on the Transitions tab, (See Fig. 2.279).

  Step 2 : Click the Play Animations command in the Slide Navigation pane, (See Fig. 2.280).

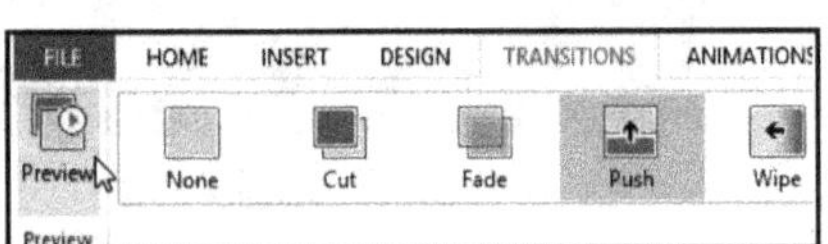
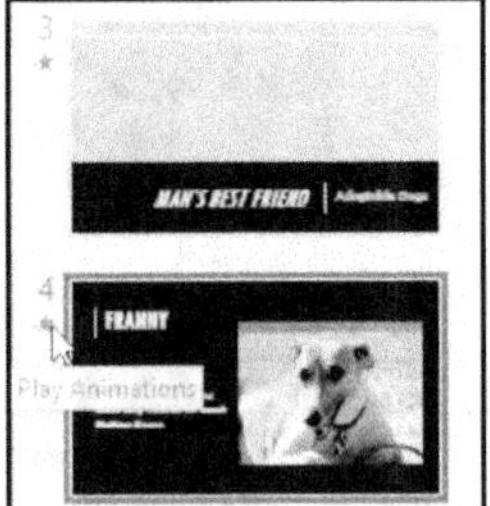

Fig. 2.279　　　　　　　　　　**Fig. 2.280**

Slide Animation:

- Slide animation effects are predefined special effects that we can add to objects on a slide.

- Animation or movement on the slide can be used to draw the audience's attention to specific content or to make the slide easier to read.

- There are several animation effects we can choose from, and they are organized into four types:

1. **Entrance:** These control how the object enters the slide. For example, with the Bounce animation the object will "fall" onto the slide and then bounce several times.

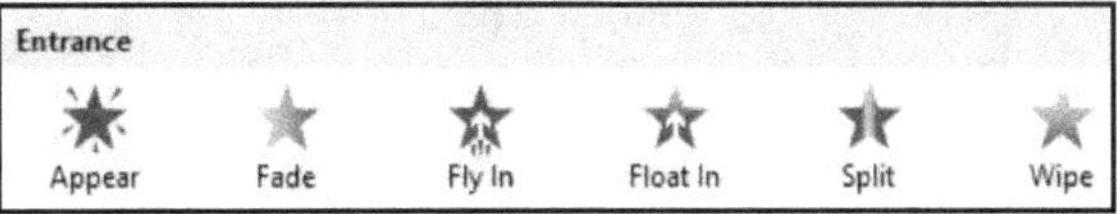

Fig. 2.281

2. **Emphasis:** These animations occur while the object is on the slide, often triggered by a mouse click. For example, we can set an object to spin when we click the mouse.

Fig. 2.282

3. **Exit:** These control how the object exits the slide. For example, with the Fade animation the object will simply fade away.

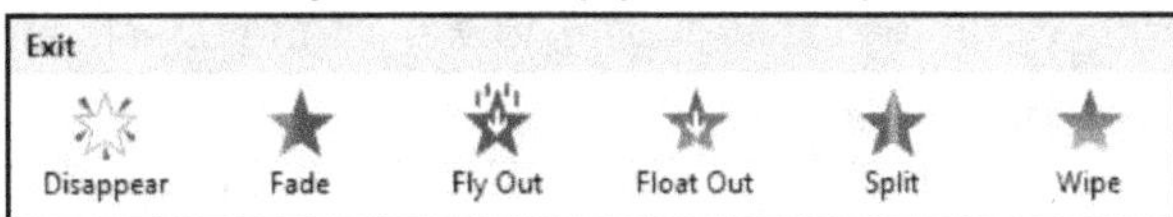

Fig. 2.283

4. **Motion Paths:** These are similar to Emphasis effects, except the object moves within the slide along a predetermined path, like a circle.

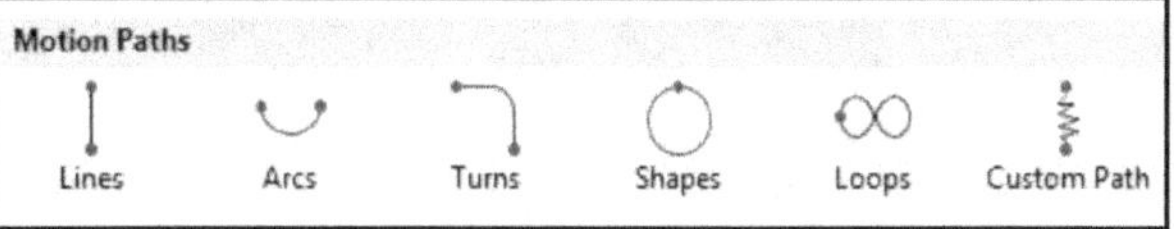

Fig. 2.284

To apply an Animation to an Object:

Step 1 : Select the object we want to animate.

Step 2 : On the ANIMATIONS tab, click the More drop-down arrow in the Animation group, (See Fig. 2.285).

Step 3 : A drop-down menu of animation effects will appear as shown in Fig. 2.286. Select the desired effect.

Step 4 : The effect will apply to the object. The object will have a small number next to it to show that it has an animation. In the Slide pane, a star symbol also will appear next to the slide.

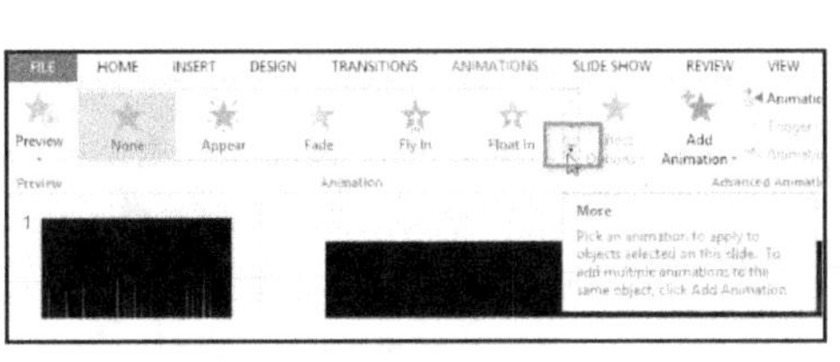

Fig. 2.285

Fig. 2.286

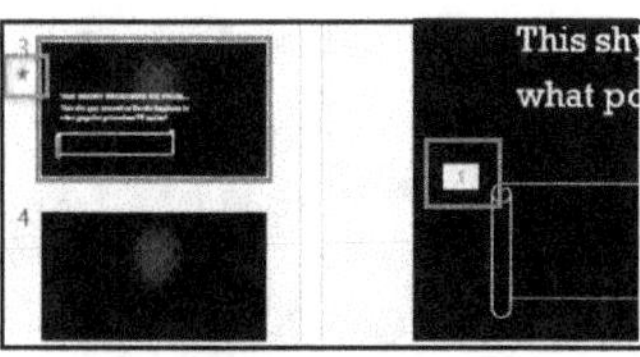

Fig. 2.287

Fig. 2.288

To Remove an Animation:

Step 1 : Select the small number located next to the animated object.

Step 2 : Press the Delete key. The animation will be deleted.

To Preview Animations:

- Any animation effects we have applied will show up when we play the slide show. However, we can also quickly preview the animations for the current slide without viewing the slide show.

Step 1 : Navigate to the slide we want to preview.

Step 2 : From the ANIMATIONS tab, click the Preview command, (See Fig. 2.289). The animations for the current slide will play.

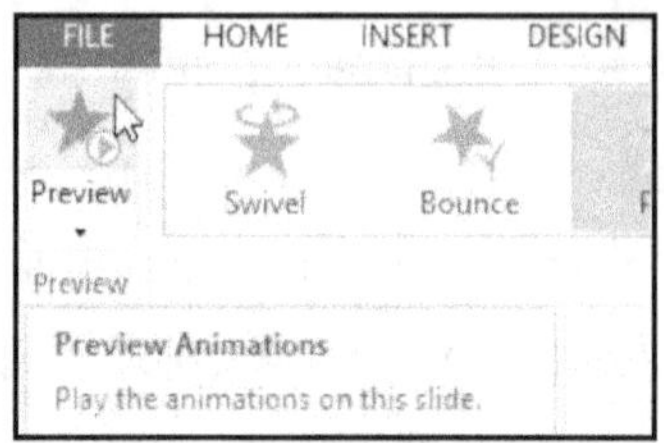

Fig. 2.289

The Animation Pane:

- It allows us to view and manage all of the effects that are on the current slide. We can modify and reorder effects directly from the Animation Pane, which is especially useful when we have several effects.

To open the Animation Pane:

Step 1 : From the ANIMATIONS tab, click the Animation Pane command, (See Fig. 2.290).

Step 2 : The Animation Pane will open on the right side of the window. It will show all of the effects for the current slide in the order in which they will appear, (See Fig. 2.291).

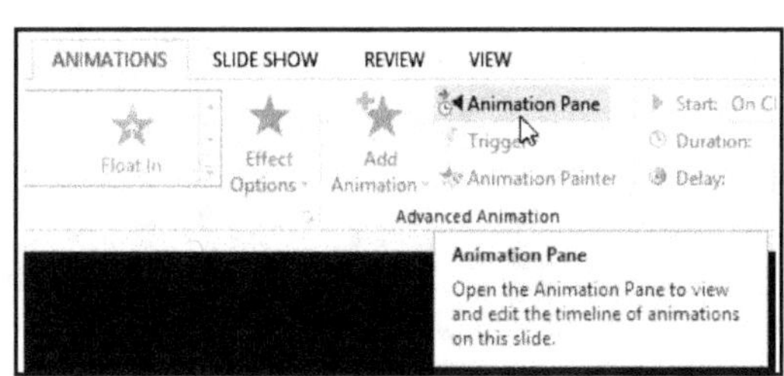

Fig. 2.290

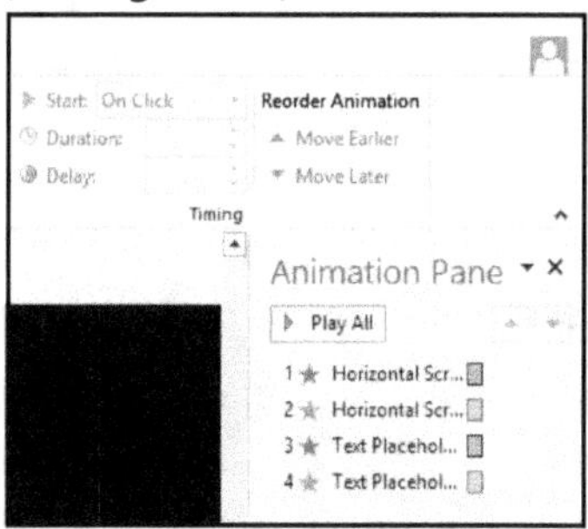

Fig. 2.291

Inserting Videos:

- PowerPoint allows us to insert a video onto a slide and play it during the presentation. This is a great way to make the presentation more engaging for the audience.

- We can even edit the video within PowerPoint and customize its appearance. For example, we can trim the video's length, add a fade in, and much more.

To Insert a Video From a File:

Step 1 : From the INSERT tab, click the Video drop-down arrow, then select Video on My PC, (See Fig. 2.292).

Step 2 : Locate and select the desired video file, then click Insert, (See Fig. 2.293).

Step 3 : The video will be added to the slide, (See Fig. 2.294).

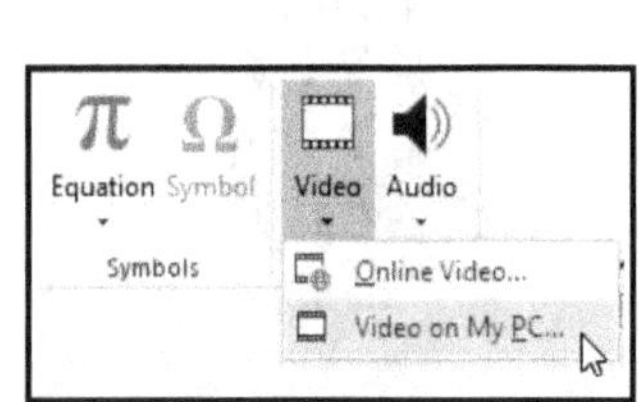

Fig. 2.292

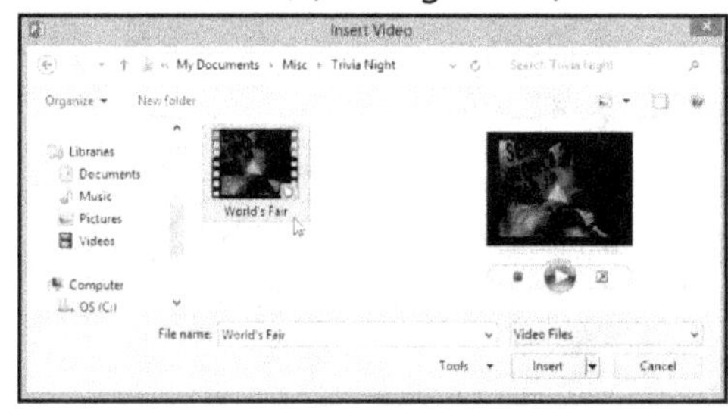

Fig. 2.293

Fig. 2.294

To Insert an Online Video:

- Some websites like YouTube allow we to embed videos into the slides.

- An embedded video will still be hosted on its original website, meaning the video itself won't be added to the file.

- Embedding can be a convenient way to reduce the file size of the presentation, but we will also need to be connected to the Internet for the video to play.

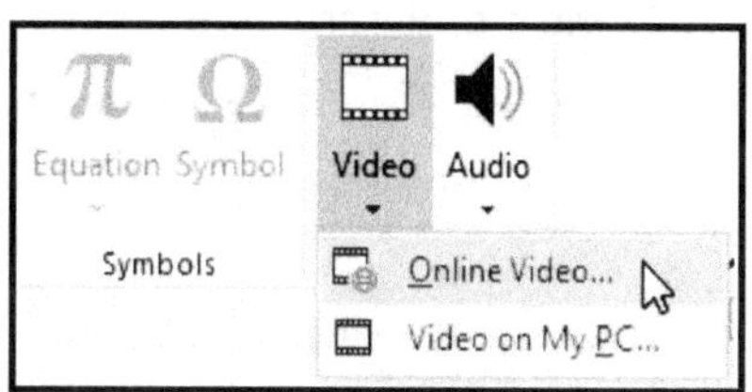

Fig. 2.295

Inserting Audio:

- PowerPoint allows us to add audio to the presentation. For example, we could add background music to one slide, a sound effect to another, and even record the own narration or commentary.

- We can then edit the audio to customize it for the presentation.

To Insert Audio from a File:

Step 1 : From the INSERT tab, click the Audio drop-down arrow, then select Audio on My PC, (See Fig. 2.296).

Step 2 : Locate and select the desired audio file, then click Insert, (See Fig. 2.297).

Step 3 : The audio file will be added to the slide.

Fig. 2.296

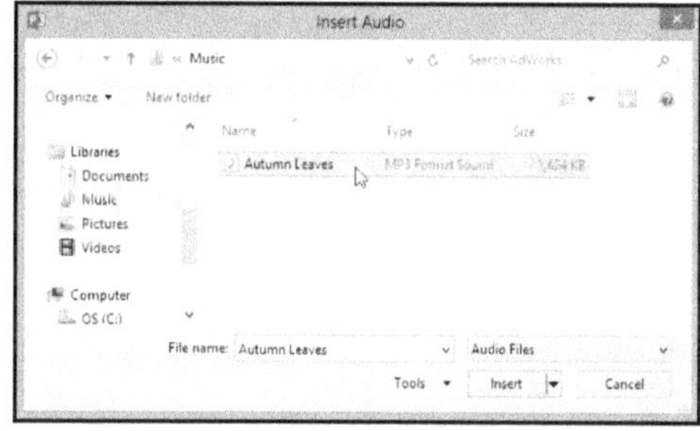

Fig. 2.297

- Sometimes we may want to record audio directly into a presentation. We can also add online audio using Online Audio …

1. What is word processing?

2. What is MS Word? How to create new document in word?

3. What is spreadsheet? Explain its uses.

4. What is presentation?

5. How to create hyperlink in word?

6. How to create table in word?

7. What is watermark? How to insert it in word? Explain with example.

8. In spreadsheet how to get external data from web?

9. Explain creation of macros in word.

10. What is lists in word? How to create it? Explain with example.

11. How to import file in Excel?

12. How to compare and merge file in word?

13. Explain formulas in Excel.

14. What is chart? How to create it in Word, Excel and PowerPoint?
15. With the help of diagram explain screen elements of Word.
16. With the help of describe explain screen elements of Excel.
17. With the help of diagram explain screen elements of PowerPoint.
18. How to insert function in Excel?

■■■

Chapter 3

Introduction to Internet

Contents

3.1 INTRODUCTION

- Over the past decade technological developments have created a global environment i.e., drawing the people around the world closer and closer together.
- During the industrial revolution, we learned to put motors to work to magnify human. In the new Information Age, we are learning to magnify brainpower by putting the power of computation whenever we need it and to provide information services on a global basis.
- Computer resources are infinitely flexible tools; networked together, allowing us to generate, exchange, share and manipulate information in an uncountable number of ways.
- The Internet as an integrating force, has combined the technology of communications and computing to provide instant connectivity and global information services to all its users at very low cost with fast speed.
- Internet is the name for a world-wide group of information resources available through a vast number Information Systems (ISs) connected through computer networking technology. The first experimental system started with four computers in 1969.
- This later evolved into a now defunct network called ARPANET (Advanced Research Project Agency NETwork) which has evolved into the global backbone which is called the Internet.
- Technically, the Internet is "an ever growing Wide Area Network (WAN) of millions of computers and computer networks across the globe which can exchange information through standard rules (protocols). Information is divided into packets which may travel through different paths to the destination address where it is reassembled into the original form."

3.2 WHAT IS INTERNET?

- The Internet is a global network of computers. The Internet is also often referred to as the Net, as a short form of network.
- Internet is a network of networks that consists of millions of private, public, academic, business, and government networks, of local to global scope, that are linked by a broad array of electronic, wireless and optical networking technologies.

- The Internet carries an extensive range of information resources and services, such as the inter-linked hypertext documents of the World Wide Web (WWW) and the e-mail.
- The Internet is a large collection of networks which run the TCP/IP (Transmission Control Protocol/Internet Protocol) protocols. They are tied together so that users of any of the users of one network can reach users on any of the other networks.
- The Internet now interconnects thousands of computer networks as shown in Fig. 3.1.

Fig. 3.1: Internet

Definition of Internet:

- "The global communication network that allows almost all computers world-wide to connect and exchange (share) information electronically".　　　　　　　**OR**
- Internet is defined as, "an information super highway, to access information over the Web".
- A global view of the Internet architecture is shown in Fig. 3.2. The Internet is essentially a network of interconnected networks. There is a hierarchical structure in this enormous mass.
- From a top-down view the Internet consists of a backbone that connects Internet Service Provider (ISP) backbones; the ISP backbones connect the backbones of various organizations; an organization's backbone is used to connect LANs; and finally, the LANs connect the hosts that are running such things as HyperText Transfer Protocol (HTTP) or mail.

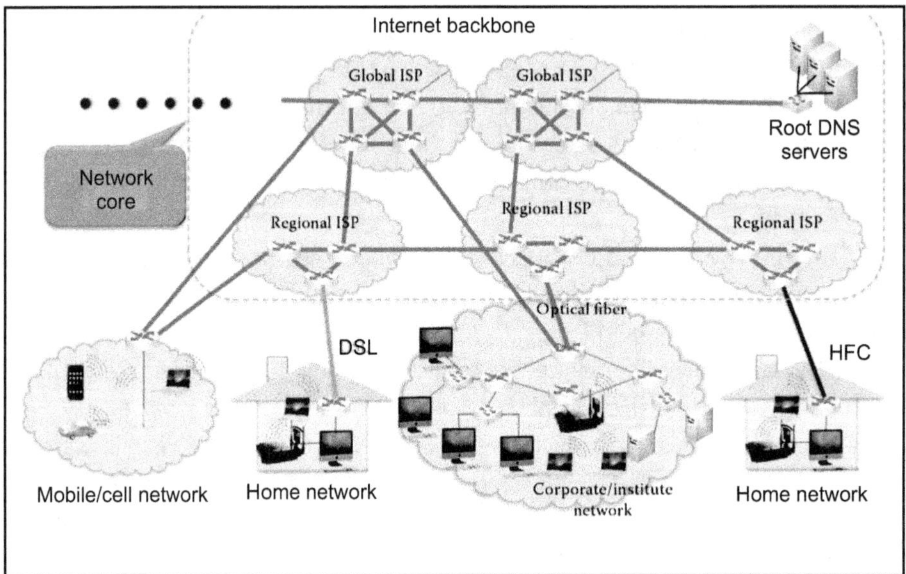

Fig. 3.2

Advantages of Internet:

1. **Global Scope and Reach:** Internet is world-wide global network. Global reach refers to a business initiative to increase the access between a company and their current and potential customers through the use of the Internet (world-wide).
2. **24×7 Availability:** Internet is always available any-time and any-where round the clock of 24×7, 7 days a week and 365 days.

3. **Information Resource:** All kinds of information is present on the Internet and easily accessed and can be searched more to get more additional knowledge. Information like educational related, government laws, market sales, stocks and shares, new creations etc., is gathered from a single place.

4. **Fast and Reliable Communication:** The main advantage of Internet is the faster communication than any other devices. It is an instant process for communication in the form of chatting, video conferencing, instant messaging, e-mails and so on.

5. **Service Provider:** In today's modern economy, Internet provides valuable services to the public. Banking sectors allow Internet banking which is online banking, movie tickets can be booked online, booking tickets to travel online, hotel reservations booking online can also be done. All these services are provided by Internet.

6. **Entertainment:** The Internet is now the most popular form of entertainment with free of cost. Movies, songs, videos, games etc. is available on Internet.

7. **E-commerce:** E-Commerce (EC) is a transaction of buying or selling (goods and services) online. Online shopping is now the latest trend in Internet world where number of products are available.

8. **Social Networking:** The social networking is the sharing of information to people across the world using Internet. Any emergency news, job vacancy, emergency news, ideas etc. can be shared in the website and the information gets passed on quickly to wide area. Also the social networking websites (Facebook and twitter) are used to easy communications.

9. **Learning:** The Internet has now become a part of education. Education like e-learning is easily carried out using Internet.

10. **Convenience Services:** The Internet has made life very convenient. With numerous online services we can now perform all the transactions online. We can book tickets (airplane or railway) for a movie, transfer funds, pay utility bills, taxes etc., right from our home.

Disadvantages of Internet:

1. **Virtual World:** The people using Internet often will forget the difference between virtual and real world. This causes the people to get depressed quickly and it leads to social isolation and obesity problems.

2. **Virus Threat:** Virus can easily be spread to the computers connected to Internet. Such virus attacks may cause computer system to crash or loss of data.

3. **Theft of Personal Information:** If we use the Internet for online banking, social networking or other services, we may risk a theft to the personal information such as name, address, credit card number etc.

4. **Pornography:** There are many pornographic sites that can be found, letting the children or adults to use Internet which indirectly affects the children/adults healthy mental life.

5. **Spamming:** Spamming refers to sending unwanted e-mails in bulk, which provide no purpose and needlessly obstruct the entire system. Such illegal activities can be very frustrating for us as it makes the internet slower and less reliable.

6. **Problem with Authenticated Information:** There are various websites that do not provide the authenticated information. This leads to misconception among many people.

3.3 COMPUTER COMMUNICATION AND INTERNET

- Communication is the activity of exchanging (sending or receiving) information using some medium like peoples, computers, smartphones and others devices.

- Computers in organizations have been used to extract and correlate information. Resource sharing, high reliability and communication can be achieved by connecting computers.

Basics for Computer Communication:

- Computer communications describes a process in which two or more computers or devices transfer (send and receive) data, instructions, and information over transmission media via a communications device(s).

- A communications device is hardware capable of transferring data from computers and devices to transmission media and vice versa. Examples of communications devices are modems, wireless access points, and routers. As shown in Fig. 3.3, some communications involve cables and wires; others are sent wirelessly through the air.

- Wired communications often use some form of telephone wiring, coaxial cable, or fiber-optic cables to send communications signals.

- The wiring or cables typically are used within buildings or underground between buildings.

- Because it is more convenient than installing wires and cables, many users go for wireless communications, which sends signals through the air or space. Examples of wireless communications technologies include Wi-Fi, Bluetooth, and so on.

- Wi-Fi uses radio signals to provide high-speed Internet and network connections to computers and devices capable of communicating via Wi-Fi. Most computers and many mobile devices, such as smartphones and portable media players, can connect to a Wi-Fi network.

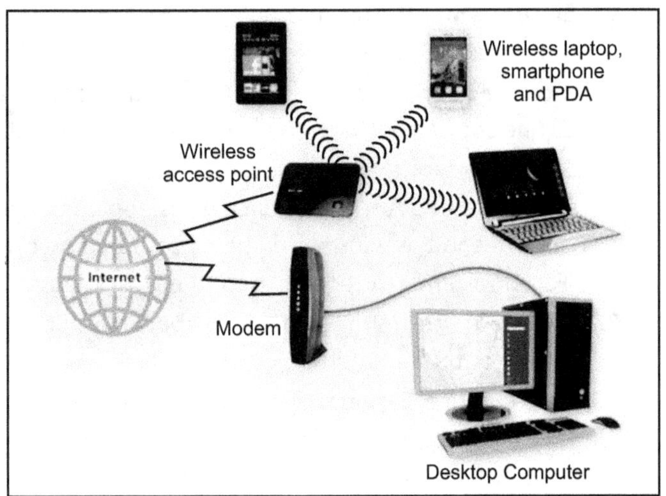

Fig. 3.3

- Bluetooth uses short-range (less than 15 feet) radio signals to enable Bluetooth-enabled computers and devices to communicate with each other.. For example, Bluetooth headsets allow us to connect a Bluetooth-enabled phone to a headset wirelessly

- Cellular radio uses the cellular network to enable high-speed Internet connections to devices with built-in compatible technology; such as smartphones. Cellular network providers use the categories 3G, 4G etc., to denote cellular transmission speeds.

- Wi-Fi and Bluetooth are both hot spot technologies. A hot spot is a wireless network that provides Internet connections to mobile computers and devices.

- Wi-Fi hot spots provide wireless network connections to users in public locations such as airports and airplanes, train stations, hotels, convention centers, schools, campgrounds, marinas, shopping malls, bookstores, libraries, restaurants, coffee shops, and more.

- Bluetooth hot spots provide location-based services, such as sending coupons or menus, to users whose Bluetooth-enabled devices enter the coverage range.

Communication through Internet:

- Communication is the most popular use of the Internet, with e-mail top most the list of all the technologies used. Some of the types of communication technologies used also include email discussion groups, usenet news, chat groups, and IRC. These are unique to networked computer environments and have come into wide popularity because of the Internet.

- Other technologies, including video and audio conferencing and Internet telephony, are also available on the Internet.

- They require more multimedia capabilities of computer systems and are more taxing of network resources than the others. They also are adaptations of other technologies to the Internet.

- Fig. 3.4 shows communication techniques used with Internet.

- Internet services allows us to access huge amount of information such as text, graphics, sound and software over the Internet.

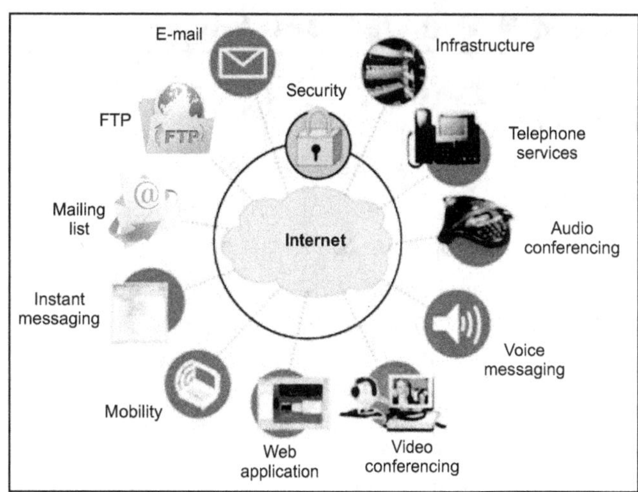

Fig. 3.4: Ways of Communication via Use of Internet

- There are various ways of communication available that offer exchange of information with individuals or groups. Some of them are explained below:

  1. **Electronic Mail:** Used to send electronic message over the Internet.

  2. **Telnet:** Used to log on to a remote computer that is attached to Internet.

  3. **Newsgroup:** Newsgroup are online areas in which users have written discussions about a particular subject. It offers a forum for people to discuss topics of common interests.

  4. **Internet Relay Chat (IRC):** Allows the people from all over the world to communicate in real-time.

  5. **Mailing Lists:** Used to organize group of internet users to share common information through e-mail.

  6. **Internet Telephony (VoIP):** VoIP is a conversation that takes place over the Internet using a phone connected to a computer, mobile device, or other device. It allows the Internet users to talk across Internet to any PC equipped to receive the call.

  7. **Instant Messaging:** Offers real time chat between individuals and group of people like Yahoo messenger, MSN messenger. Instant messaging is the real-time typed conversation with another connected user where we also can exchange photos, videos, and other content.

  8. **Chatting:** It offers real-time typed conversation among two or more people on a computers or mobile devices connected to a network.

  9. **File Transfer Protocol (FTP):** Enable the users to transfer files from the Internet.

  10. **World Wide Web (WWW):** It offers a way to access documents spread over the several servers over the Internet. These documents may contain texts, graphics, audio, video, hyperlinks. The hyperlinks allow the users to navigate between the documents.

  11. **Video Conferencing (Video teleconferencing):** It is a method of communicating by two-way video and audio transmission with help of telecommunication technologies. It is the real-time meeting between two or more geographically separated people(s) who use a network to transmit audio and video.

  12. **Voice Mail:** It allows users to leave a voice message for one or more people.

3.4 WWW AND WEB BROWSERS

- In this section we study basic concepts of WWW and Web Browsers.

1. WWW:

- WWW stands for World Wide Web.
- The World Wide Web was central to the development of the Information Age and is the primary tool billions of people use to interact on the Internet.
- WWW is an information space where documents and other web resources are identified by URLs, interlinked by hypertext links, and can be accessed via the Internet.
- The World Wide Web, or just "the Web," is a subset of the Internet. The Web consists of pages that can be accessed using a Web browser.
- A document on the Web is called a Web page and is identified by a unique address called the Uniform Resource Locator (URL).
- In other words, the World Wide Web (WWW) consists of files, called pages or web pages, which contain information and links to resources throughout the Internet.
- A web page is an electronic document written in a computer language called HTML (HyperText Markup Language).
- Web pages are also known as HTML documents. There are two types of web pages:
    (i) A static web page (sometimes called a flat page/stationary page) is a web page that is delivered to the user exactly as stored.
    (ii) A dynamic web page is a web page with web content that varies based on parameters provided by a user or a computer program. Dynamic web page example of server-side scripting like JSP, PHP and MySQL.
- A URL is also commonly referred to as a Web address. A URL is a type of Uniform Resource Identifier (URI), which is a generic term for many types of names and addresses on the Web.
- A Website is a location on the Internet where an individual, company, or organization keeps their Web pages and related files (such as text, graphic and video files). We display a Web page on the computer screen by using a program called a Web browser.
- A person user can retrieve and open a Web page in a Web browser either by entering a URL in the Web browser's address box or by clicking a hypertext link. When a user wants to access a Web page using either method, the user's Web browser sends a Web server a request for the Web page.
- A Web server is a computer that delivers Web pages. The Web server's reaction to the user's request is called the response.
- Web publishing or Online publishing, is the process of publishing content on the Internet. It includes creating and uploading websites, updating web pages, and posting blogs online and the published content may include text, images, videos, and other types of media.

2. Web Browsers:

- To view Web pages, we need a Web browser, which is a software program that requests, downloads and displays Web pages stored on a Web server.
- Web browser used to access the Internet and the WWW. It is basically used to access and view the web pages of the various websites available on the Internet.
- A web browser (a browser) is an application software for retrieving, presenting, and traversing information resources on the Internet and World Wide Web (WWW).

- A web browser can be defined as, "a software used to locate, retrieve and display content on the Internet and World Wide Web, including web pages, images, video and other files".
- Browsers are of two types as explained below:

   (i) Graphical Browsers: These browsers allows retrieval of text, images, audio, and video. Navigation is accomplished by pointing and clicking with a mouse on highlighted words and graphics. Both Netscape Navigator and Internet Explorer are graphical browsers.

   (ii) Text Browsers: These browsers provide access to the web in text-only mode. Navigation is accomplished by highlighting emphasised words on the screen with the arrow up and down keys, and then pressing the Enter key to follow the link. Lynx is an example of text-based browser.

- Today's common and popular Web Browsers are explained below:

   (i) Opera (◯): Opera is a web browser developed by Opera Software. The latest version is available for Windows, macOS, and Linux operating systems. Opera is a fast and secure browser.

   (ii) Internet Explorer (e): Internet Explorer (IE) is a web browser from Microsoft. IE was released as the default browser with Windows 95 (1995). Internet Explorer connects the computer to the Web through an Internet connection. The browser lets we view, print, and search for information on the Web.

   (iii) Mozilla Firefox ( ): Firefox is a free and open source web browser developed for Microsoft Windows, Mac OS X, and Linux (with its mobile versions available for Android, and Firefox OS) co-ordinated by Mozilla Corporation and Mozilla Foundation. It was first version 1.0 released on November 9, 2004. It is now available in about 78 languages world-wide.

   (iv) Safari ( ): Safari is web browser that was produced and developed by Apple Inc. which functions on a Mac OX, iOS, and Windows operating system.

   (v) Google Chrome ( ): Google Chrome is a freeware browser developed by Google using the WebKit layout engine. It was first release on Microsoft Windows operating system on September 2, 2008 in 43 different languages in beta version.

3.5 CREATING OWN E-MAIL ACCOUNT

- E-mail (electronic mail) is a way to send and receive messages across the Internet/Web/WWW. E-mail is a service which allows us to send the message in electronic mode over the Internet.
- E-mail is defined as, "the transmission of messages over communication networks".
- E-mail protocols are set of rules that help the client to properly transmit the information to or from the mail server. Various protocols for e-mail are listed below:

   1. **SMTP (Simple Mail Transfer Protocol):** It was first proposed in 1982. It is a standard protocol used for sending e-mail efficiently and reliably over the internet.
   2. **IMAP (Internet Mail Access Protocol):** IMAP allows the client program to manipulate the e-mail message on the server without downloading them on the local computer.
   3. **POP (Post Office Protocol):** It is generally used to support a single client. There are several versions of POP but the POP3 is the current standard.

- The popular e-mail service providers are listed below:

   1. **Gmail:** Gmail is an email service that allows users to collect all the messages. It also offers approx 7 GB of free storage.
   2. **Hotmail:** Hotmail offers free email and practically unlimited storage accessible on web.

3. **Yahoo Mail:** Yahoo Mail offers unlimited storage, SMS texting, social networking and instant messaging to boot.

4. **Mail.com and GMX Mail:** Mail.com and GMX Mail offers reliable mail service with unlimited online storage.

Creating Gmail Account:

- It is as easy to create a new Gmail email account as it is free. To set up a new Gmail email account follow the following steps:

Step 1 : Visit Create the Google Account for Gmail, (See Fig. 3.5).

Step 2 : Enter the first and last names under Name, (See Fig. 3.6).

Step 3 : Type the desired username under Choose the username, (See Fig. 3.6).

Note: Your Gmail e-mail address will be that user name followed by "@gmail.com"; if the Gmail username is "example", for instance, the Gmail address will be "example@gmail.com".

Step 4 : If Gmail lets we know that the desired username is not available, enter a different desired name under Choose the username or click one of the proposals under Available:

Step 5 : Type the desired password for the Gmail account under both Create a password and Confirm the password, (See Fig. 3.6).

Step 6 : Select and enter the birth date under Birthday.

Step 7 : Choose the gender under Gender, (See Fig. 3.6).

Step 8 : Optionally, enter the mobile phone number under Mobile phone for account verification and authorization.

Step 9 : Optionally, enter an existing e-mail address under Your current email address if we want to be able to recover a lost password with it.

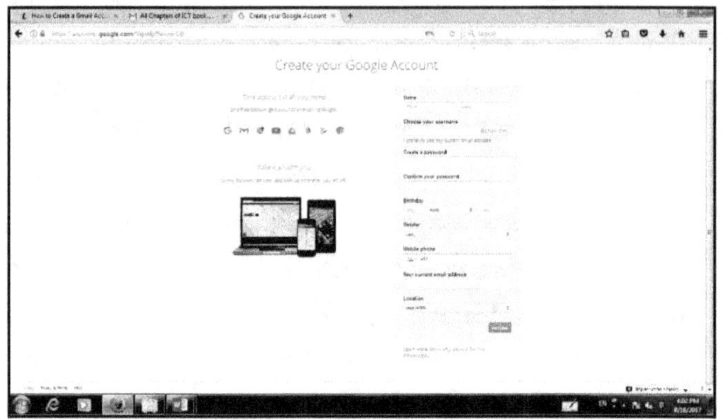

Fig. 3.5

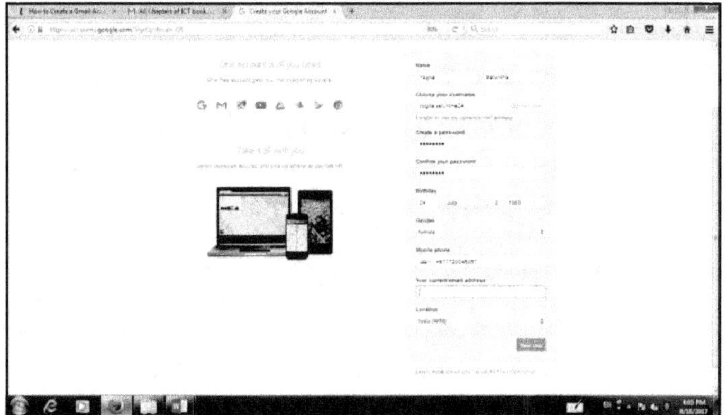

Fig. 3.6

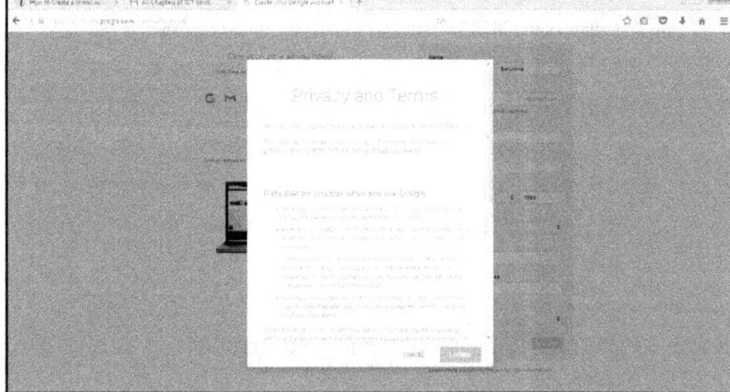

Fig. 3.7

Step 10 : If prompted, type the characters in the captcha picture under Prove you are not a robot.

Step 11 : Select the country or location under Location, (See Fig. 3.6).

Step 12 : Click Next step, (See Fig. 3.6).

Step 13 : Examine Google's terms for serving Gmail and the Gmail privacy policy.

Step 14 : Click I AGREE, (See Fig. 3.7).

Step 15 : Now click Continue to Gmail, (See Fig. 3.8).

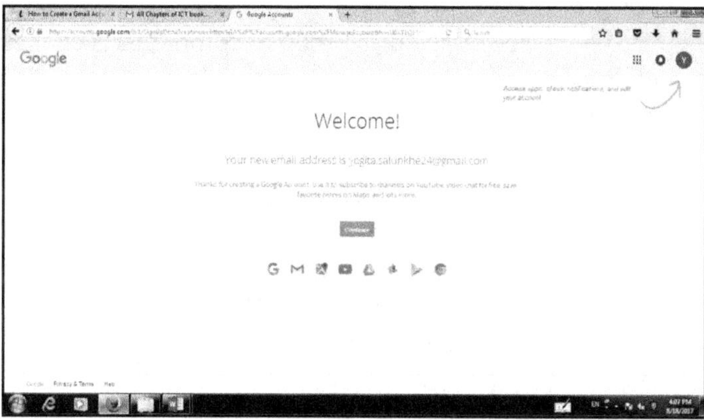

Fig. 3.8

- The Inbox of New created gmail account is shown in Fig. 3.9.
- To Sign out in current account, click Sign out Button as shown in Fig. 3.10.

Fig. 3.9

Fig. 3.10

3.6 NETWORKING AND TYPES

- A network is a collection of computers and devices connected together, often wirelessly via communications devices and transmission media.
- A group of computers and other devices/peripherals connected together is called a network. The best-known computer network is the Internet.
- In simple words, a computer network is a group of interconnected computing devices. The interconnected computers can share resources, which called networking.

Definition of Computer Network:

- A computer network can be defined as, "an interconnected collection of autonomous computers and computing devices".   **OR**
- A computer network is, "an interconnection of computers and computing equipments (devices) like printers, scanner etc., using either wires or radio waves (wireless) made to share hardware and software resources".
- A network is a set of devices often referred to as nodes like computer, printer, or any other device capable of sending and receiving data connected by media links as shown in Fig. 3.11.

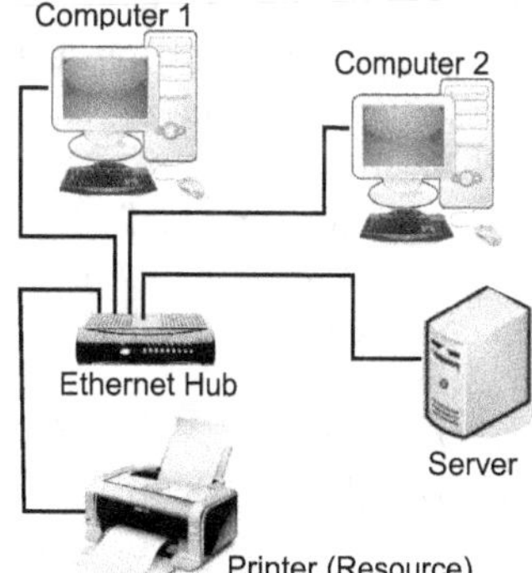

Fig. 3.11: Typical Network

- A network is either a peer-to-peer network (also called a workgroup) or a server-based network (also called a client/server network).
- In a peer-to-peer network, a group of computers is connected together so that users can share resources and information. There is no central location for authenticating users, storing files, or accessing resources.
- In a server-based network, the server is the central location where users (clients) share and access network resources.

Types of Computer Networks:

- Computer networks fall into three classes regarding the size, distance and the structure namely LAN (Local Area Network), MAN (Metropolitan Area Network), WAN (Wide Area Network).

1. Local Area Network (LAN):

- LANs are privately-owned networks covering a small geographical area, (less than 1 km) like a home, office, or groups of buildings.
- LANs are widely used to connect personal computers and workstations to share resources like printers and exchange information.
- Fig. 3.12 shows a typical LAN.
- Early LAN had data rates in the 4 to 16 mbps range. Today, LAN's speeds are normally 100 to 1000 mbps. Wireless LANs are the newest evolution in LAN technology.

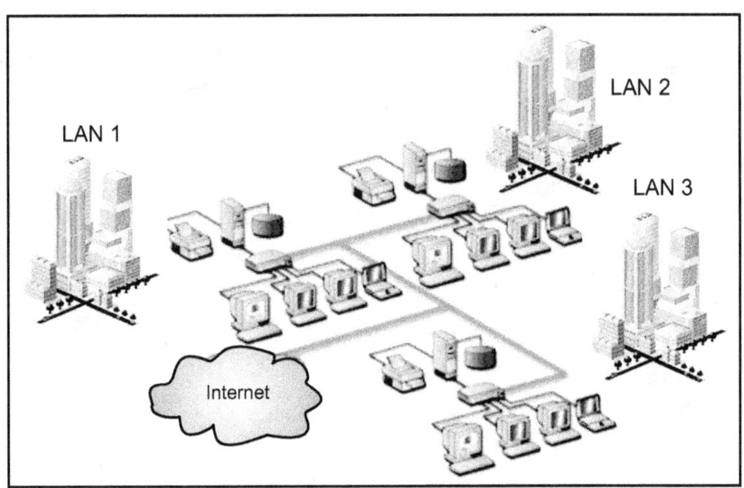

Fig. 3.12: Local Area Network (LAN)

Advantages of LAN	Disadvantages of LAN
(i) The reliability of LAN is high because the failure of one computer in the network does not effect the functioning for other computers.	(i) Used for small geographical areas (less than 1 km).
(ii) Addition of new computer to network is easy and simple i.e., easy to install.	(ii) Limited computers are connected in LAN (1 to 20).
(iii) High rate of data transmission is possible.	

2. Metropolitan Area Networks (MAN):

- The term metropolitan relates to city area. It means MAN covers comparatively large area than LAN but restricted upto city or a town.
- If a network spanning a physical area larger than a LAN but smaller than a WAN, such as a city or a town then, this network is called as Metropolitan Area Network (MAN).
- Geographical area for MAN lies between 1 km to 16 km.
- ATM (Asynchronous Transfer Modes) FDDI (Fiber Distributed Data Interface) etc., are the technologies is used in MAN.

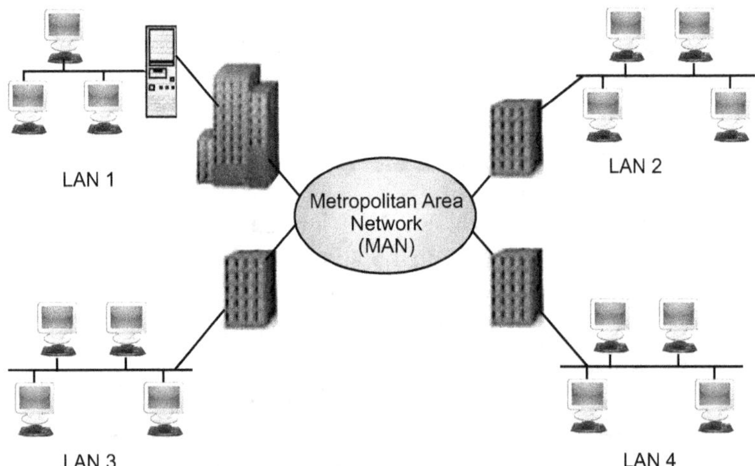

Fig. 3.13: Metropolitan Area Network (MAN)

Advantages of MAN	Disadvantages of MAN
(i) MAN spans large geographical area than LAN (1 to 16 kms). (ii) MAN offers centralized management of data.	(i) Cost is high. (ii) Speed is slow.

3. Wide Area Network (WAN):

- A WAN provides long distance transmission of data, (voice, image, audio and video information) over large geographical areas that may comprise a country, a state or even the whole world.

- The Internet is popular example of WAN, which spans the World today.

- Fig. 3.14 shows a typical WAN.

- WAN are commonly connected either through the Internet or special arrangements made with phone companies or other service providers.

- WAN may use advanced technologies like Frame Relay.

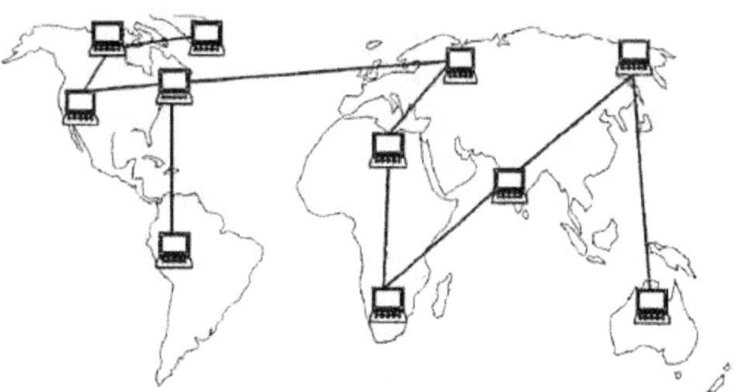

Fig. 3.14: WAN

Advantages of WAN	Disadvantages of WAN
(i) WAN covers a large geographical area. (ii) Using WAN messages can be sent very quickly.	(i) WANs are expensive. (ii) Slow in speed than LAN and MAN.

Differentiation between LAN, MAN and WAN:

Sr. No.	Parameter	LAN	WAN	MAN
1.	Stand for	Local Area Network.	Wide Area Network.	Metropolitan Area Network.
2.	Area covered	Covers small area i.e. within the building (less than 1 km).	Covers large geographical area, like country, state etc.	Covers larger area than LAN and smaller than WAN like city or town or campus.
3.	Error rates	Lowest.	Highest.	Moderate.
4.	Transmission speed	High.	Low.	Moderate.
5.	Equipment cost	Uses inexpensive equipment.	Uses most expensive equipment.	Uses moderately expensive equipment.
6.	Example	Offices, Cyber Café.	Internet.	ATM, FDDI etc.
7.	Data transfer rate	High.	Low.	Moderate.
8.	Set-up cost	Low.	High.	Moderate.

Practice Questions

1. What is internet?
2. What is WWW and Web browser?
3. Explain the term LAN in detail.
4. Explain the term MAN in detail.
5. Explain the term WAN in detail.
6. What is computer network? Enlist types of networks.

7. Define the following in terms:
   (i) Internet
   (ii) WWW
   (iii) Web browser
   (iv) Network.
8. Explain computer communication and internet in detail.
9. How to create email account?
10. Compare LAN, WAN and MAN.

∎∎∎

Chapter 4

Introduction to HTML and Software

Contents

4.1 INTRODUCTION

- Today software is everywhere, such as on to the mobile phones, tablets, notebooks, computer everything runs on software.

- Now-a-days software plays an essential role in our daily life in the last decades. It is behind banking, government, education, medical etc., systems. In our increasingly complex and technology-driven world, industry is demanding for quality softwares.

- Software is a general term used to describe a collection of computer programs, procedures and documentation that perform some tasks (operations) on a computer system.

- Software is responsible for controlling, integrating and managing the hardware components of a computer and for accomplishing specific tasks. In other words, software tells the computer what to do and how to do it.

- For example, the software instructs the hardware what to display on the user's screen, what kinds of input to take from the user, and what kinds of output to generate.

- Computer software can be defined as, "a series of instructions that directs a computer to perform specific tasks or operations".

- Fig. 4.1 shows the various types of computer softwares. Software can be categorized as system software and application software as explained in Chapter 1, Section 1.3.2, Point (2).

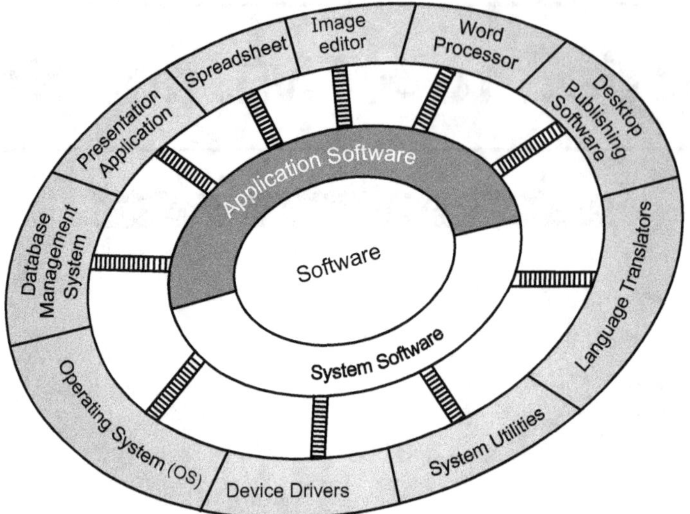

Fig. 4.1: Categories of Computer Softwares

- HTML is a standard markup language used for creating web pages and web applications. In this chapter we study HTML in detail.

4.2 INTRODUCTION TO HTML

- HTML stands for HyperText Markup Language. HTML is the standard markup language for creating web pages and web applications.
- HTML was invented in 1990 by a scientist called Tim Berners-Lee. HTML is the World Wide Web's core markup language. Without HTML, the World Wide Web (WWW) will not exist.
- In HTML **HyperText** simply means "Text within Text". A text has a link within it, is a hypertext. Hypertext refers to the way in which web pages (HTML documents) are linked together. When we click a link in a web page, we are using hypertext. **Markup Language** describes how HTML works. With a markup language, we simply "mark up" a text document with tags that tell a Web browser how to structure it to display.
- A markup language is a set of markup tags and the tags describes document content.
- Since, the early days of the web, there have been many versions of HTML, they are given below:

Versions of HTML	Year
HTML	1991
HTML+	1993
HTML 2.0	1995
HTML 3.2	1997
HTML 4.01	1999
XHTML 1.0	2000
HTML5	2012
XHTML5	2013

- To create the source document, we need an HTML Editor. An HTML editor is a computer program for creating and editing HTML, the markup of a web page.
- There are two main types of HTML editors i.e. text and WYSIWYG (What We See Is What We Get) editors.

- A **text editor** is a type of program used for editing plain text files. A text-based HTML editor includes Notepad on Windows, GNU Emacs on UNIX/Linux, or SimpleText on the Macintosh.
- **WYSIWYG editors** provide an editing interface which resembles how the page will be displayed in a web browser like Macromedia Dreamweaver.

4.2.1 Features of HTML

- Various features of HTML are given below:
  1. HTML is a very **easy and simple language**. It can be easily understood and modified.
  2. It is a **markup language** so it provides a flexible way to design web pages along with the text.
  3. It is **platform-independent** because it can be displayed on any platform like Windows, Linux and Macintosh etc.
  4. It facilitates the programmer to add Graphics, Videos, and Sound to the web pages which makes it more **attractive and interactive**.
  5. It is very easy to make **effective presentation** with HTML because it has a lot of formatting tags.
  6. It support **scripting languages** to support dynamic web applications (e.g., forms that react to user input).

4.2.2 Advantages of HTML

- Various advantages of HTML are listed below:
  1. HTML is highly flexible and user-friendly.
  2. It is an open technology that supports almost all the Web browser and platforms like Windows, Macintosh etc.
  3. HTML is efficient and reliable. We can create the Web page in order to advertise and promote products and services.
  4. It is easily understandable and does not require long time training.
  5. It provides search engine compatible to the Web sites.

4.2.3 HTML Page Structure

- Fig. 4.2 shows page structure of HTML. A web page is also known as HTML page or HTML document.
- HTML files are text files featuring semantically tagged elements. HTML filenames are suffixed with ".htm or .html" file extensions.
- A web page is marked by an opening <html> tag and a closing </html> tag and is divided into the following three major sections:

  1. **Comment Section (optional):** This section contains comments about the web page. It is important to include comments that tell us that is going on in the web page.
  2. **Head Section (optional):** The head section is defined with a starting <head> tag and closing </head> tag. This section usually contains a title for the Web page as shown in the Fig. 4.2.
  3. **Body Section:** The body section comes after the head section. The body section contains the entire information about the web page and its behaviour. A web page outline containing these three sections and the opening and closing HTML tags is illustrated in Fig. 4.2.

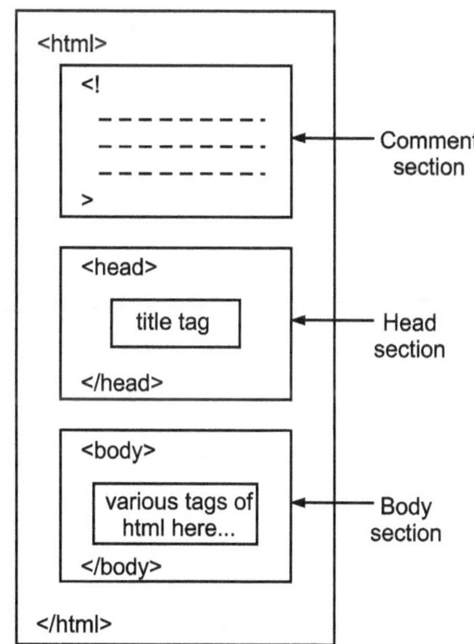

Fig. 4.2: Web Page Structure in HTML

4.2.4 Working of HTML (Creating and Loading HTML Pages)

- Creating an HTML document is very easy and simple. To begin coding HTML we need only two things i.e., a text editor and a web browser.
- Notepad is the most basic of text editor and we will probably code a fair amount of HTML with it.
- Follow the following steps for creating a web page:
  - **Step 1 :** Open Notepad or another text editor and type the following code given in   following example.
  - **Step 2 :** Then, in the File menu, choose Save As to save the program. In the Save as Type option box, choose All Files. Name the file Program1.1.htm or Program1.1.html then Click Save.
  - **Step 3 :** Then go to the folder which we saves the HTML file. Then Double Click on  file which displays the output.

Example for creating a html document:

```
<!DOCTYPE html>
<head>
<title>
</title>
<body>
<h1>Nirali Prakashan</h1>
<p>Textbook Publication Firm.</p>
</body>
</head>
</html>
```

Output:

> # Nirali Prakashan
>
> Textbook Publication Firm.

- In above Example:
  - The !DOCTYPE declaration defines the document type. The <!DOCTYPE> declaration helps the browser to display a web page correctly.
  - The text between <html> and </html> describes the web page.
  - The text between <body> and </body> is the visible page content.
  - The text between <h1> and </h1> is displayed as a heading.
  - The text between <p> and </p> is displayed as a paragraph contents.

4.2.5 HTML Tags

- HTML documents are simply a text file made up of HTML elements and these elements are defined using HTML tags.
- Tags contain elements which provide instructions for how information will be processed or displayed on a web page. HTML is made up of different tags and attributes.
- The tag is an HTML command that shows the layout or displays the desired output of a whole or part of the Web page.
- HTML tags are labels we use to mark up the beginning and end of an element. All tags have the same format and they begin with a less-than sign "<" and end with a greater-than sign ">" also known as angle brackets.
- Generally speaking, there are two kinds of tags i.e., a opening tag like <html> and a closing tag like </html>. A closing tag must include forward slash "/" and it controls the appearance, layout and flow of the web page.

- The information within an element's opening and closing tags is its content. A tag contains three parts i.e. element (identification of tag), attribute and value.

  Syntax of Tag: `<tag_name>content................. </tag_name>`

- HTML tags tells, the browser which elements to present and how to present them. Where the element appears is determined by the order in which the tags appear. HTML tags sometimes require additional information to be supplied to them.

- The additional information supplied to an HTML tag is known as **attributes** of a tag.

- Attributes are written immediately following the tag, separated by a space like <font face="arial"> Welcome </font>. Multiple attributes can be associated with a tag, also separated by a space like <font face = "arial" size = 12> Welcome </font>. The face and size are multiple attributes of the <font> tag.

Types of Tags:

- HTML tags can be of two types:

  1. **Paired Tags:** In paired tag, first tag is called the opening tag like <b> and the second tag is called the closing tag like </b>. Paired tag is also called as container tag. In other words, a tag is said to a paired tag if it along with a companion tag or closing tag appears at the end.

  2. **Singular Tags:** The second type of tag is the singular tag, which is also known as a stand-alone tag or empty tag. The stand-alone tag does not have companion tag or closing tag. For example <br> tag will insert a line break. This tag does not require any companion tag or a closing tag.

- There are two methods or styles of formatting characters in HTML as explained below:

  1. **Logical Styles:** The logical styles inform the browser what kind of text to present. The browser takes care of how to present it. For example the <em> tag tells emphasis must be given.

  2. **Physical Styles:** The physical styles tags explicitly informs the browser how the characters must be shown like bold, italics etc.

4.2.5.1 Basic HTML Tags

- HTML contains following basic tags:

  1. **<!DOCTYPE> Tag:** The <!DOCTYPE> tag is used for specifying which language and version the document is using. This is referred to as the Document Type Declaration (DTD). The <!DOCTYPE> declaration must go right at the top of the page, before any other HTML code.

  2. **<html> Tag:** The <html> tag represents the root of an HTML document. The <html> tag is the container that contains all other HTML elements (except for the <!DOCTYPE> tag which is located before the opening HTML tag). The <html> tag tells the browser that this is an HTML document.

     Syntax: `<html>..........................</html>`

Attribute	Description
1. Manifest	This attribute specifies the address of the document's application cache manifest and the value must be a valid URL.

  3. **<title> Tag:** The <title> tag is used for declaring the title, or name, of the HTML document. The title is usually displayed in the browser's title bar (at the top) or the <title> tag defines a title in the browser toolbar. It is also displayed in browser bookmarks and search results.

     Syntax: `<title>..........................</title>`

  4. **<head> Tag:** The <head> tag is used for indicating the head section of the HTML document. The <head> tag is a container for all the head elements and must include a title for the document, and can include scripts, styles, meta information, and so on.

     Syntax: `<head>............</head>`

Example for basic HTML tags:

```
<!DOCTYPE html>
<html>
<head>What is meant by Web Technologies?
<title>Web Technologies</title>
<body>
<p>Web technologies related to the interface between web servers and their
clients. This information includes markup languages, programming interfaces
and languages, and standards for document identification and display.</p>
<p>Parts of Web Technologies:<br>
1. HyperText Markup Language (HTML) is the main markup language for
   displaying web pages and other information that can be displayed in a web
   browser.<br>
2. Cascading Style Sheets (CSS) is a style sheet language used for describing
   the presentation semantics (the look and formatting) of a document written
   in a markup language.<br>
3. JavaScript (JS) is an open source client-side scripting language commonly
   implemented as part of a web browser in order to create enhanced user
   interfaces and dynamic websites.</p>
</body>
</head>
</html>
```

Output:

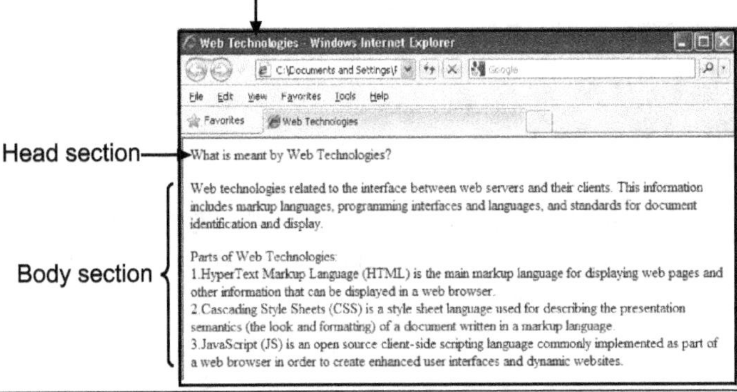

5. **<body> Tag:** The <body> tag defines the document's body. The body tag is placed between the </head> and the </html> tags. The <body> tags contains all the contents of an HTML document, such as text, hyperlinks, images, tables, lists, etc.

   Syntax: `<body>.................</body>`

Attributes:

Attribute	Value	Description
1. alink	color	This attribute specifies the color of an active link in a document.
2. background	URL	This attribute specifies a background image for a document.
3. bgcolor	color	This attribute specifies the background color of a document.
4. link	color	This attribute specifies the color of unvisited links in a document.
5. text	color	This attribute specifies the color of the text in a document.
6. vlink	color	This attribute specifies the color of visited links in a document.

Example for <body> tag:

```
<!DOCTYPE html>
<html>
<head>
<title>HTML <body> tag</title>
<body bgcolor = "orange">
<p>Document content goes... here.</p>
</body>
</head>
</html>
```

Output:

> Document content goes... here.

4.2.5.2 Text Formatting Tags

- The HTML tags are used for formatting text are called as text formatting tags. Following are the text formatting tags used in HTML.

Tag	Description
1. <b>	The <b> tag defines bold text. Anything that appears in a <b>...</b> element is displayed in bold. **Syntax:** <b>......</b>
2. <em>	The <em> tag is used for indicating emphasis. The <em> tag surrounds the word/term being emphasised. **Syntax:** <em>......</em>
3. <i>	The content of the <i> tag is usually displayed in italic. The <i> tag can be used to indicate a technical term, a phrase from another language, a thought etc. **Syntax:** <i>......</i>
4. <small>	The <small> tag defines smaller text (and other side comments). The content of the <small> element is displayed one font size smaller than the rest of the text surrounding it. **Syntax:** <small>......</small>
5. <strong>	The <strong> tag is used for indicating strong importance for its contents. The strong tag surrounds the emphasized word/phrase. **Syntax:** <strong>......</strong>
6. <sub>	The <sub> tag defines subscript text. Subscript text appears half a character below the baseline. Subscript text can be used for chemical formulas, like H_2O. **Syntax:** _{......}
7. <sup>	The <sup> tag defines superscript text. Superscript text appears half a character above the baseline like 10^5. **Syntax:** ^{......}
8. <ins>	The <ins> tag defines a text that has been inserted into a document. **Syntax:** <ins>......</ins>
9. <del>	The <del> tag defines text that has been deleted from a document. **Syntax:** <del>......</del>
10. <u>	The <u> tag usually results in the text being underlined. Anything that appears in a <u>...</u> element is displayed with underline, **Syntax:** <u>......</u>
11. <strike>	Anything that appears in a <strike> tag is displayed with strikethrough, which is a thin line through the text like, ~~strikethrough.~~ **Syntax:** <strike>......</strike>
12. <big>	The content of the <big> element is displayed one font size larger than the rest of the text surrounding it. **Syntax:** <big>......</big>

Example for text formatting tags:

```
<!DOCTYPE html>
<head>
<title> HTML Text formatting Example</title>
</head>
<body>
<p>Nirali Prakashan</p>
<font size="4">Wecome to <b>Nirali Prakashan.</b> <br/></font>
<font size="2"><i>Nirali Prakashan</i> is a textbook publication
firm.<br/></font>
<font face="verdana" color="black">Nirali Prakashan <u>publised more than
3500 book titles.</u></font>
</body>
</html>
```

Output:

> Nirali Prakashan
>
> **Wecome to Nirali Prakashan.**
> *Nirali Prakashan* is a textbook publication firm.
> Nirali Prakashan <u>publised more than 3500 book titles.</u>

Example for subscript and superscript tags:

```
<!DOCTYPE html>
<head>
<title> HTML sub  and sup Tags example</title>
</head>
<body>
<h3> Mathematical Formula</h3>
<p>2<sup>10</sup>+ LOG<sub>10</sup><sup>5</sup></p>
</body>
</html>
```

Output:

> ## Mathematical Formula
>
> $2^{10}+ LOG_{10}{}^{5}$

- **HTML "Computer Output" or "Code Formatting" Tags:** HTML normally uses variable letter size and spacing. This is not we want when displaying computer code for this purpose Computer Code Elements (tags) are used like <kbd>, <samp>, <code> etc. and all these tags are displayed in fixed letter size and spacing.
- Following table shows various types of computer output tags:

Tag	Description
1. `<code>`	The `<code>` tag is used for indicating a piece of code. The code tag surrounds the code being marked up. **Syntax:** `<code>.........</code>`
2. `<kbd>`	`<kbd>`tag defines keyboard input text or keystroke. **Syntax:** `<kbd>.........</kbd>`
3. `<samp>`	`<samp>`tag defines sample computer code. Syntax is: **Syntax:** `<samp>.........</samp>`

contd. ...

4.	`<var>`	It defines a variable. **Syntax:** `<var>`………`</var>`

5.	`<pre>`	The `<pre>` tag is used for indicating preformatted text. The code tag surrounds the code being marked up. Browsers normally render `<pre>` text in a fixed-pitched font, with whitespace in tact, and without word wrap.
		Syntax: `<pre>`………`</pre>`

	Attributes	Value	Description
	1. `width`	`number`	This attribute specifies the maximum number of characters per line.

Example for `<pre>` tag:

```
<!DOCTYPE html>
<head>
<title>HTML <pre> tag Example</title>
<body>
<pre>This text has
been formatted using
    the HTML pre tag. The brower should display all white space
as it was entered.
</pre>
<pre width="30">
Text in a pre tag is displayed in a fixed-width
font, and it preserves both       spaces      and
line breaks
</pre>
</body>
</head>
</html>
```

Output:

```
This text has
been formatted using
    the HTML pre tag. The brower should
       display all white space
as it was entered.

Text in a pre tag is d i s p l a y e d i n a f i x e d - w i d t h
font, and it preserves both       spaces      and
line breaks
```

- **HTML Citations, Quotations and Definition Tags:** Following table shows various tags in this category:

	Tag	Description
1.	`<abbr>`	The `<abbr>` tag is used for indicating an abbreviation, like "WWW" or "NATO". **Syntax:** `<abbr>`…………`</abbr>`
2.	`<address>`	The `<address>` tag defines the contact information for the author/owner of a document or an article. **Syntax:** `<address>`…………`</address>`
3.	`<bdo>`	This tag defines the text direction. The `<bdo>` tag is used to override the current text direction. **Syntax:** `<bdo dir=" ">`…………`</bdo>`

		Attribute	Value	Description
		1. `dir`	`ltr` `rtl`	It specifies the text direction of the text inside the `<bdo>` element

contd. …

4.	`<blockquote>`	This tag defines a section that is quoted from another source. The HTML <blockquote> tag is used for indicating long quotations (i.e. quotations that span multiple lines).
5.	`<q>`	The <q> tag defines a short quotation. Browsers normally insert quotation marks around the quotation using <q>tag. **Syntax:** <q>............</q> <table><tr><th>Attribute</th><th>Value</th><th>Description</th></tr><tr><td>1. cite</td><td>URL</td><td>Specifies the source URL of the quote</td></tr></table>
6.	`<cite>`	The <cite> tag is used for indicating a citation. **Syntax:** <cite>............</cite>
7.	`<dfn>`	The HTML <dfn> tag is used for indicating a definition. The <dfn> tag surrounds the word/term being defined. **Syntax:** <def>............</def>

Example for HTML citations, quotations and definition tags:

```html
<!DOCTYPE html>
<title> HTML <dfn> <strong> <cite> <code> tags example</title>
<head></head>
<body>
<p>Nirali Prakashan</p>
This is <strong>important</strong>. It <strong>really is important. <strong>
And this is even more important!</strong></strong>
<p><dfn>Definition</dfn>: To define the meaning of a word, phrase orterm.</p>
<p>Have we heard that <dfn><abbr title="HyperText Markup Language">
HTML</abbr></dfn> is the predominant markup language for Web pages?</p>
According to <cite title="HTML & XHTML: The Definitive Guide. Published by
Nirali Prakashan">Chuck Musciano and Bill Kennedy</cite>, the HTML cite tag
actually exists!<br>
To create a new array, type the following: <code>var faq = newArray(3)</code>
</body>
</html>
```

Output:

Nirali Prakashan

This is **important**. It **really is important. And this is even more important!**

Definition: To define the meaning of a word, phrase or term.

Have you heard that *HTML* is the predominant markup language for Web pages?

According to *Chuck Musciano and Bill Kennedy*, the HTML cite tag actually exists!
To create a new array, type the following: `var faq = new Array(3)`

4.2.5.3 Block Level Tags

- Most HTML elements (tags) are defined as block level elements or as inline elements. Block level elements normally start (and end) with a new line when displayed in a browser.

    1. **<!--...--> Comment Tag:** The comment tag is used to insert comments in the source code. Comments are not displayed in the browsers. Comments can be inserted into the HTML code to make it more readable and understandable. Comments are ignored by the browser and are not displayed.

Example for comment tag:

```
<!DOCTYPE html>
<title> HTML comment </title>
<!-- The level 4 heading goes here -->
<h4>How to comment out the code.</h4>
<h3>!--...-- tag used for comment and they are simply there for the
programmer's benefit.</h3>
<!-- The text goes here -->
<p>This code demonstrates the HTML code to hide the comments.</p>
</html>
```

Output:

How to comment out your code.

!--.....-- tag used for comment and they are simply there for
the programmer's benefit.

This code demonstrates the HTML code to hide your comments.

2. **Heading <h1> to <h6> Tags:** The HTML <h1> to <h6> tags are used to define HTML headings. <h1> defines the most important heading, while <h6> defines the least important heading.

   Syntax:

```
<h1>............</h1>
<h2>............</h2>
<h3>............</h3>
<h4>............</h4>
<h5>............</h5>
<h6>............</h6>
```

Attribute	Value	Description
1. align	left center right justify	This attribute specifies the alignment of a heading.

Example for HTML heading tags:

```
<!DOCTYPE html>
<head>
<title> HTML headings example</title>
<body>
<h1 align="center">This is heading 1 Nirali Prakashan</h1>
<h2 align="left">This is heading 2 Nirali Prakashan </h2>
<h3>This is heading 3 Nirali Prakashan </h3>
<h4 align="right">This is heading 4 Nirali Prakashan </h4>
<h5 align="left">This is heading 5 Nirali Prakashan </h5>
<h6 align="center">This is heading 6 Nirali Prakashan </h6>
</body>
</head>
</html>
```

Output:

> **This is heading 1 Nirali Prakashan**
>
> **This is heading 2 Nirali Prakashan**
>
> **This is heading 3 Nirali Prakashan**
>
> This is heading 4 Nirali Prakashan
>
> **This is heading 5 Nirali Prakashan**
>
> This is heading 6 Nirali Prakashan

3. **<p> Tag:** HTML documents are divided into paragraphs. The HTML <p> tag is used for defining a paragraph.

    Syntax: <p>............</p>

Attribute	Value	Description
1. align	left right center justify	This attribute specifies the alignment of the text within a paragraph.

4. **
 Tag:** The HTML <br> tag is used for specifying a line break. The <br> tag is an empty tag. In other words, it has no end tag.

    Syntax:

Example for <p> and <br> tags:

```html
<!DOCTYPE html>
<head>
<title> HTML paragraph example</title>
<body>
<p>A technical definition of the World Wide Web is all the resources and
users on the Internet that are using the Hypertext Transfer Protocol (HTTP).
<br>W3C (World Wide Web Consortium), an international consortium of companies
involved with the Internet and the Web.<br> HTML is a markup language. A
markup language is a set of markup tags and the tags describes document
content </p>
<p align="center">HTML stands for Hyper Text Markup Language.
</p>
<p>Hypertext <br> Text displayed on screen which can be accessed usually by a
mouse click or key press. Apart from text, it can display other cool stuff
like images, tables, frames etc.
</p>
<p align="left">Markup Language <br>  Language which is used to structure
data and present it to the user.
</p>
<p align="left">HTML is a language for describing web pages.</p>
</body>
</head>
</html>
```

Output:

```
A technical definition of the World Wide Web is all the resources and users on the
Internet that are using the Hypertext Transfer Protocol (HTTP)
W3C (World Wide Web Consortium), an international consortium of companies involved
with the Internet and the Web.
HTML is a markup language. A markup language is a set of markup tags and the tags
describes document content

                    HTML stands for Hyper Text Markup Language

Hypertext
Text displayed on screen which can be accessed usually by a mouse click or key press.
Apart from text, it can display other cool stuff like images, tables, frames etc.

Markup Language
Language which is used to structure data and present it to the user.

HTML is a language for describing web pages.
```

5. **<center> Tag:** The <center> tag is used to center-align text.

6. **Non breaking Spaces:** Suppose we were to use the phrase "14 Angry Men." Here we would not want a browser to split the "14" and "Angry" across two lines. A good example of this technique appears in the movie "14 Angry Men." In cases where we do not want the client browser to break text, we should use a nonbreaking space entity ( ) instead of a normal space. For example, when coding the "14 Angry Men" paragraph, we would use something similar to the following code:

```
<p>A good example of this technique appears in the movie "14 
Angry Men."</p>
```

7. **<div> Tag:** The <div> tag defines a division or a section in an HTML document. The <div> tag is used to group block-elements to format them with CSS.

   Syntax: `<div>............</div>`

Attribute	Value	Description
1. align	left right center justify	This attribute specifies the alignment of the content inside a <div> element.

Example for <div> tag:

```
<!DOCTYPE html>
<title> HTML div example</title>
<head></head>
<body>
<div style="background-color:orange;text-align:center">
<p>Nirali Prakashan</p>
</div>
<div style="background-color:pink;text-align:center">
<p>Pragati Group</p>
</div>
</body>
</html>
```

Output:

```
                    Nirali Prakashan

                    Pragati Group
```

8. **<span> Tag:** The <span> tag is used for grouping and applying styles to inline elements. The <span> tag provides a way to add a hook to a part of a text or a part of a document. The <span> tag provides no visual change by itself. The difference between the <span> tag and the <div> tag is that the <span> tag is used with inline elements while the <div> tag is used with block-level content.

Syntax: `<span>............</span>`

Example for <span> tag:

```
<!DOCTYPE html>
<head> span tag </head>
<title> HTML span tag example</title>
<body>
<p>The <span style="color:red">span tag is used with  inline  elements</span>
and  the  <span  style="color:purple">div  tag</span>  is  used  with  block-level
content.</p>
</body>
</html>
```

Output:

> span tag
>
> The span tag is used with inline elements **and the div tag is used with block-level content.**

9. **<hr> Tag:** Horizontal rules are used to visually break up sections of a document. The HTML <hr> tag is used for specifying a horizontal rule in an HTML document. The <hr> tag creates a line from the current position in the document to the right margin and breaks the line accordingly. The <hr> tag is used to separate content (or define a change) in an HTML page. The intention behind the <hr> tag is that it indicates a paragraph-level thematic break.

Syntax: `<hr/>`

Attribute	Value	Description
1. align	left center right	This attribute specifies the alignment of a <hr> element.
2. noshade	noshade	This attribute specifies that a <hr> element should render in one solid color (noshaded), instead of a shaded color.
3. size	pixels	This attribute specifies the height of a <hr> element.
4. width	pixels %	This attribute specifies the width of a <hr> element.

Example for <hr> tag:

```
<!DOCTYPE html>
<head>
<title> HTML <hr> example</title>
</head>
<body>
<p>This is paragraph one and should be on top</p>
<hr align="right" width="50%">
<p>This is paragraph two and should be at middle</p>
<hr size="7">
<p>This is paragraph two and should be at bottom</p>
<hr width="80%" size="12" noshade>
</body>
</html>
```

Output:

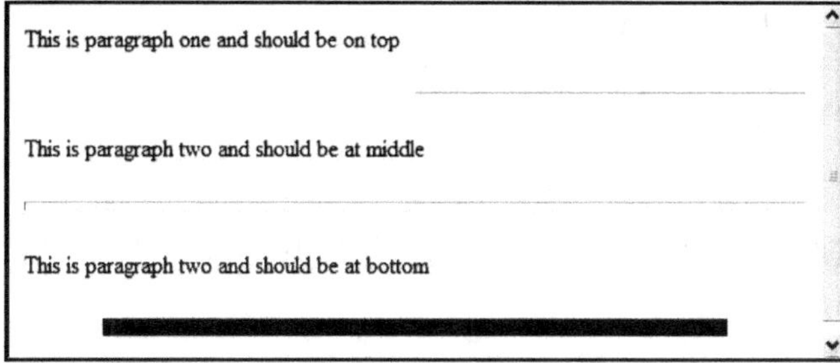

10. <font> Tag: Fonts play very important role in making a website more user friendly and increasing content readability. The <font> tag specifies the font face, font size, and font color of text.

Syntax: `<font>............</font>`

Attribute	Value	Description

1. `color`	`rgb(x,x,x)`
	`#xxxxxx`
	`colorname`
		This attribute specifies the color of text.

2. `face`	`font_family`	This attribute specifies the font of text.
3. `size`	`number`	This attribute specifies the size of text.

Example for <font> tag:

```html
<!DOCTYPE html>
<head>
<title> HTML font tag example</title>
</head>
<body>
<p><h2>Nirali Prakashan</h2></p>
<font size="4" color="red">Wecome to Nirali Prakashan. <br/></font>
<font size="2" color="blue">Nirali Prakashan is a textbook publication firm.<br/></font>
<font face="verdana" color="black">Nirali Prakashan publised more than 3500 book titles.</font>
</body>
</html>
```

Output:

> **Nirali Prakashan**
>
> Wecome to Nirali Prakashan.
> Nirali Prakashan is a textbook publication firm.
> Nirali Prakashan publised more than 3500 book titles.

4.2.6 Hyperlinks

- The real power of HTML is its ability to link one document to the other document. A link is a connection from one web resource (documents) to another.

- A webpage can contain various links that take we directly to other pages and even specific parts of a given page. These links are known as hyperlinks.

- A hypertext link, or hyperlink or link, contains a reference to a specific Web page that we can click to open that Web page.

- Hyperlinks allow visitors to navigate between Web sites by clicking on words, phrases, and images. Thus we can create hyperlinks using text or images available on a webpage.

<a> Anchor Tag:

- The HTML <a> tag is used for creating a hyperlink to another web page. The <a> tag can be used in two ways:
  1. To create a link to another document, by using the href attribute, and
  2. To create a bookmark inside a document, by using the name attribute.
- The <a> tag defines a hyperlink, which is used to link from one page to another. By default, links will appear as follows in all browsers:
  1. An unvisited link is underlined and blue,
  2. A visited link is underlined and purple, and
  3. An active link is underlined and red.

Syntax: `<a href="Document URL">............</a>`

Attributes	Value	Description
1. charset	char_encoding	This attribute specifies the character-set of a linked document.
2. coords	coordinates	This attribute specifies the coordinates of a link.
3. href	URL	This attribute specifies the URL of the page the link goes to.
4. hreflang	language_code	This attribute specifies the language of the linked document.
5. media	media_query	This attribute specifies what media/device the linked document is optimized for.
6. name	section_name	This attribute specifies the name of an anchor.
7. rel	alternate author bookmark help license next nofollow noreferrer prefetch prev search tag	This attribute specifies the relationship between the current document and the linked document.
8. rev	text	This attribute specifies the relationship between the linked document and the current document.
9. shape	default rect circle poly	This attribute specifies the shape of a link.
10. target	_blank _parent _self _top framename	This attribute specifies where to open the linked document.
11. type	MIME_type	This attribute specifies the MIME type of the linked document.

Example for <a> tag:

```
<!DOCTYPE html>
<body>
<p>
<a href="default.asp">HTML</a> This is a link to a
page on this website.
</p>
<p>
<a href="http://www.google.com/">Google</a> This is a link to a website on
the World Wide Web.
</p>
</body>
</html>
```

Output:

> How to create hyperlinks
>
> HTML This is a link to a page on this website.
>
> Google This is a link to a website on the World Wide Web.

Concept of URL:

- Each and every web page has a unique address, called a Uniform Resource Locator (URL) that identifies its location on the Internet.

Parts of URL:

- URL has several parts i.e., the protocol, the domain name, the directory or folder, and the filename and its extension. Below is an example of URL and the corresponding parts.

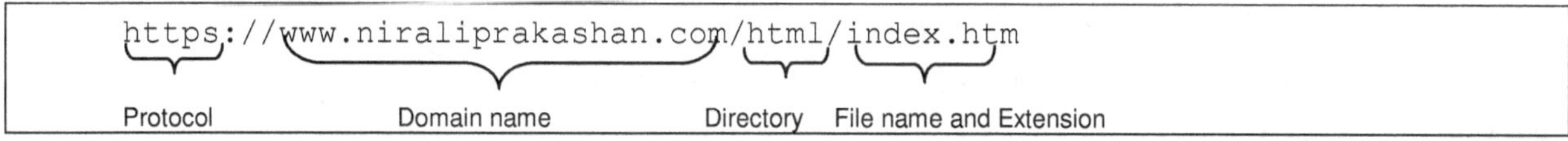

- The first part of URL is called protocol, in the URL above the protocol being used is the https protocol which enables the web browsing. HTTP protocol allows the computers in the world wide web to talk to each other. It provides a set of instructions for precise information exchange.

- The domain name is the second part of URL. It is a text name that corresponds to the IP address of the server that serves the web site. Text name is easier to remember that is why domain name is used instead of the IP address. In our example the domain name is www.niraliprakashan.com.

- There are domain names that ends with .com (for commercial), .edu (for education), .org (for organization), .info (for information) and many more. Another part of the URL identifies the directory name or the folder name. It is the actual location of the file we want to access.

- When creating links make sure that the directory exists and contains the file we want to link to. In the given example the name of the directory or folder is html.

- A directory may consist of one or more sub directories. The last part of the URL is the file name is index.html. In creating links, make sure that the file exists and make sure to use the appropriate file extension such as .htm or .html.

- We may address a URL in one of the following two ways:

  1. **Absolute:** An absolute URL is the complete address of a resource. For example, `https://www.niraliprakashan.com/html/html_text_links.htm`

2. **Relative:** A relative URL indicates where the resource is in relation to the current page. Given URL is added with the `<base>` element to form a complete URL. For example, `/html/html_text_links.htm`. There are two types of relative URLs:

   (i) In **Document Relative URLs**, the document address is given in relation with the originating document.

   (ii) In **Server Relative URLs**, the document address is given in relation with the server on which the document is present.

Examples for linking:

1. **Linking to a Page Section:** We can create a link to a particular section of a page by using name attribute. Here, we will create three links with-in this page itself. First create a link to reach to the top of this page. Here, is the code we have used for the title heading HTML Text Links:

   ```
   <h1>HTML Text Links <a name="top"></a></h1>
   ```

   Now we have a place where we can reach. To reach to this place use the following code with-in this document anywhere:

   ```
   <a href="/html/html_text_links.htm#top">Go to the Top</a>
   ```

   This will produce following link and we try using this link to reach to the top of this page:

   ```
   Go to the Top
   ```

2. **Setting Link Colors:** We can set colors of our links, active links and visited links using link, alink and vlink attributes of `<body>` tag. But it is recommended to use CSS to set colors of links, visited links and active links.

   Following is the example we have used for our example.

   ```
   a:link    {color:#900B09; background-color:transparent}

   a:visited {color:#900B09; background-color:transparent}

   a:active  {color:#FF0000; background-color:transparent}

   a:hover   {color:#FF0000; background-color:transparent}
   ```

3. **Use an image as a link:** We can also use image as link, following example shows this.

Example for image as a link:

```
<!DOCTYPE html>
<body>
<p>Create a link of an image:
<a href="default.asp">
<img src="smiley.gif" alt="HTML tutorial" width="32" height="32"/>
   </a></p>
<p>No border around the image, but still a link:
<a href="default.asp">
<img border="0" src="smiley.gif" alt="HTML tutorial" width="32" height="32"/>
</a></p>
</body>
</html>
```

Output:

4. **Example shows link to another part of a document (on the same page).**

```html
<!DOCTYPE html>
<body>
<p>
<a href="#C4">See also Chapter 1.</a><br/>
<a href="#C4">See also Chapter 2.</a><br/>
<a href="#C4">See also Chapter 3.</a><br/>
<a href="#C4">See also Chapter 4.</a><br/>
<a href="#C4">See also Chapter 5.</a><br/>
<a href="#C4">See also Chapter 6.</a><br/>
<a href="#C4">See also Chapter 7.</a><br/>
</p>
<h2>Chapter 1</h2>
<p>This chapter explains web</p>
<h2>Chapter 2</h2>
<p>This chapter explains web site</p>
<h2>Chapter 3</h2>
<p>This chapter explains HTML</p>
<h2><a name="C4">Chapter 4</a></h2>
<p>This chapter explains ASP</p>
<h2>Chapter 5</h2>
<p>This chapter explains Javascript</p>
<h2>Chapter 6</h2>
<p>This chapter explains CSS</p>
<h2>Chapter 7</h2>
<p>This chapter explains JQuery</p>
</body>
</html>
```

Output:

5. **Following example shows link to a mail message (will only work if we have mail created).**

```html
<!DOCTYPE html>
<body>
<p>
This is an email link:
<a href="mailto:someone@example.com?Subject=Hello%20again">
Send Mail</a>
</p>
```

```
<p>
<b>Note:</b> Spaces between words should be replaced by %20 to ensure that
the browser will display the text properly.
</p>
</body>
</html>
```

Output:

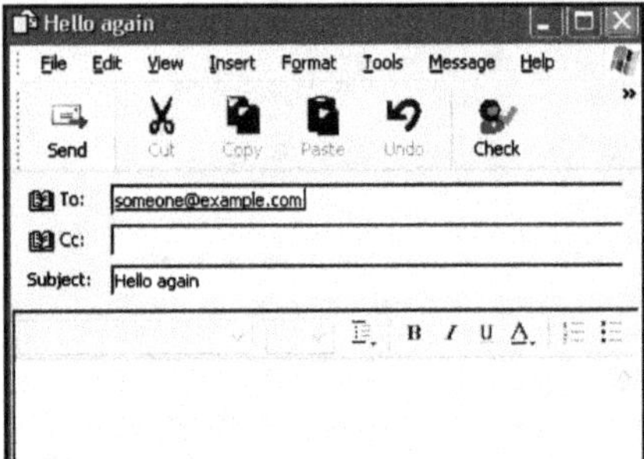

This is an email link: Send Mail

Note: Spaces between words should be replaced by %20 to ensure that the browser will display the text properly.

When we click, Send Mail then following screen displays:

4.2.7 Images

- Images are used in many ways to enhance the look of a Web page and make it more interesting and colorful.

- Images can be used to add background color, to help organize a Web page, to help clarify a point being made in the text, or to serve as links to other webpages.

- It is true that one single image is worth than thousands of words. So as a web developer we should have clear understanding on how to use images in the web pages. An image (from Latin: imago) is an artifact that depicts visual perception, for example a two-dimensional picture, that has a similar appearance to some subject—usually a physical object or a person, thus providing a depiction of it.

- There are basically two types of graphic programs i.e., **Bitmap (or raster) images** are stored as a series of tiny dots called pixels. Each pixel is actually a very small square that is assigned a color, and then arranged in a pattern to form the image. When we zoom in on a bitmap image we can see the individual pixels that make up that image. Bitmap graphics can be edited by erasing or changing the color of individual pixels using a program such as Adobe Photoshop. **Vector images** are not based on pixel patterns, but instead use mathematical formulas to draw lines and curves that can be combined to create an image from geometric objects such as circles and polygons. Vector images are edited by manipulating the lines and curves that make up the image using a program such as Adobe Illustrator.

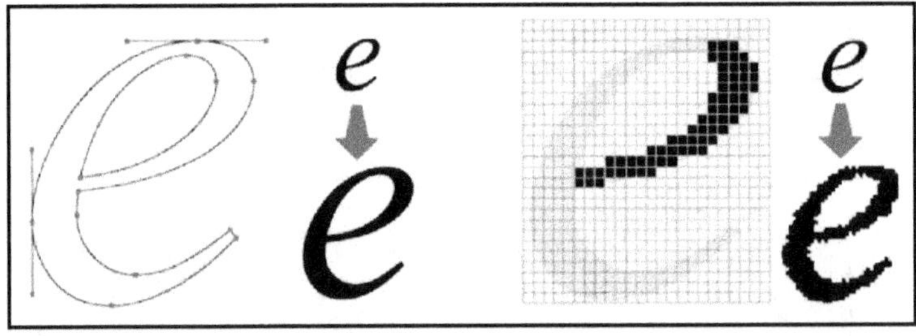

(a) Vector Image   (b) Bitmap Image

Fig. 4.3

Types of Images:

- Image file formats are standardized means of organizing and storing digital images. A filename extension is an identifier specified as a suffix to the name of a computer file.
- The extension indicates a characteristic of the file contents or its intended use. A file extension is typically delimited from the filename with a full stop " . " (period) like .bmp (Bitmap image file), .exe (executable file), .txt (text file) and so on.
- Some common types of images file formats that supported by HTML are explained below:
    1. **JPEG (Joint Photographic Experts Group):** JPEG is the most popular among the image formats used on the web. JPEG files are very 'lossy', meaning so much information is lost from the original image when we save it in JPEG file. JPEG files usually have a filename extension of .jpg or .jpeg.
    2. **GIF (Graphic Interchange Format):** GIF images are limited to 256 colors; they are cross-platform, which means any computer can view them. GIF files are compressed which makes them small in file size but not in dimension. GIF files unlike JPEG files do not lose quality in compression. GIF files have the .gif extension.
    3. **BMP (Bitmap):** BMP is the standard Windows image format on DOS and Windows compatible computers. The BMP format supports RGB (Red, Green, Blue) indexed-colors, grayscale, and Bitmap color modes. BMP files have the .bmp extension. A bitmap image is made up of pixels or bits (binary digits) of information arranged on a grid. Each bit can be visualized as a dot. The number of pixels per unit of measurement, example, ppi (pixels per inch) or dpi (dots per inch) determines the resolution of the image.
    4. **PDF (Portable Document Format):** PDF is used by Adobe Acrobat. PDF files can represent both vector and bitmap graphics and can contain electronic document search and navigation features such as electronic links. The Photoshop PDF format supports RGB, indexed-colors, CMYK (Cyan, Magenta, Yellow and Black), grayscale and Bitmap. This file has .pdf extension.
    5. **Targa**: Format is designed for systems using the true vision video board and is commonly supported by MS-DOS color applications. The Targa format supports 32 bit RGB, grayscale, and 16 bit and 24 bit RGB files without alpha channels. While saving an RGB image in this format, we can choose a pixel depth. Targa files have the .tga extension.
    6. **TIFF (Tagged-Image File Format):** TIFF is used to exchange files between applications and computer platforms. Virtually all paint programs, image editing, and page layout applications support TIFF file format. TIFF files have the .tif or .tiff extension.
    7. **PNG (Portable Network Graphics):** PNG Pronounced "ping" was developed as an alternative to GIF. PNG files support 24-bit images and produces background transparency without jagged edges. Some older versions of Web browsers may not support PNG images. Like GIF and JPEG files, PNG files are cross-platform and compressed. PNG files can have more colors than GIF files and also compress smaller. PNG files have the .png extension.

Image <img> Tag:

- In HTML, images are defined with the <img> tag. The <img> tag is empty, which means that it contains attributes only, and has no closing tag. We will insert any image in our web page by using <img> tag.

Syntax:

```
<img src="image URL" attr_name="attr_value"... more attributes />
```

Attributes	Value	Description
1. `alt`	`text`	This attribute specifies an alternate text for an image.
2. `src`	`URL`	This attribute specifies the URL of an image.
3. `align`	`top` `bottom` `middle` `left` `right`	This attribute specifies the alignment of an image according to surrounding elements.

contd. ...

4.	border	pixels	This attribute specifies the width of the border around an image.
5.	crossoriginNew	anonymous use-credentials	This attribute allow images from third-party sites that allow cross-origin access to be used with canvas.
6.	height	pixels	This attribute specifies the height of an image.
7.	hspace	pixels	This attribute specifies the whitespace on left and right side of an image.
8.	ismap	ismap	This attribute specifies an image as a server-side image-map.
9.	longdesc	URL	This attribute specifies the URL to a document that contains a long description of an image.
10.	src	URL	This attribute specifies the URL of an image.
11.	usemap	#mapname	This attribute specifies an image as a client-side image-map.
12.	vspace	pixels	This attribute specifies the whitespace on top and bottom of an image.
13.	width	pixels	This attribute specifies the width of an image.

Example for <img> tag:

```
<!DOCTYPE html>
<head>
<meta name="GENERATOR" content="MICROSOFT FRONTPAGE 4.0">
<meta name="PROGID" content="FRONTPAGE.EDITOR.DOCUMENT">
<title>IMAGE ALIGNMENT</title>
</head>
<body>
<center>THIS EXAMPLE IS FOR IMAGE ALIGNMENT:</center>
<p><img src = "http://images3.wikia.nocookie.net/__cb20120720034705/christmas
 specials/images/8/80/Mickey_Mouse.jpg" width="90" height="80" align ="TOP"
border = 1> HI, I AM MICKEY, A TOP TEXT</p>
<p><img src = "http://images3.wikia.nocookie.net/__cb20120720034705/christmas
 specials/images/8/80/Mickey_Mouse.jpg" width="90" height="80" align ="MIDDLE"
border = 1> HI, I AM MICKEY, A MIDDLE TEXT</p>
<p><img src = "http://images3.wikia.nocookie.net/__cb20120720034705/christmas
 specials/images/8/80/Mickey_Mouse.jpg" width="90" height="80" align ="BOTTOM"
border = 1> HI, I AM MICKEY, A BOTTOM TEXT</p>
</body>
</html>
```

Output:

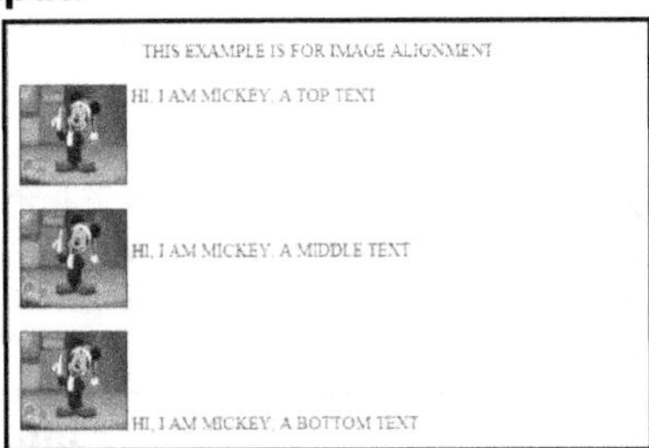

Example for image alignment:

```
<!DOCTYPE html>
<head>
<meta name="GENERATOR" content="MICROSOFT FRONTPAGE 4.0">
<meta name="PROGID" content="FRONTPAGE.EDITOR.DOCUMENT">
<title>HI</title>
</head>
<body>
<p><img  src  =  "http://images.free-extras.com/pics/m/mickey_mouse-1104.jpg"
width="90" height="80" align ="LEFT" border = 1> HI, I AM MICKEY, A TEXT
AFTER IMAGE<br>YOU CAN WRITE A WHOLE PARAGRAPH<br> ALSO AND CAN SEE IT AS
<b>LEFT ALIGNMENT </b> </p> <br>
<p><img  src  =  "http://images.free-extras.com/pics/m/mickey_mouse-1104.jpg"
width="90" height="80" align ="RIGHT" border = 1> HI, I AM MICKEY, A TEXT
BEFORE  IMAGE  YOU  CAN  WRITE  IN  <b>RIGHT ALIGNMENT</b>  SAME  THAT  OF  LEFT
ALIGNMENT</p>
</body>
</html>
```

Output:

Example for image spacing:

```
<!DOCTYPE html>
<head>
<meta name="GENERATOR" content="MICROSOFT FRONTPAGE 4.0">
<meta name="PROGID" content="FRONTPAGE.EDITOR.DOCUMENT">
<title>IMAGE SPACING</title>
</head>
<body>
<p><img src = "http://www.umnet.com/pic/diy/screensaver/8/mickey-mouse 87422.
 jpg" border=1 width="90" height="80">
<img src = http://www.umnet.com/pic/diy/screensaver/8/mickey-mouse-87422.jpg
 border=1 width="90" height="80"> NO VERTICLE OR HORIZONTAL SPACING
</p>
<p>
<img src = http://www.umnet.com/pic/diy/screensaver/8/mickey-mouse-87422.jpg
 width = "90" height = "80" border=1 HSPACE = "20" VSPACE = "20" >
<img  src  =  "http://www.umnet.com/pic/diy/screensaver/8/mickey-mouse-87422.
 jpg" width = "90" height = "80" border=1 HSPACE ="20" VSPACE = "20" > 20
 PIXELS VERTICLE AND HORIZONTAL SPACING FROM EACH SIDE
</p>
</body>
</html>
```

Output:

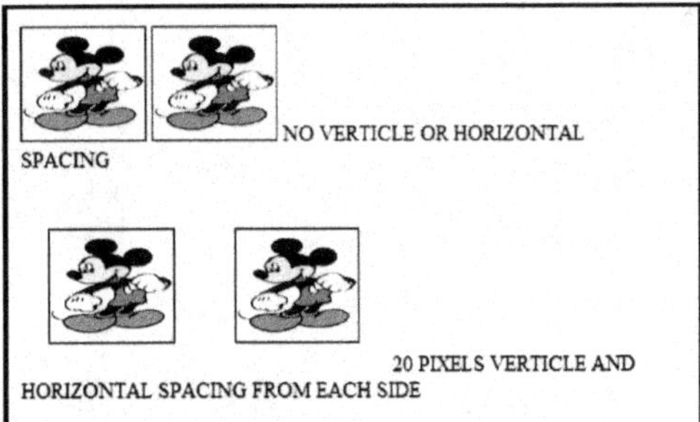

- **Image as a Link:** Anything can be a link i.e., text or images. To make an image into a link we simply put the image tag inside the tag for a link. The tag would look like this:

```
<a href="http://www.anycartoonsite.com"><img src="z:\book\mikey.jpg"></a>
```

Example for image as link:

```
<!DOCTYPE html>
<head>
<meta http-equiv="CONTENT-TYPE" CONTENT="TEXT/HTML; CHAR SET = WINDOWS-1252">
<meta name="GENERATOR" CONTENT="MICROSOFT FRONTPAGE 4.0">
<meta name="PROGID" CONTENT="FRONTPAGE.EDITOR.DOCUMENT">
<title>IMAGELINK</title>
</head>
<body>
IMAGE LINK EXAMPLE
<p>
<a  href  =  "HTTP://WWW.YAHOO.COM"  ><img  src  =  "http://images4.wikia.
 nocookie.net/__cb20120710211159/cartoons/images/8/80/Mickey_Mouse.jpg"
 border = "0"  width ="90" height = "80"> </a>
</p>
</body>
</html>
```

Output:

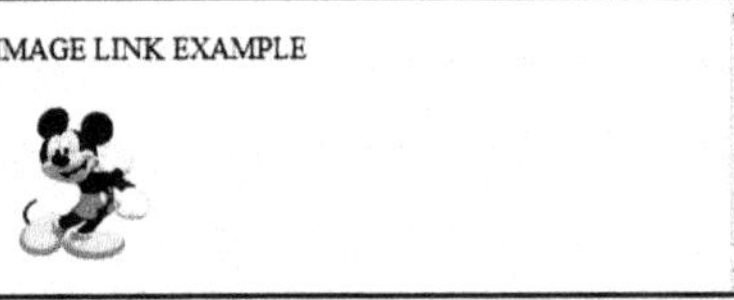

4.2.8 Tables

- Tables are just like spreadsheets and they are made up of rows and columns.
- The HTML table model allows to arrange data - text, preformatted text, images, links, forms, form fields, other tables etc. into rows and columns of cells.
- Tables are used on websites for two major purposes:
  1. The obvious purpose of arranging information in a table.
  2. The less obvious - but more widely used - purpose of creating a page layout with the use of hidden tables.
- Various tags in HTML for tables are explained below:
  1. **<table> Tag:** The <table> tag defines an HTML table. An HTML table consists of the <table> element and one or more <tr>, <th>, and <td> elements. The <tr> element defines a table row, the <th> element defines a table header, and the <td> element defines a table cell.

     Syntax: <table>............</table>

Attributes	Value	Description
1. align	left center right	This attribute specifies the alignment of a table according to surrounding text.
2. bgcolor	rgb(x,x,x) #xxxxxx colorname	This attribute specifies the background color for a table.
3. border	pixels	This attribute specifies the width of the borders around a table.
4. cellpadding	pixels	This attribute specifies the space between the cell wall and the cell content.
5. cellspacing	pixels	This attribute specifies the space between cells.
6. frame	void above below hsides lhs rhs vsides box border	This attribute specifies which parts of the outside borders that should be visible.
7. rules	none groups rows cols all	This attribute specifies which parts of the inside borders that should be visible.
8. summary	text	This attribute specifies a summary of the content of a table.
9. width	pixels %	This attribute specifies the width of a table.

Example for <table> tag:

```
<!DOCTYPE html>
<body>
<p>
Each table starts with a table tag.
Each table row starts with a tr tag.
Each table data starts with a td tag.
</p>
<h4>One column:</h4>
<table border="1">
<tr>
   <td>100</td>
</tr>
</table>
<h4>One row and three columns:</h4>
<table border="1">
```

```
<tr>
   <td>100</td>
   <td>200</td>
   <td>300</td>
</tr>
</table>
<h4>Two rows and three columns:</h4>
<table border="1">
<tr>
   <td>100</td>
   <td>200</td>
   <td>300</td>
</tr>
<tr>
   <td>400</td>
   <td>500</td>
   <td>600</td>
</tr>
</table>
</body>
</html>
```

Output:

Each table starts with a table tag. Each table row starts with a tr tag. Each table data starts with a td tag.

One column:

100

One row and three columns:

100 200 300

Two rows and three columns:

100 200 300
400 500 600

2. **<th> Tag:** Table heading can be defined using <th> tag or the <th> tag defines a header cell in an HTML table. An HTML table has two kinds of cells:

   (i) Header Cells: Contains header information (created with the <th> element), and

   (ii) Standard Cells: Contains data (created with the <td> element).

   The text in <th> elements are bold and centered by default. The text in <td> elements are regular and left-aligned by default.

   Syntax: `<th>............</th>`

Attributes	Value	Description
1. abbr	text	This attribute specifies an abbreviated version of the content in a cell.
2. align	left right center justify char	This attribute aligns the content in a cell.

contd. ...

3.	axis	category_name	This attribute specifies categorizes cells.
4.	bgcolor	rgb(x,x,x) #xxxxxx colorname	This attribute specifies the background color of a cell.
5.	char	character	This attribute specifies aligns the content in a cell to a character.
6.	charoff	number	This attribute sets the number of characters the content will be aligned from the character specified by the char attribute.
7.	colspan	number	This attribute sets the number of columns a cell should span.
8.	height	pixels %	This attribute sets the height of a cell.
9.	nowrap	nowrap	This attribute specifies that the content inside a cell should not wrap.
10.	rowspan	number	This attribute sets the number of rows a cell should span.
11.	scope	col colgroup row rowgroup	This attribute defines a way to associate header cells and data cells in a table.
12.	valign	top middle bottom baseline	This attribute specifies vertical aligns the content in a cell.
13.	width	pixels %	This attribute specifies the width of a cell.

Example of <th> tag.

```
<!DOCTYPE html>
<body>
<h4>Table headers:</h4>
<table border="1">
<tr>
  <th>Name</th>
  <th>Telephone</th>
  <th>Telephone</th>
</tr>
<tr>
  <td>Bill Gates</td>
  <td>555 77 854</td>
  <td>555 77 855</td>
</tr>
</table>
<h4>Vertical headers:</h4>
<table border="1">
<tr>
  <th>First Name:</th>
  <td>Bill Gates</td>
</tr>
```

```
<tr>
   <th>Telephone:</th>
   <td>555 77 854</td>
</tr>
<tr>
   <th>Telephone:</th>
   <td>555 77 855</td>
</tr>
</table>
</body>
</html>
```

Output:

Table headers:

| Name | Telephone | Telephone |
| Bill Gates | 555 77 854 | 555 77 855 |

Vertical headers:

First Name:	Bill Gates
Telephone:	555 77 854
Telephone:	555 77 855

3. **<tr> Tag:** The <tr> tag defines a row in an HTML table. A <tr> element contains one or more <th> or <td> elements.

    Syntax: `<tr>............</tr>`

Attributes	Value	Description

1. `align`	`right`
	`left`
	`center`
	`justify`
	`char`
		This attribute aligns the content in a table row.

2. `bgcolor`	`rgb(x,x,x)`
	`#xxxxxx`
	`colorname`
		This attribute specifies a background color for a table row.

3. `char`	`character`	This attribute aligns the content in a table row to a character.
4. `charoff`	`number`	This attribute sets the number of characters the content will be aligned from the character specified by the char attribute.

5. `valign`	`top`
	`middle`
	`bottom`
	`baseline`
		This attribute vertical aligns the content in a table row.

4. **<td> Tag:** The <td> tag defines a standard cell in an HTML table. The <td> tag is used to mark up individual cells inside a table row.

    Syntax: `<td>............</td>`

Attribute	Value	Description
1. `abbr`	`text`	This attribute specifies an abbreviated version of the content in a cell.
2. `align`	`left` `right` `center` `justify` `char`	This attribute aligns the content in a cell.
3. `axis`	`category_name`	This attribute categorizes cells.
4. `bgcolor`	`rgb(x,x,x)` `#xxxxxx` `colorname`	This attribute specifies the background color of a cell.
5. `char`	`character`	This attribute aligns the content in a cell to a character.
6. `charoff`	`number`	This attribute specifies sets the number of characters the content will be aligned from the character specified by the char attribute.
7. `colspan`	`number`	This attribute specifies the number of columns a cell should span.
8. `headers`	`header_id`	This attribute specifies one or more header cells a cell is related to.
9. `height`	`pixels` `%`	This attribute specifies sets the height of a cell.
10. `nowrap`	`nowrap`	This attribute specifies that the content inside a cell should not wrap.
11. `rowspan`	`number`	This attribute specifies sets the number of rows a cell should span.
12. `scope`	`col` `colgroup` `row` `rowgroup`	This attribute specifies defines a way to associate header cells and data cells in a table.
13. `valign`	`top` `middle` `bottom` `baseline`	This attribute specifies vertical aligns the content in a cell.
14. `width`	`pixels` `%`	This attribute specifies the width of a cell.

Example of <td> tag:

```
<!DOCTYPE html>
<head>
<body>
<table border="1">
<tr>
<th>Name</th>
<th>Salary</th>
</tr>
```

```
<tr>
<td>Ramesh Salunkhe</td>
<td>50,000</td>
</tr>
<tr>
<td>Amar Salvi</td>
<td>70,000</td>
</tr>
</table>
</body>
</head>
</html>
```

Output:

Name	Salary
Ramesh Salunkhe	50,000
Amar Salvi	75,000

5. **<caption> Tag:** The <caption> tag defines a table caption. The <caption> tag is used to provide a caption for a table. This caption can either appear above or below the table. This can be indicated with the align attribute.

   Syntax: `<caption>............</caption>`

   The <caption> tag must be inserted immediately after the <table> tag. We can specify only one caption per table.

Attribute	Value	Description
1. align	left right top bottom	This attribute defines the alignment of a caption.

Example for <caption> tag:

```
<!DOCTYPE html>
<body>
<table border="1">
  <caption>Monthly Savings</caption>
  <tr>
    <th>Month</th>
    <th>Savings</th>
  </tr>
  <tr>
    <td>January</td>
    <td>Rs. 1000</td>
  </tr>
  <tr>
    <td>February</td>
    <td> Rs. 1500</td>
  </tr>
</table>
</body>
</html>
```

Output:

Cellpadding and Cellspacing in a Table:

- There are two attribiutes called cellpadding and cellspacing which we will use to adjust the white space in our table cell. Cellspacing defines the width of the border, while cellpadding represents the distance between cell borders and the content within.

Example for cellspacing:

```
<!DOCTYPE html>
<body>
<h4>Without  cellspacing:</h4>
<table border="1">
<tr>
  <td>First</td>
  <td>Row</td>
</tr>
<tr>
  <td>Second</td>
  <td>Row</td>
</tr>
</table>
<h4>With cellspacing="0":</h4>
<table border="1" cellspacing="0">
<tr>
  <td>First</td>
  <td>Row</td>
</tr>
<tr>
  <td>Second</td>
  <td>Row</td>
</tr>
</table>
<h4>With cellspacing="10":</h4>
<table border="1" cellspacing="10">
<tr>
  <td>First</td>
  <td>Row</td>
</tr>
<tr>
  <td>Second</td>
  <td>Row</td>
</tr>
</table>
</body>
</html>
```

Output:

```
Without cellspacing:

First    Row
Second Row

With cellspacing="0":

First    Row
Second Row

With cellspacing="10":

First    Row

Second   Row
```

Example of cellpadding:

```html
<!DOCTYPE html>
<body>
<h4>Without cellpadding:</h4>
<table border="1">
<tr>
  <td>First</td>
  <td>Row</td>
</tr>
<tr>
  <td>Second</td>
  <td>Row</td>
</tr>
</table>
<h4>With cellpadding:</h4>
<table border="1"
cellpadding="10">
<tr>
  <td>First</td>
  <td>Row</td>
</tr>
<tr>
  <td>Second</td>
  <td>Row</td>
</tr>
</table>
</body>
</html>
```

Output:

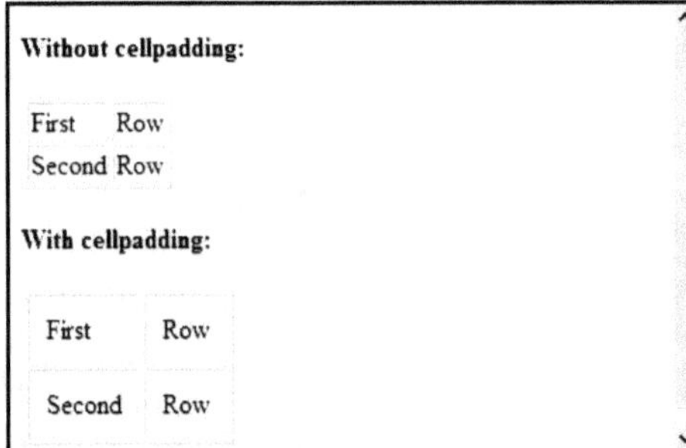

Colspan and Rowspan Attributes in Table: We will use colspan attribute if we want to merge two or more columns into a single column. Similar way we will use rowspan if we want to merge two or more rows.

Example for rawspan and colspan:

```
<!DOCTYPE html>
<head>
<title>Practice HTML table spans</title>
</head>
<body>
<p>Merge two or more rows or columns and then see the result:</p>
<table border="1">
<tr>
<th>Column 1</th>
<th>Column 2</th>
<th>Column 3</th>
</tr>
<tr><td rowspan="2">Row 1 Cell 1</td>
<td>Row 1 Cell 2</td><td>Row 1 Cell 3</td></tr>
<tr><td>Row 2 Cell 2</td><td>Row 2 Cell 3</td></tr>
<tr><td colspan="3">Row 3 Cell 1</td></tr>
</table>
</body>
</html>
```

Output:

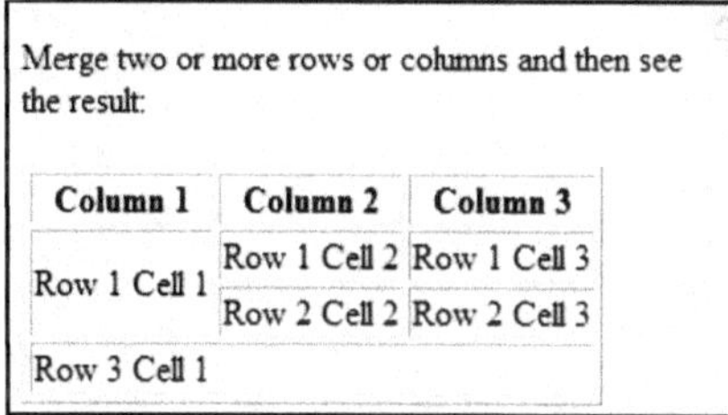

Nested Tables: We can use one table inside another table is called as nesting of tables (or nested tables). Not only tables we can use almost all the tags inside table data tag <td>. Following is the example of using another table and other tags inside a table cell.

Example for nested tables:

```
<!DOCTYPE html>
<head>
<title>HTML Nested table</title>
</head>
<body>
<table border="1">
<tr>
<td>
    <table border="1">
    <tr>
    <th>Name</th>
    <th>Salary</th>
    </tr>
```

```
      <tr>
      <td>Ramesh Raman</td>
      <td>5,700</td>
      </tr>
      <tr>
      <td>Shabbir Hussein</td>
      <td>7,500</td>
      </tr>
      </table>
   </td>
   <td>
      <ul>
      <li>This is another cell</li>
      <li>Using list inside this cell</li>
      </ul>
   </td>
   </tr>
   <tr>
   <td>Row 2, Column 1</td>
   <td>Row 2, Column 2</td>
   </tr>
   </table>
   </body>
   </html>
```

Output:

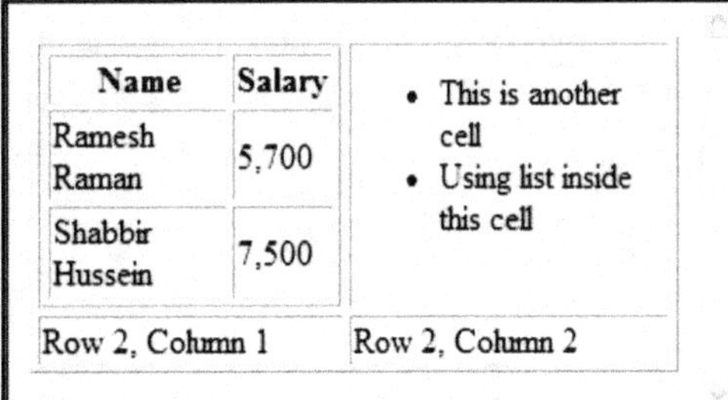

4.2.9 Lists

- In HTML, we can list out our items, subjects or menu in the form of a list. HTML gives us three different types of lists as explained below:

  1. **Unordered Lists:** An unordered list is a collection of related items that have no special order or sequence. The most common unordered list, we will find on the Web is a collection of hyperlinks to other documents. Unordered list is created by using <ul> tag. Each item in the list is marked with a butllet. The bullet itself comes in three flavors: squares, discs, and circles. The default bullet displayed by most web browsers is the traditional full disc.

  2. **Ordered Lists:** The typical browser formats the contents of an ordered list just like an unordered list, except that the items are numbered instead of bulleted. The numbering starts at one and is incremented by one for each successive ordered list element tagged with <li> tag. Ordered list is created by using <ol> tag. Each item in the list is marked with a number.

  3. **Definition Lists:** HTML also support a list style entirely different from the ordered and unordered lists we have discussed so far - definition lists. Like the entries we find in a dictionary or encyclopedia,

complete with text, pictures, and other multimedia elements, the Definition List is the ideal way to present a glossary, list of terms, or other name/value list. Definition List makes use of following tags:

<dl> - Defines the start of the list

<dt> - A term.

<dd> - Term definition.

</dl> - Defines the end of the list.

- Various lists of tags are explained below:

  1. **<meta> Tag:** Meta elements are typically used to specify page description, keywords, author of the document, last modified, and other metadata. The <meta> tag always goes inside the head element. Metadata is data (information) about data. The <meta> tag is used for declaring metadata for the HTML document. Metadata will not be displayed on the page, but will be machine parsable.

     Syntax: `<meta name = string content = string>`

Attributes	Value	Description
1. `charset`	`character_set`	This attribute specifies the character encoding for the HTML document.
2. `content`	`text`	This attribute specifies gives the value associated with the http-equiv or name attribute.

3. `http-equiv`	`content-type`
	`default-style`
	`refresh`
		This attribute specifies provides an HTTP header for the information/value of the content attribute.

4. `name`	`application-name`
	`author`
	`description`
	`generator`
	`keywords`
		This attribute specifies a name for the metadata.

5. `scheme`	`format/URI`	This attribute specifies a scheme to be used to interpret the value of the content attribute.

  o Metadata can include document description, keywords, author etc. It can also be used to refresh the page or set cookies. The <meta> tag is placed between the opening/closing <head> tags.

Example for <meta> tag:

```html
<!DOCTYPE html>
<head>
<title>Meta Refresh Example</title>
<meta http-equiv="refresh"
content="5;url=/html_5/tags/html_meta_tag_example.cfm" />
</head>
<body style="background-color:#ff9900;">
<p>Watch me redirect to another page in 5 seconds...</p>
</body>
</html>
```

Output:

2. **<li> List Tag:** The <li> tag defines a list item. The <li> tag is used in ordered lists (<ol>), unordered lists (<ul>), and in menu lists (<menu>).

Syntax: `<li>.........</li>`

Attributes	Value	Description

1. `type`	`1`
	`A`
	`a`
	`I`
	`i`
	`disc`
	`square`
	`circle`
		This attribute specifies which kind of bullet point will be used.

2. `value`	`number`	This attribute specifies the value of a list item.

3. **<ul> Unordered List Tag:** An unordered list starts with the <ul> tag. Each list item starts with the <li> tag. The <ul> tag defines an unordered (bulleted) list.

Syntax: `<ul>.........</ul>`

Attributes	Value	Description
1. `compact`	`compact`	This attribute specifies that the list should render smaller than normal.

2. `type`	`disc`
	`square`
	`circle`
		This attribute specifies the kind of marker to use in the list.

Example for <ul> tag:

```
<!DOCTYPE html>
<body>
<head><h3>An Unordered List Example</h3>
<h4>Disc bullets list:</h4>
<ul type="disc">
 <li>Apples</li>
 <li>Bananas</li>
 <li>Lemons</li>
 <li>Oranges</li>
</ul>
<h4>Circle bullets list:</h4>
<ul type="circle">
 <li>Apples</li>
 <li>Bananas</li>
 <li>Lemons</li>
 <li>Oranges</li>
</ul>
<h4>Square bullets list:</h4>
<ul type="square">
 <li>Apples</li>
 <li>Bananas</li>
 <li>Lemons</li>
 <li>Oranges</li>
</ul>
</body>
</head>
</html>
```

Output:

```
An Unordered List Example

Disc bullets list:

   • Apples
   • Bananas
   • Lemons
   • Oranges

Circle bullets list:

   o Apples
   o Bananas
   o Lemons
   o Oranges

Square bullets list:

   ▪ Apples
   ▪ Bananas
   ▪ Lemons
   ▪ Oranges
```

- **Nested List:** One list inside another list is known as nesting of list or nested list. We can nested unordered list as follows:

```
<!DOCTYPE html>
<body>
<h4>A nested List:</h4>
<ul>
  <li>Coffee</li>
  <li>Tea
    <ul>
    <li>Black tea</li>
    <li>Green tea
      <ul>
      <li>China</li>
      <li>Africa</li>
      </ul>
    </li>
    </ul>
  </li>
  <li>Milk</li>
</ul>
</body>
</html>
```

Output:

```
A nested List:

   • Coffee
   • Tea
        o Black tea
        o Green tea
             ▪ China
             ▪ Africa
   • Milk
```

4. **<ol> Ordered List Tag:** An ordered list starts with the <ol> tag. Each list item starts with the <li> tag. Each item in the list is marked with a number.

   Syntax: `<ol>............</ol>`

Attributes	Value	Description
1. compact	compact	This attribute specifies that the list should render smaller than normal.
2. reversed	reversed	This attribute specifies that the list order should be descending like 9,8,7...
3. start	number	This attribute specifies the start value of an ordered list.
4. type	1 A a I i	This attribute specifies the kind of marker to use in the list.

Example for <ol> tag:

```html
<!DOCTYPE html>
<head>
<body>
<head><h3>An Ordered List Example</h3>
<h4>Numbered list:</h4>
<ol>
 <li>Apples</li>
 <li>Bananas</li>
 <li>Lemons</li>
</ol>
<h4>Letters list:</h4>
<ol type="A">
 <li>Apples</li>
 <li>Bananas</li>
 <li>Lemons</li>
</ol>
<h4>Lowercase letters list:</h4>
<ol type="a">
 <li>Apples</li>
 <li>Bananas</li>
 <li>Lemons</li>
</ol>
<h4>Roman numbers list:</h4>
<ol type="I">
 <li>Apples</li>
 <li>Bananas</li>
 <li>Lemons</li>
 </ol>
<h4>Lowercase Roman numbers list:</h4>
<ol type="i">
 <li>Apples</li>
 <li>Bananas</li>
 <li>Lemons</li>
 </ol>
</body>
</head>
</html>
```

Output:

```
An Ordered List Example

Numbered list:

    1.  Apples
    2.  Bananas
    3.  Lemons

Letters list:

    A.  Apples
    B.  Bananas
    C.  Lemons

Lowercase letters list:

    a.  Apples
    b.  Bananas
    c.  Lemons

Roman numbers list:

    I.    Apples
    II.   Bananas
    III.  Lemons

Lowercase Roman numbers list:

    i.    Apples
    ii.   Bananas
    iii.  Lemons
```

5. **Definition Lists:** A definition list is a list of items, with a description of each item. The <dl> tag defines a definition list. <dl> tag is used to provide a list of items with associated definitions. Every item should be put in a dt and its definition goes in the dd directly following it. This list is typically rendered without bullets of any kind. The definition list is the ideal way to present a glossary, list of terms, or other name/value list.

   Syntax: `<dl>............</dl>`

   Definition List makes use of following two tags:

   (i) **<dt>:** The <dt> tag is used inside dl. It marks up a term whose definition is provided by the next dd. The dt tag may only contain text-level markup.

   (ii) **<dd>:** The <dd> tag is used inside a dl definition list to provide the definition of the text in the dt tag. It may contain block elements but also plain text and markup.

Example for definition list:

```html
<!DOCTYPE html>
<body>
<h4>A Definition List:</h4>
<dl>
   <dt>Coffee</dt>
   <dd>Black hot drink</dd>
   <dt>Milk</dt>
   <dd>White cold drink</dd>
</dl>
</body>
</html>
```

Output:

```
A Definition List:

Coffee
        Black hot drink
Milk
        White cold drink
```

4.2.10 Frames

- The term frames is used as shorthand for, "a technique to display multiple documents at once". This technique is only used when the browser is running on a graphical display.

- Frames allow us to split the browser window into multiple windows that can display different pages. This will make the navigation much easier for the visitor.

- With frames, we can display more than one HTML document in the same browser window. Each HTML document is called a frame, and each frame is independent of the others.

- A collection of frames in the browser window is known as a frameset.

- The window is divided up into frames in a similar pattern to the way tables are organized into rows and columns. The simplest of framesets might just divide the screen into two rows, while a complex frameset could use several rows and columns.

- Frames are used to display more than one document in the same window. Frames are defined in <frameset>, and inside it we have <frame> which defines the location of the web page to load into the frame. See the example given below to understand it clearly.

Example for Vertical frames:

```
<!DOCTYPE html>
<frameset cols="50%,30%,20%">
<frame src="frame1.html" />
<frame src="frame2.html" />
<frame src="frame3.html" />
</frameset>
</html>
```

Output:

This is frame1	This is frame2	This is frame3

Example for Horizontal frames:

```
<!DOCTYPE html>
<frameset rows="50%,30%,20%">
<frame src="frame1.html" />
<frame src="frame2.html" />
<frame src="frame3.html" />
</frameset>
</html>
```

Output:

This is frame1
This is frame2
This is frame3

1. **<frame> Tag:** The <frame> tag indicates what goes in each frame of the frameset. The <frame> element is always an empty element, and therefore should not have any content, although each <frame> element should always carry one attribute, src, to indicate the page that should represent that frame. The <frame> tag defines one particular window (frame) within a <frameset> and each <frame> in a <frameset> can have different attributes, such as border, scrolling, the ability to resize, etc.

   Syntax: <frame>........</frame>

Attributes	Value	Description
1. `frameborder`	`0` `1`	This attribute specifies whether or not to display a border around a frame.
2. `longdesc`	`URL`	This attribute specifies a page that contains a long description of the content of a frame.
3. `marginheight`	`pixels`	This attribute specifies the top and bottom margins of a frame.
4. `marginwidth`	`pixels`	This attribute specifies the left and right margins of a frame.
5. `name`	`name`	This attribute specifies the name of a frame.
6. `noresize`	`noresize`	This attribute specifies that a frame cannot be resized.

7. `scrolling`	`yes`
	`no`
	`auto`
		This attribute specifies whether or not to display scrollbars in a frame.

8. `src`	`URL`	This attribute specifies the URL of the document to show in a frame.

2. **<frameset> Tag:** The <frameset> tag holds one or more frame elements. Each frame element can hold a separate document. The <frameset> tag replaces the <body> element in frameset documents. The <frameset> tag defines how to divide the window into frames. Each frameset defines a set of rows or columns. If we define frames by using rows then horizontal frames are created. If we define frames by using columns then vertical frames are created. The values of the rows/columns indicate the amount of screen area each row/column will occupy. Each frame is indicated by <frame> tag and it defines what HTML document to put into the frame.

Syntax:
```
<frameset cols/rows="   ">
     <frame src=  "            ">
     .
     .
     .
     </frameset>
```

Attributes	Value	Description

1. `cols`	`pixels`
	`%`
	`*`
		This attribute pecifies the number and size of columns in a frameset.

2. `rows`	`pixels`
	`%`
	`*`
		This attribute specifies the number and size of rows in a frameset.

Example for frameset element defining rows and columns with vertical:
```
<!DOCTYPE html>
<head>
<meta name="GENERATOR" content="Microsoft FrontPage 4.0">
<meta name="ProgId" content="FrontPage.Editor.Document">
```

```
<title>Frameset Element Defining Rows and Columns</title>
</head>
<frameset rows="64,*,64">
<frame src="z:\comp.html">
<frameset cols="150,*">
<frame src="z:\about.html">
<frame src="z:\dept1.html">
</frameset>
<frame src="z:\it.html">
</frameset>
</html>
```

Output:

Example for frames horizontal:

```
<!DOCTYPE html>
<head>
<meta name="GENERATOR" content="Microsoft FrontPage 4.0">
<meta name="ProgId" content="FrontPage.Editor.Document">
<title>FrameSet Element with rows and columns</title>
</head>
<frameset cols="150,*">
<frame name="contents" scrolling= "no" noresize src="z:\comp.html">
<frameset rows="20%,*">
    <frame name="rtop" src="z:\it.html">
    <frame name="main" src="z:\etc.html">
</frameset>
</frameset>
</html>
```

Output:

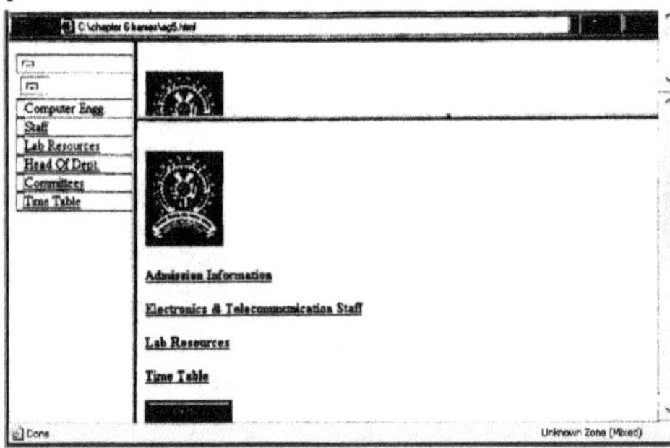

Code of frame defining src, name, noresize and scrolling attributes:

```html
<!DOCTYPE html>
<head>
<meta name="GENERATOR" content="Microsoft FrontPage 4.0">
<meta name="ProgId" content="FrontPage.Editor.Document">
<title>example of frame src,name,scrolling</title></head>
<frameset rows="160,*,160">
<frame name="banner" scrolling="no" noresize src="z:\comp.html">
<frame name="content" scrolling=yes noresize src="z:\etc.html">
<frame scrolling="no" noresize src="z:\about.html">
</frameset>
</html>
```

Output:

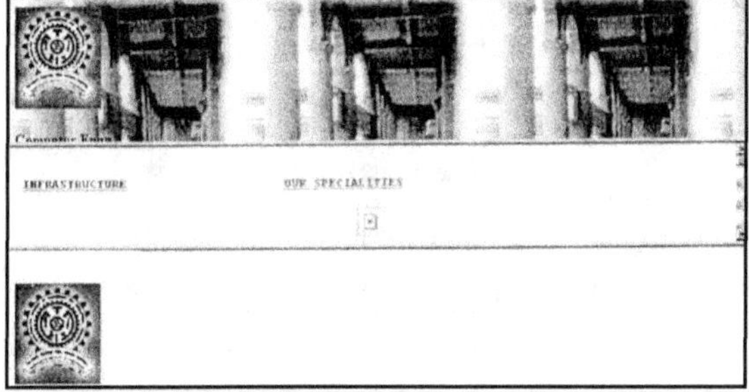

3. **<noframes> Tag:** If a user is using any old browser or any browser which does not support frames then <noframes> element should be displayed to the user.

 Syntax: `<noframes>............</noframes>`

 The noframes element contains content that should only be rendered when frames are not displayed. noframes is typically used in a frameset document to provide alternate content for browsers that do not support frames or have frames disabled.

Example for <noframes> tag:

```html
<!DOCTYPE html>
<head>
<meta name="generator" content="Microsoft FrontPage 4.0">
<meta name="ProgId" content="FrontPage.Editor.Document">
<title>a frameset document which has a noframes alternative</title>
</head>
<frameset rows="*,*">
<frame src="z:\comp.html" name=foo>
<frame src="z:\etc.html" name=bar>
<noframes>
<body>
this is the noframes alternative section.
any block-level html element may be used here.
</body>
</noframes>
</frameset>
</html>
```

Output:

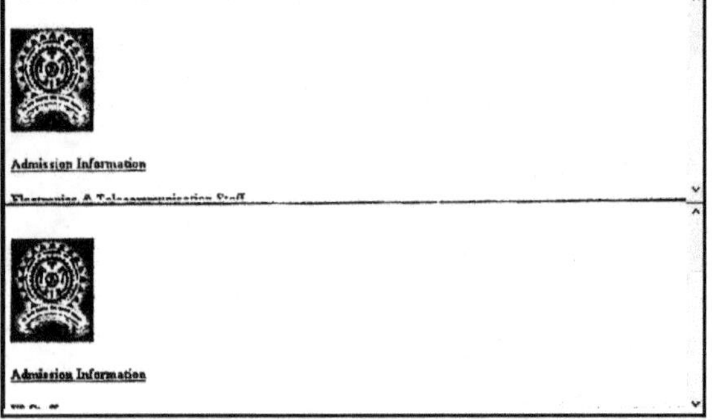

4. **<iframe> Tag:** The <iframe> tag specifies an inline frame. An inline frame is used to embed another document within the current HTML document. The <iframe> tag defines a rectangular region within the document in which the browser displays a separate document, including scrollbars and borders. HTML5 has added some new attributes, and several HTML 4.01 attributes are removed from HTML5.

Syntax: `<iframe> ...... </iframe>`

Attributes	Value	Description
1. height	`pixels`	This attribute specifies the height of an <iframe>.
2. name	`text`	This attribute specifies the name of an <iframe>.

3. sandbox	`allow-forms`
	`allow-pointer-lock`
	`allow-popups`
	`allow-same-origin`
	`allow-scripts`
	`allow-top-navigation`
		This attribute enables an extra set of restrictions for the content in an <iframe>.

4. src	`URL`	This attribute specifies the address of the document to embed in the <iframe>.
5. srcdoc	`HTML_code`	This attribute specifies the HTML content of the page to show in the <iframe>
6. width	`pixels`	This attribute specifies the width of an <iframe>.

Example:

```
<!DOCTYPE html>
<body>
<iframe src="http://www.pragati.com">
  <p>Your browser does not support iframes.</p>
</iframe>
</body>
</html>
```

Output:

4.2.11 Forms

- Forms are an essential part of the Internet, as they provide a way for websites to capture information from users and to process requests.
- HTML forms are one of the main points of interaction between a user and a website. They allow users to send data to the web site. Most of the time that data is sent to the web server, but the web page can also intercept it to use it on its own.
- A form is a group of controls that the user interacts with and sends the result to specific files as designed by an application developer. A form can contain input elements like text fields, checkboxes, radio-buttons, submit buttons and more. A form can also contain select lists, textarea, fieldset, legend, and label elements. The <form> tag is used to create an HTML form.

Syntax:

```
<form>
.
.
.
input elements ...
.
.
.
</form>
```

	Attributes	Value	Description
1.	accept	file_type	This attribute specifies a comma-separated list of file types that the server accepts (that can be submitted through the file upload).
2.	accept-charset	character_set	This attribute specifies the character encodings that are to be used for the form submission.
3.	action	URL	This attribute specifies where to send the form-data when a form is submitted.
4.	autocomplete	on off	This attribute specifies whether a form should have autocomplete on or off.
5.	enctype	application/x-www-form-urlencoded multipart/form-data text/plain	This attribute specifies how the form-data should be encoded when submitting it to the server (only for method="post").
6.	method	get post	This attribute specifies the HTTP method to use when sending form-data.
7.	name	text	This attribute specifies the name of a form.
8.	novalidate	novalidate	This attribute specifies that the form should not be validated when submitted.
9.	target	_blank _self _parent _top	This attribute specifies where to display the response that is received after submitting the form. Target to open the given URL: _blank :The target URL will open in a new window. _self :The target URL will open in the same frame as it was clicked. _parent:The target URL will open in the parent frameset. _top :The target URL will open in the full body of the window.

- Users interact with forms through named controls. A control's "control name" is given by its name attribute. The <form> element can contain one or more of the form elements: <input>, <textarea>, <button>, <select>, <option>, <fieldset>, <label> etc.

<input> Tag:

- The most important form element is the input tag. <input> tag defines an input control. The input tag is used to select user information.
- An input tag can vary in many ways, depending on the type attribute. An <input> tag can be of type text field, checkbox, password, radio button, submit button, and more.
- <input> tag are used within a <form> element to declare input controls that allow users to input data. An input field can vary in many ways, depending on the type attribute.

Syntax: `<input type= " ">`

Attributes:

Attributes	Value	Description
1. accept	audio/* video/* image/* MIME_type	This attribute specifies the types of files that the server accepts (only for type="file").
2. align	left right top middle bottom	This attribute specifies the alignment of an image input (only for type="image").
3. alt	text	This attribute specifies an alternate text for an image (only for type="image").
4. checked	checked	This attribute specifies that an <input> element should be preselected when the page loads (for type="checkbox" or type="radio").
5. disabled	disabled	This attribute specifies that an <input> element should be disabled.
6. maxlength	number	This attribute specifies the maximum number of characters allowed in an <input> element.
7. name	name	This attribute specifies the name of an <input> element.
8. readonly	readonly	This attribute specifies that an input field should be read-only.
9. size	number	This attribute specifies the width, in characters, of an <input> element.
10. src	URL	This attribute specifies the URL of the image to use as a submit button (only for type="image").
11. type	button checkbox file hidden image password radio reset submit text	This attribute specifies the type of <input> element.
12. value	text	This attribute specifies the value of an <input> element.

Example of form using various controls:

```html
<!DOCTYPE html>
<body>
   <head><h3><b> EX. FORM FOR RADIO BUTTON, CHECKBOXES & PULL DOWN
                                           MENU</b></h3>
</head>
<form action="nextpage.html" method ="post">
<h3>Enter the information:</h3>
<b><i>Enter the name:</b>
<input type=text size="25"><br>
<b><i>Your age is in between:</b><br>
   <input type="radio" name="radios" value="radio1" checked>15-20
                                           yrs<br>
<input type="radio" name="radios" value="radio2">20-25 yrs<br>
<input type="radio" name="radios" value="radio3">25 yrs and above<br>
<br><b><i>
Your Hobbies
<input type=checkbox name="option" value="read">Reading novels <br>
<input type="checkbox" name="option" value="write" checked> Writing <br>
<input type="checkbox" name="option" value="sing">Singing <br>
<br><br><b><i> Enter the city: </b>
<select>
<option value=pune> Pune </option>
<option value=mumbai> Mumbai</option>
<option value=a'nagar> Ahmednagar </option>
</select><br><br>
<input type="submit" name="button" value="Submit data">
</form>
</body>
</html>
```

Output:

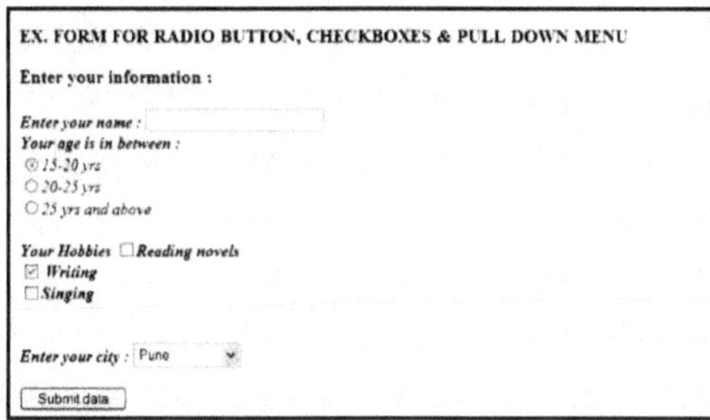

4.2.12 Embedding Multimedia

- The word multimedia consists of two words i.e., multi and media. The word multi means many and the word media (plural of medium) are the means through which information is shared. There are different mediums of sharing information like sound, text, image, graphics, animation or video.

- Multimedia represents information through a variety of media. Multimedia is a combination of more than one media like text, graphics, images, audio, or video, which is used for presenting, sharing, and disseminating the information. HTML embed multimedia defines how inserting of sound, music and video files to the website is performed in the HTML.

- HTML supports following Video file formats:

File Format	File Extension
MPEG (Moving Pictures Expert Group)	.mpg .mpeg
AVI (Audio Video Interleave)	.avi
WMV (Windows Media Video)	.wmv
QuickTime	.mov
RealVideo	.rm .ram
Flash	.swf .flv
WebM	.webm
MPEG-4 or MP4	.mp4

- HTML supports following Audio file formats:

File Format	File Extension
MIDI (Musical Instrument Digital Interface)	.mid .midi
RealAudio	.rm .ram
WMA (Windows Media Audio)	.wma
AAC (Advanced Audio Coding)	.aac
WAV	.wav
MP3	.mp3
MP4	.mp4

- Tags used in HTML to embedded multimedia explained below:

1. **<embed> Tag:** Sometimes, we need to add music or video into the web page. The easiest way to add video or sound to the web site is to include the special HTML tag called <embed>. This tag causes the browser itself to include controls for the multimedia automatically provided browser supports <embed> tag and given media type. The <embed> tag allows us to add Multimedia like sound, music and video files to the web pages.

   Syntax: `<embed>.........</embed>`

Attribute	Description
1. `align`	This attribute determines how to align the object. It can be set to either center, left or right.
2. `autostart`	This boolean attribute indicates if the media should start automatically. You can set it either true or false.
3. `loop`	This attribute specifies if the sound should be played continuously (set loop to true), a certain number of times (a positive value) or not at all (false).
4. `playcount`	This attribute specifies the number of times to play the sound. This is alternate option for loop if we are usiong IE.

contd. ...

5. `hidden`	This attribute specifies if the multimedia object should be shown on the page. A false value means no and true values means yes.
6. `width`	Width of the object in pixels.
7. `height`	Height of the object in pixels.
8. `name`	A name used to reference the object.
9. `src`	URL of the object to be embedded.
10. `volume`	This attribute controls volume of the sound. Can be from 0 (off) to 100 (full volume).

Example for <embed> tag:

```
<!DOCTYPE html>
<head>
<title>HTML embed Tag</title>
</head>
<body>
<embed src="/html/yourfile.swf" width="200" height="200" >
    <noembed><img src="yourimage.gif" alt="Alternative Media">
</noembed>
</embed>
</body>
</html>
```

Output:

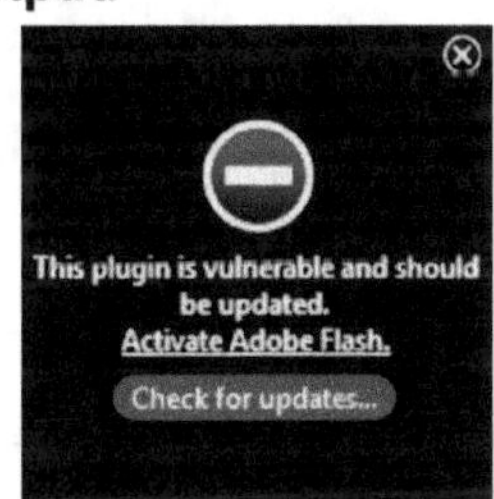

- We can also include a <noembed> tag for the browsers which don't recognize the <embed> tag. We could, for example, use <embed> to display a movie of the choice, and <noembed> to display a single JPG image if browser does not support <embed> tag.

2. **<source> Tag:** The HTML <source> tag is used to specify multiple media resources on media elements such as <audio> and <video>. The <source> tag allows us to specify alternative video/audio files which the browser may choose from, based on its media type or codec support.

Syntax: `<source> ...... </source>`

Attribute	Value	Description
1. `media`	`media_query`	This attribute specifies the type of media resource.
2. `src`	`URL`	This attribute specifies the URL of the media file.
3. `type`	`media_type`	This attribute specifies the media type of the media resource.

Example for <source> tag:

```
<!DOCTYPE html>
<body>
<audio controls>
<source src="horse.ogg" type="audio/ogg">
```

```
<source src="horse.mp3" type="audio/mpeg">
Your browser does not support the audio element.
</audio>
<p><strong>Note:</strong> The source tag is not supported in Internet
Explorer 8 and earlier versions.</p>
</body>
</html>
```

Output:

3. **<audio> Tag:** The HTML <audio> tag is used to specify audio on an HTML document.

   Syntax: `<audio> ......... </audio>`

Attribute	Value	Description
1. autoplay	autoplay	This attribute specifies that the audio will start playing as soon as it is ready.
2. controls	controls	This attribute specifies that audio controls should be displayed (such as a play/pause button etc).
3. loop	loop	This attribute specifies that the audio will start over again, every time it is finished.
4. muted	muted	This attribute specifies that the audio output should be muted.
5. preload	auto metadata none	This attribute specifies if and how the author thinks the audio should be loaded when the page loads.
6. src	URL	This attribute specifies the URL of the audio file.

Example for <audio> tag:

```
<!DOCTYPE html>
<title>My Example</title>
<audio src="/music/good_enough.mp3" controls>
<p>If we are reading this, it is because the browser does not support the
audio element.</p>
</audio>
</html>
```

Output:

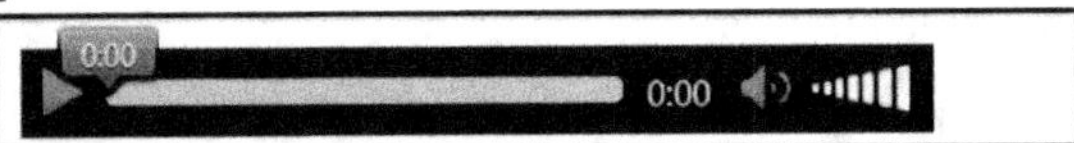

4. **Background Audio:** We can use HTML <bgsound> tag to play a soundtrack in the background of the webpage. This tag is supported by Internet Explorer only and most of the other browsers ignore this tag. It downloads and plays an audio file when the host document is first downloaded by the user and displayed. The background sound file also will replay whenever the user refreshes the browser. This tag is having only two attributes loop and src. Both these attributes have same meaning as explained above.

   Syntax: `<bgsound src=" ">.........</bgsound>`

Example for <bgsound> tag:

```
<!DOCTYPE html>
<head>
<title>HTML embed Tag</title>
</head>
<body>
<bgsound src="/html/yourfile.mid">
    <noembed><img src="yourimage.gif" ></noembed>
</bgsound>
</body>
</html>
```

5. **<video> Tag:** The <video> tag is used to specify video on an HTML document. The <video> tag specifies video, such as a movie clip or other video streams.

   Syntax: `<video>.........</video>`

	Attribute	Value	Description
1.	autoplay	autoplay	This attribute specifies that the video will start playing as soon as it is ready.
2.	controls	controls	This attribute specifies that video controls should be displayed (such as a play/pause button etc).
3.	height	pixels	This attribute sets the height of the video player.
4.	loop	loop	This attribute specifies that the video will start over again, every time it is finished.
5.	muted	muted	This attribute specifies that the audio output of the video should be muted.
6.	poster	URL	This attribute specifies an image to be shown while the video is downloading, or until the user hits the play button.
7.	preload	auto metadata none	This attribute specifies if and how the author thinks the video should be loaded when the page loads.
8.	src	URL	This attribute specifies the URL of the video file.
9.	width	pixels	This attribute sets the width of the video player.

Example for <video> tag:

```
<!DOCTYPE html>
<body>
<video width="320" height="240" controls>
  <source src="movie.mp4" type="video/mp4">
  <source src="movie.ogg" type="video/ogg">
  Your browser does not support the video tag.
</video>
<p><strong>Note:</strong> The video tag is not supported in Internet Explorer
8 and earlier versions.</p>
</body>
</html>
```

Output:

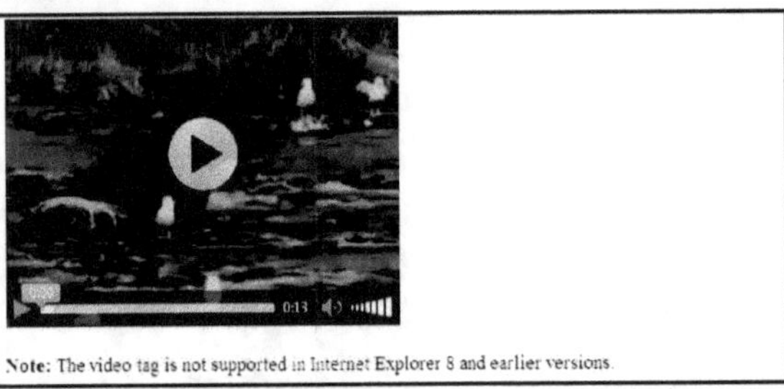

6. **<object> Tag:** The <object> tag defines an embedded object within an HTML document. Use this element to embed multimedia (like audio, video, Java applets, ActiveX, PDF, and Flash) in the web pages.

Syntax: `<object>.........</object>`

Attribute	Value	Description
1. `form`	`form_id`	This attribute specifies one or more forms the object belongs to.
2. `height`	`pixels`	This attribute specifies the height of the object.
3. `name`	`name`	This attribute specifies a name for the object.
4. `type`	`media_type`	This attribute specifies the media type of data specified in the data attribute.
5. `usemap`	`#mapname`	This attribute specifies the name of a client-side image map to be used with the object.
6. `width`	`pixels`	This attribute specifies the width of the object.

Example for <object> tag:

```
<!DOCTYPE html>
<html>
<body>
<object width="400" height="400" data="helloworld.swf">
</object>
</body>
</html>
```

Output:

4.3 CYBER SECURITY

- The Internet is proliferating in an exponential way all over the world. It has the potential to change the way people live. With only a few mouse clicks, people can follow the news, look up facts, buy goods and services and communicate with others from around the world which needed security.

- Cyber security refers generally to the ability to control access to networked systems and the information they contain.
- Where cyber security controls are effective, cyberspace is considered a reliable, resilient, and trustworthy digital infrastructure. Where cyber security controls are absent, incomplete, or poorly designed, cyberspace is considered the wild west of the digital age.
- Cyber security is a collection of defensive technologies (hardware/software), processes and practices designed to protect networks, computers, programs and information from attack, damage or unauthorized access in order to secure systems that are connected to the Internet.
- Cyber security protects against threats using defensive measures, including information assurance, computer systems, and applications hardening, malware protection, access control, information infrastructure protection, and network security.
- A threat is a potential violation of security. The violation need not actually occur for there are to be threat.
- The fact that the violation might occur means that those actions that could cause it to occur must be guarded against (or prepared for). Those action are called attacks. Those who execute such actions, or cause them to be executed, are called attackers.
- Attacks on the Internet information infrastructure originate from the corners like virus, spyware etc., of the world and can be absolutely devastating. The attacks on hosts are generated through malware and can easily gain unauthorized access to critical information.
- The Denial of Service (DoS) attack in which legitimate users are denied access to resources. These DoS attacks will typically exhaust the server's memory and processing, capacity and,/or exhaust the link band-width.
- A threat is defined as, a "potential for violation of security, which exists when there is a circumstance, capability, action or event that could breach security and cause harm i.e., a threat is a possible danger that might exploit a vulnerability".
- Malware comes in following distinct categories:
  1. **Spyware** records keystrokes and other crucial activities and uploads this information to a collection site.
  2. A **virus** provides illegal access to a host's resources, infects it, e.g., through an email attachment, and may contain spyware, Trojans or worms.
  3. A host can be infected through a **worm** by simply passively receiving an object that executes itself and then actively propagates to other hosts.
  4. **Trojans** that may be contained in spyware, a virus or a worm provide a backdoor for illegal access to a host.
  5. **Rootkits** are malware that is hidden in a host's file system and very difficult to detect.
- Computer security is the protection of computer systems from the theft or damage to their hardware, software or information, as well as from disruption or misdirection of the services they provide.
- Cyber security refers to the body of technologies, processes, and practices designed to protect networks, devices, programs, and data from attack, damage, or unauthorized access. Cyber security may also be referred to as information technology security.
- Confidentiality, integrity, and availability addresses the security **objectives** that are specific to information as explained below:
  1. **Confidentiality** refers to a system's capability to limit dissemination of information to authorized use.

2. **Availability** refers to the timely delivery of functional capability.

3. **Integrity** refers to ability to maintain the authenticity, accuracy, and provenance of recorded and reported information.

- Above information security goals applied to information even before they were on computers, but the advent of cyberspace has changed the methods by which the goals are achieved.

Importance of Cyber Security:

- Cyber security is important because government, military, corporate, financial, education and medical organizations collect, process, and store unprecedented amounts of data on computers and other devices.

- A significant portion of that data can be sensitive information, whether that be intellectual property, financial data, personal information, or other types of data for which unauthorized access or exposure could have negative consequences.

- Organizations transmit sensitive data across networks and to other devices in the course of doing businesses, and cyber security describes the discipline dedicated to protecting that information and the systems used to process or store it.

Definition of Cyber Security:

- Cyber security or information technology security are, "the techniques of protecting computers, networks, programs and data from unauthorized access or attacks that are aimed for exploitation". **OR**

- Information security is defined as, "the protection of information and information systems against unauthorised access or modification of information, whether in storage, processing, or transit, and against denial of service to authorised users". Information security includes those measures necessary to detect, document, and counter such threats. **OR**

- Cyber security refers, "to the preventative techniques used to protect the integrity of networks, programs and data from attack, damage, or unauthorized access".

- Generally, the term cyber security policy refer to directives designed to maintain cyber security. Cyber security policy is illustrated in Fig. 4.4.

- An information protection policy is the documentation of enterprise wide decisions on handling and protecting information.

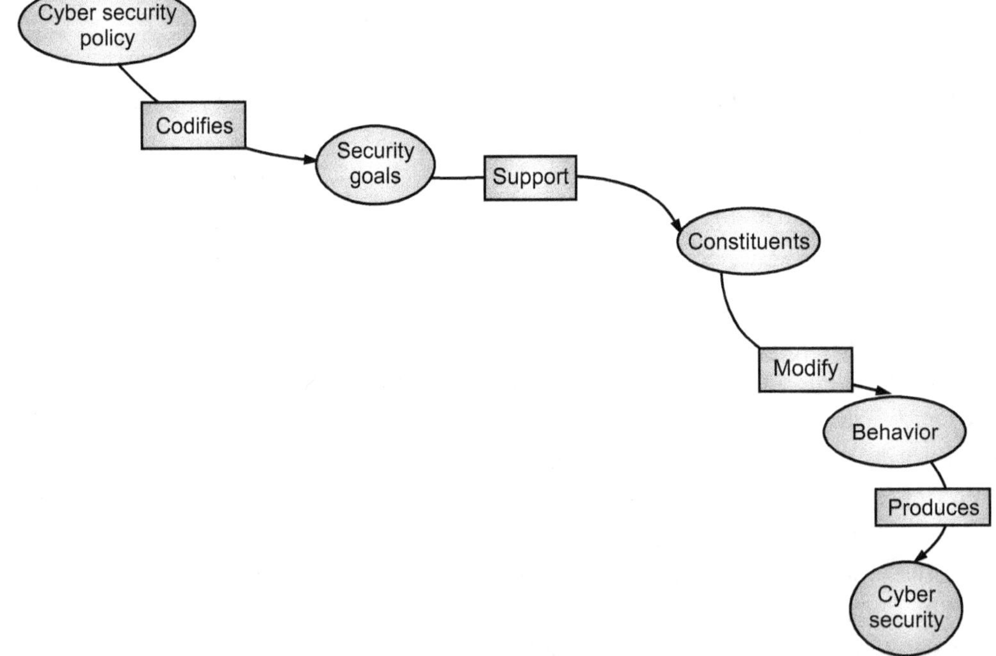

Fig. 4.4: Cyber Security Policy

- Logical security policies consists of software safeguards for an organization's systems, including user identification and password access, authenticating, access rights and authority levels.
- Physical security policies deal with hardware as a physical asset and with the protection of physical assets from harm or theft.
- Securing the physical location of computers and the computers themselves is important because a breach of physical security can result in a loss of information.
- Fig. 4.5 shows cyber security management life cycle.

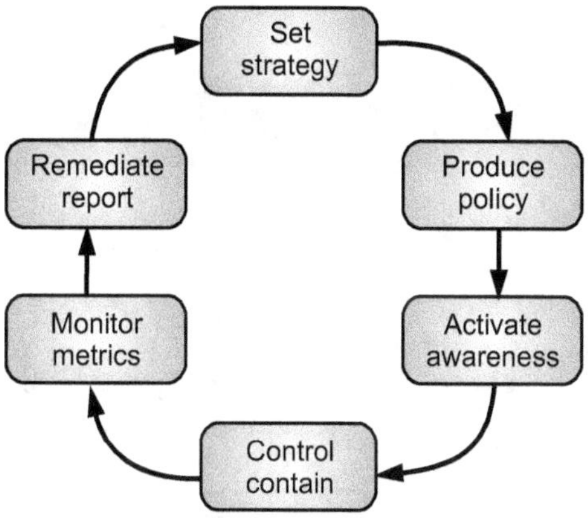

Fig. 4.5: Cyber Security Management Life Cycle

- Major areas covered in cyber security are explained below:

1. **Application security** encompasses measures or counter-measures that are taken during the development life-cycle to protect applications from threats that can come through flaws in the application design, development, deployment, upgrade or maintenance. Some basic techniques used for application security are: Input parameter validation, User/Role Authentication and Authorization, Session management, parameter manipulation and exception management, and Auditing and logging.

2. The purpose of **information protection/security** is to protect the valuable resources of an organization, such as information, hardware and software. Information security protects information from unauthorized access to avoid identity theft and to protect privacy. Major techniques used to cover this are: Identification, authentication amd authorization of user and Cryptography.

3. **Disaster recovery** planning is a process that includes performing risk assessment, establishing priorities, developing recovery strategies in case of a disaster. Any business should have a concrete plan for disaster recovery to resume normal business operations as quickly as possible after a disaster.

4. **Network security** includes activities to protect the usability, reliability, integrity and safety of the network. Effective network security targets a variety of threats and stops them from entering or spreading on the network. Network security compo-nents include anti-virus, anti-spyware, firewall, to block unauthorized access to the network, Intrusion Prevention Systems (IPS), to identify fast-spreading threats, Virtual Private Networks (VPNs), to provide secure remote access.

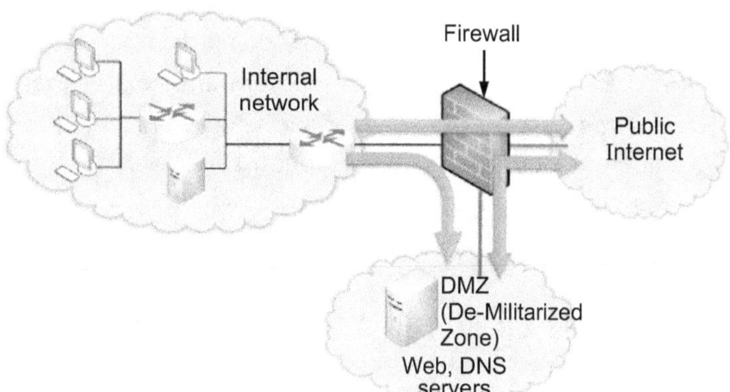

Fig. 4.6: Firewall Protection of Network for an Organization

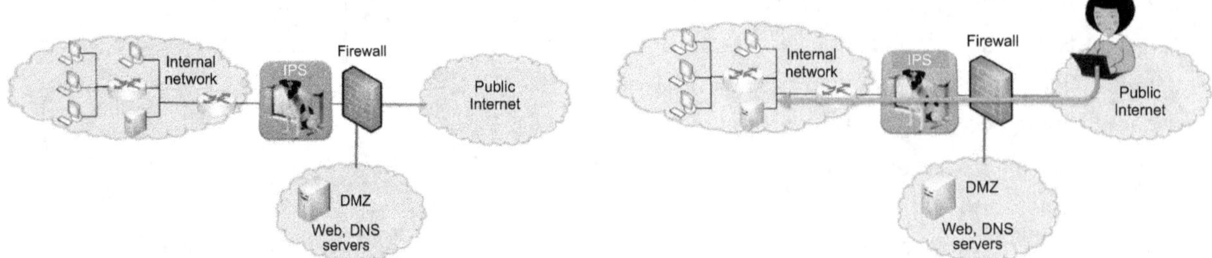

Fig. 4.7: Placement of IPS Protection System	**Fig. 4.8: VPN Computer Security**

4.4 COMPUTER VIRUS

- A virus is a program that can make copies of itself in order to 'infect' other computer programs.
- A computer virus is a type of malicious software program ("malware") that, when executed, replicates by reproducing itself (copying its own source code) or infecting other computer programs by modifying them.
- Infecting computer programs can include as well, data files, or the "boot" sector of the hard drive. When this replication succeeds, the affected areas are then said to be "infected" with a computer virus.
- Virus attaches itself to files stored on floppy disks, USBs, e-mail attachments and hard disks.

Definition of Computer Virus:

- A computer virus is, "a program that reproduces its own code by attaching itself to other programs and using the resources of the computer without knowing about it". The virus code is executed (runs) when the infrared program is executed.	**OR**
- A computer virus is, "a malware or a malicious automated code which replicates itself from one computer to another and infects the operation of the computer machine".

Types of Computer Viruses:

1. **Direct Action Viruses:** The main purpose of this virus is to replicate and take action when it is executed. When a specific condition is met, the virus will go into action and infect files in the directory or folder that it is in and in directories that are specified in the AUTOEXEC.BAT file. Example include Vienna virus.

2. **Boot Virus:** This type of virus affects the boot sector of a floppy or hard disk. This is a crucial part of a disk, in which information on the disk itself is stored together with a program that makes it possible to boot (start) the computer from the disk. Examples include Polyboot.B, AntiEXE etc.

3. **Polymorphic Virus:** Polymorphic viruses encrypt or encode themselves in a different way (using different algorithms and encryption keys) every time they infect a system. Examples include Elkern, Marburg, Satan Bug Tuareg etc.

4. **Directory Virus:** Directory viruses change the paths that indicate the location of a file. By executing a program (file with the extension .EXE or .COM) which has been infected by a virus, which are unknowingly running the virus program, while the original file and program have been previously moved by the virus. Example include Dir-2 virus.

5. **Macro Virus:** Macro viruses infect files that are created using certain applications or programs that contain macros. Examples Relax, Melissa.A, Bablas, O97M/Y2K etc.

6. **Overwrite Viruses:** Virus of this kind is characterized by the fact that it deletes the information contained in the files that it infects, rendering them partially or totally useless once they have been infected. Examples include Way, Trj.Reboot, Trivial.88.D etc.

7. **Network Virus:** Network viruses rapidly spread through a Local Network Area (LAN), and sometimes throughout the Internet. Generally, network viruses multiply through shared resources, i.e., shared drives and folders.

8. **Worms:** A worm has the ability to self-replicate, and can lead to negative effects on the system. Examples of worms include PSWBugbear.B, Lovgate.F, Trile.C, Sobig.D, Mapson etc.

9. **Trojans or Trojan Horses:** Unlike viruses they do not reproduce by infecting other files, nor do they self-replicate like worms. Example include Trojan.Vundo.

10. **Memory Resident Viruses:** This type of virus is a permanent which dwells in the RAM memory. From there it can overcome and interrupt all of the operations executed by the system like corrupting files and programs that are opened, closed, copied, renamed etc. Examples include Randex, CMJ, Meve, MrKlunky etc.

Preventing Measures against Computer Viruses:

1. Install anti-virus software like AVG anti-virus, Norton anti-virus, MaAfee anti-virus etc., and also ensure that scan and update them regularly.

2. Do not open strange e-mails, especially ones with attachments which might be any of the malware.

3. Ensure that the operating system and any program is upto date with latest updates.

4. When downloaded from internet, always check what to install. Do not simply click OK to dismiss pop-up windows.

5. Make backups of all software (including operating system).

6. Never double-click on an executable attachment that arrives via e-mail, unless it comes from a trustworthy source. Most anti-virus software automatically scans attachments as they are loaded.

7. Avoid downloading programs from unknown sources (such as sites on the Internet).

Practice Questions

1. What is meant by HTML?

2. What is tag? Enlist types of tags with explanation.

3. Explain Basic, block and text formatting HTML tags.

4. What are the styles used in HTML?

5. What is meant by link? Explain with example.

6. What is image? Enlist various image file formats.

7. What is table? How to create in HTML? Explain with example.

8. What is list in HTML? How to create it? Explain with example.

9. Which basic tags are required in every HTML page?

10. How will we use different header tags in the HTML page?

11. Write the HTML codes for the following:

    I have :

    - One car
    - One Motorcycle
    - One Bicycle

    I also have:

    o One Bookshelf
    o One Computer
      ▪ Laptop
      ▪ Notebook
      ▪ Tab
    o One CD-Player

12. How can we open a link in a new browser window?

13. Enlist types of lists in HTML. Explain them with example.
14. how to insert multimedia in HTML?
15. Create a Email link to write a mail to the friend to invite him/her for the Birthday party, describing the venue, out and time.
16. Develop a web page to give information about the college.
17. Write short note on: (i) Frame, (ii) Form, (iii) Table.
18. What is cyber security? Enlist its objectives.
19. What is computer virus? Enlist their types with example.
20. Define the terms: (i) Cyber security, (ii) Virus.

■■■

Chapter 5

Information Technology

Contents

5.1 INTRODUCTION

- The world today is witnessing a new kind of revolution i.e., the Information Revolution - ushered in by technology. This revolution is far more sweeping than any other revolution in history in its reach and influence, bringing fundamental changes in all aspects of our life.
- Information Technology (IT) is concerned with all forms of tools, techniques, and technology applied for transmitting, storing, processing and disseminating information.
- IT provides the means for collecting, storing, encoding, processing, analyzing, transmitting, receiving and printing text, graphic, audio or video information.
- The Information Technology Association of America (ITAA) defines IT as, "the study, design, development, implementation, support or management of Computer-Based Information Systems (CBISs), particularly software applications and computer hardware".   **OR**
- Information Technology (IT) may be defined as, "the technology which is used to acquire, store, organize, and process data to a form which can be used in specified applications and disseminate the processed data".

5.2 CURRENT IT TOOLS

- The computer used as a tool that can perform so many different tasks/operations that it is perhaps the most versatile tool ever made.
- For examples, to the accountant, computers balance books, analyze profits and losses, and prepare tax reports, to the factory worker, computers control manufacturing machines and track production and so on.
- The computer can do a wide variety of tasks/operations (like accountancy) because it can be programmed. It is a machine specifically designed to follow instructions.
- Because of the computer's programmability, it doesn't belong to any single profession. Computers are designed to do whatever job their programs, or software, tell them to do.
- The widespread use of computers in the workplace began in the early 1990s. Since then, computer-based tools have been developed.

- In general the techniques are the methods used in performing functions or tasks/operations. Tools are implements used to perform the functions or tasks or operations.
- Techniques and tools work hand in hand to improve quality and productivity.
- A tool is usually given a name denoting the type of work performed. Tools can be equipment, computer software, or a standardized, codified series of steps to accomplish a specific operation/task.
- The current tools in IT are described below:

1. **Office Tools:**

- The office tools are computer software programs which allow a user to create specific office tasks such as letters, memos, reports etc., and easily as opposed to creating the same items by hand.
- Microsoft Office is a common office software package today. Some tools are given below:

   (i) Word Processing. Word Processing is a software program that creates documents using text and/or graphics. Examples include Microsoft Word, Corel WordPerfect, Google Docs, WordPad and so on.

   (ii) Spreadsheets. Spreadsheets quickly organize numerical information and allows the creator to input formulas into the spreadsheet for easy calculation. Examples include Microsoft Excel, Corel Quattro Pro, Google Sheets and so on.

   (iii) Databases: Databases allow the user to save collections of information in one easily accessible place. Examples include Microsoft Access, Corel Paradox, Oracle, MySQL and so on.

   (iv) Presentations: The presentation software is used to display the information in the form of slide show. Examples include Microsoft PowerPoint, OpenOffice Impress and so on.

2. **Photo Editing Tools:**

- Photo editing tools are computer software programs which allow a user to creating, editing, viewing, storing, retrieving photographs and images. Some common tools are given below:

   (i) Microsoft Paint: Microsoft Paint is one of the oldest graphic illustrators. Paint (formerly Paintbrush), commonly known as Microsoft Paint, is a simple computer graphics app that has been included with all versions of Microsoft Windows.

   (ii) Adobe Photoshop: Adobe Photoshop is the industry-standard image editing software, and is used worldwide by photographers and graphic designers to perfect their digital images. Adobe Photoshop lets us to enhance, retouch, and manipulate photographs and other images. Photoshop allows us to transform our images to the workings of our imagination and showcase them for the world to see.

   (iii) CorelDraw: It is a vector graphics editor developed and marketed by Corel Corporation. It is also the name of Corel's Graphics Suite, which bundles CorelDraw with bitmap image editor Corel Photo-Paint as well as other graphics-related programs.

   (iv) Adobe Illustrator: Adobe Illustrator is a vector graphics editor developed and marketed by Adobe Systems. Illustrator is a vector-based imaging program.

3. **Communication Tools:**

- Communication is the process of transmission of information ideas or thoughts from a sender to a receiver.
- The Internet IT tool plays a significant role in communication. Information can be sent to an individual or a group of persons through following various forms:

   (i) Mail (Electronic Mail): Even with all the modern methods of communication, regular postal mail is still one of the most powerful tools for individuals or a business. It adds a personal touch, it used for delivering secure documents, contracts and so on.

   (ii) Smartphones: A smartphone is a cell phone offering advanced capabilities with computer like functionality. A smartphone can incorporate advanced features like e-mail, Internet and e-book reader capabilities and include a full keyboard or an external keyboard.

(iii) Social Networking Sites: A social network is a social structure made of individuals or organizations that are tied together by common interests, often like a community. Internet-based social networking occurs through a variety of websites like MySpace, Facebook, Twitter etc., that allow users to share content and interact around similar interests.

(iv) Online Chat Tools: Online chat can refer to any kind of communication over the Internet, but is primarily meant to refer to direct one-on-one chat or text-based group chat using tools such as instant messengers. It is commonly used in place of e-mail when there is a need to communicate live. Chat tools can be used both for internal and external communication and can be placed on a website so customers can talk to a customer service person in real-time. GoogleTalk and Skype are the most common chart tool today.

(v) Video and Web Conferencing: Video conferencing transmits and receives images and voice in real-time. Web conferencing adds another dimension - it allows us to share documents and applications.

(vi) Cloud Computing: The Cloud has become a bit of a buzzword in business recently, with its offer to share resources and data on-demand.

4. Video Editing Tools:

- Video editing tools are the software which is an application program which handles the post-production video editing of digital video sequences on a computer Non-Linear Editing (NLE) System. Some common Video editing tools/software are listed below:

(i) Adobe Premiere Pro: Adobe Premiere Pro is the cross-platform, uber-popular timeline based video editor that's long set the standard for video editing software.

(ii) VEGAS Pro: It is a video editing software package for Non-Linear Editing (NLE).

(iii) Windows Movie Maker: Windows Movie Maker is a fairly simple video editing program that comes installed on most Windows computers.

5.3 SOCIAL NETWORKING

- Social networking is the way the 21st century communicates now. A social networking service (also Social Networking Site, (SNS) or social media) is an online platform which people use to build social networks or social relations with other people who share similar personal or career interests, activities, backgrounds or real-life connections.

- Social networking is defined as, "the use of Internet-based social media programs to make connections with friends, family, classmates, customers and clients."

- The variety of stand-alone and built-in social networking services currently available online introduces challenges of definition; however, there are some common features:
  1. Social networking services are Internet-based applications.
  2. User-Generated Content (UGC) is the lifeblood of SNS organisations.
  3. Users create service-specific profiles for the site or app that are designed and maintained by the SNS organization.
  4. Social networking services facilitate the development of online social networks by connecting a user's profile with those of other individuals or groups.

- A popular definition social media is, "involves user generated online text, image, audio, and video content that are delivered via Web platforms and tools. The media is used primarily for social interactions and conversations such as sharing opinions, experiences, insights, and perceptions and to collaborating online".

- Social Networking Services (SNSs), such as LinkedIn and Facebook, provide and host a Web space for people to build their homepages for free. SNSs also provide basic support tools for conducting different activities and allow many vendors to provide apps.

- The following are examples of social network services:
    1. **Flickr.com:** Users share and comment on photos.
    2. **Facebook.com:** The most visited social network website.
    3. **LinkedIn.com:** The major enterprise-oriented social network.
    4. **YouTube.com and metacafe.com:** Users can upload and view video clips.
    5. **Google+:** A business-oriented social network.
    6. **MySpace.com:** Facilitates socialization and entertainment for people of all ages.
    7. **Pinterest.com:** Provides a platform for organizing and sharing images.

Fig. 5.1: Logos of various Social Network Sites

Advantages of Social Networking:

1. **Establishing Connection with People(s):** The main advantage of social networking site is that it helps in establishing connection with people, friends and relatives.
2. **Real-time Information Sharing:** Many social networking sites incorporate an instant messaging feature, which lets people exchange information in real-time via a chat.
3. **Cost Effective:** Social media is cost-efficient because of signing up and creating a profile is free of cost.
4. **Information Spreads Incredibly Fast:** Breaking news and other important information can spread like wildfire on social media sites. Important things like recalls, storm information or missing children are all communicated and taken seriously very quickly without any cost.
5. **Worldwide (Global) Connectivity:** Social Media has made the prospect of global connectivity just a mouse click away. The social networking websites can be accessed from any part of the globe. Using social media sites like Facebook, Twitter, LinkedIn, Pinterest etc., people(s) from anywhere can connect with anyone overall world.

Disadvantages of Social Networking:

1. **Causing Major Relationship Problems:** Online social interactions with social networking have not only been starting new relationships, but ending many others. Social networking puts trust to the limit.
2. **Illegal Crime and Virus Attacks:** Social networking sites can be used to commit scams, spread computer viruses or defraud people by stealing their identity.
3. **Time Waster:** Social networking can also be tremendous time wasters. Whether it is in work or personal time, people can sometimes spend hours on social networking sites without any obvious benefits.
4. **Privacy Issues:** There is privacy issues associated with social networking. It is not always clear who is able to read the profile and personal information.

5.4 MOBILE COMPUTING

- In today's computing world, different technologies have emerged like mobile computing. These have grown to support the existing computer networks all over the world.
- Mobile computing refers the computational tasks performed by a mobile user with his smart phone.
- Mobile computing is human–computer interaction by which a computer is expected to be transported during normal usage, which allows for transmission of data, voice and video.
- Mobile Computing is a technology that allows transmission of data, voice and video via a computer or any other wireless enabled device without having to be connected to a fixed physical link.

Principles of Mobile Computing:

1. **Portability:** Facilitates movement of device(s) within the mobile computing environment.
2. **Connectivity:** Ability to continuously stay connected with minimal amount of lag/downtime, without being affected by movements of the connected nodes
3. **Social Interactivity:** Maintaining the connectivity to collaborate with other users, at least within the same environment.
4. **Individuality:** Adapting the technology to suit individual needs.

- Mobile computing can be defined as, "a computing environment of physical mobility".
- Mobile computing is the technology used for transmitting voice and data through small, portable devices using wireless enabled networks.
- Mobile computing system is a distributed system, which is connected via a wireless network for communication. The clients or the nodes possess mobility and the ability to provide computing at anytime, anywhere.
- Mobile computing involves mobile communication, mobile hardware, and mobile software.

1. Mobile Hardware:

- Mobile hardware includes mobile devices or device components that receive or access the service of mobility.
- Mobile Hardware is a small and portable computing device with the ability to retrieve and process data. Smartphones, handheld and wearable devices fall under mobile hardware. These devices typically have an Operating System (OS) embedded in them and able to run application software on top of it.
- These devices are equipped with sensors, full-duplex data transmission and have the ability to operate on wireless networks such as IR, WiFi, and Bluetooth.

2. Mobile Software:

- Mobile Software is the software program which is developed specifically to be run on mobile hardware. This is usually the operating system like android, windows mobile etc., in mobile devices.
- These operating systems provide features such as touchscreen, cellular connectivity, Bluetooth, Wi-Fi, GPS mobile navigation, camera, video camera, speech recognition, voice recorder, music player, near field communication and sensors.

3. Mobile Communication:

- The mobile communication in this case, refers to the infrastructure put in place to ensure that seamless and reliable communication goes on. These would include devices such as protocols, services, bandwidth, and portals necessary to facilitate and support the stated services.
- Mobile communication refers to the exchange of data and voice using existing wireless networks. The data being transferred are the applications including File Transfer (FT), the interconnection between Wide Area Networks (WANs), facsimile (fax), electronic mail, access to the internet and the World Wide Web. The wireless networks utilized in communication are IR, Bluetooth, WLANs, Cellular and satellite communication system.
- The mobile computing functions can be logically divided into following major segments (See Fig. 5.2).

1. **User with device:** The user device, this could be a fixed device like desktop computer in office or a portable device like mobile phone. Examples includes, laptop computers, desktop computers, fixed telephone, mobile phones, digital TV with set-top box, palmtop computers, pocket PCs, two way pagers, handheld terminals, etc.

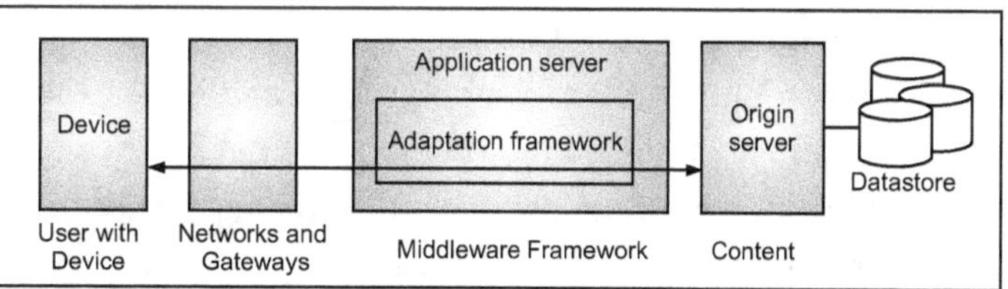

Fig. 5.2: Mobile Computing Functions

2. **Network:** Whenever a user is mobile, he/she will be using different networks at different places at different time. Examples include GSM, CDMA, iMode, Ethernet, Wireless LAN, Bluetooth etc.

3. **Gateway:** This is required to interface different transport bearers. These gateways convert one specific transport bearer to another transport bearer. Examples includes From a fixed phone (with voice interface [we access a service by pressing different keys on the telephone]).

4. **Middleware:** This is more of a function rather than a separate visible node. In the present context middleware handles the presentation and rendering of the content on a particular device. It will also handle the security and personalization for different users.

5. **Content:** This is the domain where the origin server and content is. This could be an application, system or even an aggregation of systems. The content can be mass market, personal or corporate content. Origin server will have some means to accessing the database and the storage devices.

Advantages of Mobile Computing:

- Mobile computing has changed the complete landscape of our day-to-day life. Following are the major advantages of mobile computing:

  1. **Location Flexibility:** This has enabled users to work from anywhere as long as there is a connection established.

  2. **Saves Time:** The time consumed or wasted while travelling from different locations or to the office and back, has been slashed. One can now access all the important documents and files over a secure channel or portal and work as if they were on their computer.

  3. **Enhanced Productivity:** Users can work efficiently and effectively from whichever location they find comfortable. This in turn enhances their productivity level.

  4. **Entertainment:** Video and audio recordings can now be streamed on-the-go using mobile computing. It is easy to access a wide variety of movies, educational and informative material.

  5. **Streamlining of Business Processes:** Business processes are now easily available through secured connections. Looking into security issues, adequate measures have been put in place to ensure authentication and authorization of the user accessing the services.

Disadvantages of Mobile Computing:

  1. **Range and Bandwidth:** Mobile Internet access is generally slower than direct cable connections. High speed network wireless LANs are inexpensive but have very limited range.

  2. **Security Standards:** Security is a major concern while concerning the mobile computing standards on the fleet. One can easily attack the VPN through a huge number of networks interconnected through the line.

  3. **Transmission Interferences:** Weather, terrain, and the range from the nearest signal point can all interfere with signal reception. Reception in tunnels, some buildings, and rural areas is often poor.

5.5 CLOUD COMPUTING

- Cloud computing is currently the buzzword in IT industry, and many are curious to know what cloud computing is and how it works.

- Cloud computing refers to applications and services that run on a distributed network using virtualized resources and accessed by common Internet protocols and networking standards.

- Cloud computing provides access of the applications as utilities over the Internet. It allows us to create, configure, and customize the business applications online.

- Cloud computing is a computing infrastructure and software model for enabling ubiquitous access to shared pools of configurable resources (e.g., computer networks, servers, storage, applications and services), which can be rapidly provisioned with minimal management effort, often over the Internet.

- Cloud computing allows users, and enterprises, with various computing capabilities to store and process data in either a privately owned cloud, or on a third-party server located in a data center in order to make data accessing mechanisms more efficient and reliable.

- The NIST (National Institute of Standards and Technology) of U.S. defines cloud computing as "a model for enabling ubiquitous, convenient, on-demand network access to a shared pool of configurable computing resources (e.g., networks, servers, storage, applications, and services) that can be rapidly provisioned and released with minimal management effort or service provider interaction."   **OR**

- Cloud computing is defined as, "a blend of computing concepts that include a huge number of computers associated through a real-time communication network (internet)".   **OR**

- Rimal defined cloud computing as, "a model of service delivery and access where dynamically scalable and virtualized resources are provided as a service over the Internet".

- Cloud computing is a center point for the most highly impactful technologies such as mobile Internet, automation of knowledge work, the Internet of Things (IoT), and big data.

What is Cloud?

- The term Cloud refers to a Network or Internet. In other words, we can say that Cloud is something, which is present at remote location. Cloud can provide services over public and private networks, i.e., WAN, LAN or VPN.

- Applications such as e-mail, web conferencing, Customer Relationship Management (CRM) execute on cloud.

- The basic cloud computing model is shown in Fig. 5.3. Servers, storage, applications and services are accessed via a common network. They are shared between organizations and accessed by users or applications.

- The users may be members of the organizations working on-the premises, remote workers, customers or members of the general public.

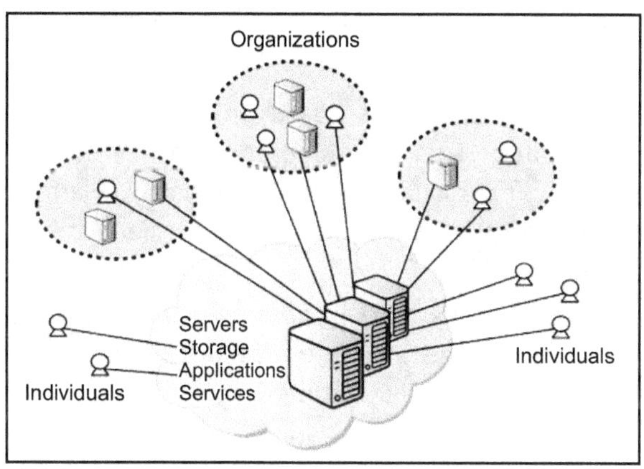

Fig. 5.3: Basic Cloud Computing Model

What is Cloud Computing?

- Cloud computing refers to manipulating, configuring, and accessing the hardware and software resources remotely. It offers online data storage, infrastructure, and application.

- Cloud computing offers platform independency, as the software is not required to be installed locally on the PC. Hence, the Cloud Computing is making our business applications mobile and collaborative.

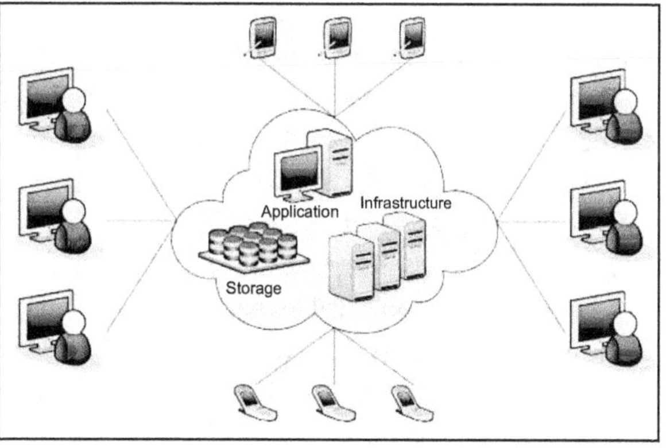

Fig. 5.4: Cloud Computing

Basic Concepts of Cloud Computing:

- There are certain services and models working behind the scene making the cloud computing feasible and accessible to end users. The working models for cloud computing are Deployment Models and Service Models.

1. Deployment Models:

- Deployment models define the type of access to the cloud, i.e., how the cloud is located? Cloud can have any of the four types of access i.e., Public, Private, Hybrid, and Community as shown in Fig. 5.5.

(i) The **public cloud** allows systems and services to be easily accessible to the general public. Public cloud may be less secure because of its openness.

(ii) The **private cloud** allows systems and services to be accessible within an organization. It is more secured because of its private nature.

(iii) The **community cloud** allows systems and services to be accessible by a group of organizations.

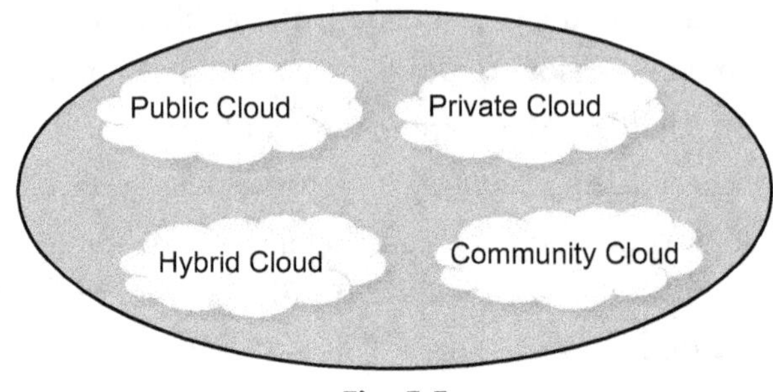

Fig. 5.5

(iv) The **hybrid cloud** is a mixture of public and private cloud, in which the critical activities are performed using private cloud while the non-critical activities are performed using public cloud.

2. Service Models:

- Cloud computing is based on service models. These are categorized into three basic service models which are: Infrastructure-as–a-Service (IaaS), Platform-as-a-Service (PaaS) and Software-as-a-Service (SaaS) as shown in Fig. 5.6.

- The IaaS is the most basic level of service. Each of the service models inherit the security and management mechanism from the underlying model.

- IaaS provides access to fundamental resources such as physical machines, virtual machines, virtual storage, etc. PaaS provides the runtime environment for applications, development and deployment tools, etc. SaaS model allows to use software applications as a service to end-users.

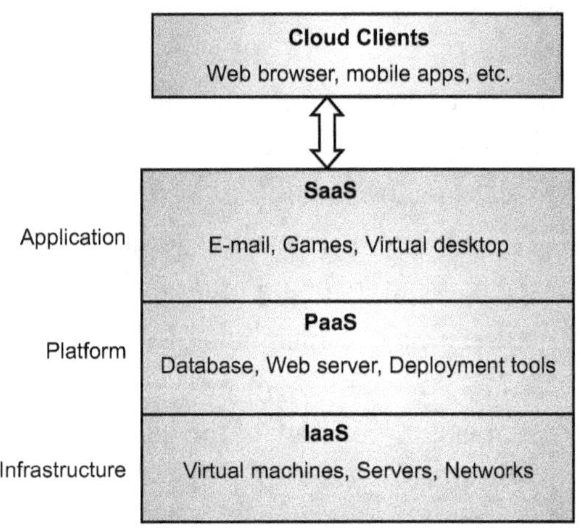

Fig. 5.6

Characteristics of Cloud Computing:

- There are four key characteristics of cloud computing. They are shown in Fig. 5.7.
- The characteristics in Fig. 5.7 of cloud computing are explained below:

  1. **Broad Network Access:** Cloud computing is completely web based, it can be accessed from anywhere and at any time.

  2. **Resource Pooling:** Cloud computing allows multiple tenants to share a pool of resources. One can share single physical instance of hardware, database and basic infrastructure.

  3. **Rapid Elasticity:** It is very easy to scale the resources vertically or horizontally at any time. Scaling of resources means the ability of resources to deal with increasing or decreasing demand. The resources being used by customers at any given point of time are automatically monitored.

  4. **Measured Service:** In this service cloud provider controls and monitors all the aspects of cloud service. Resource optimization, billing, and capacity planning etc. depend on it.

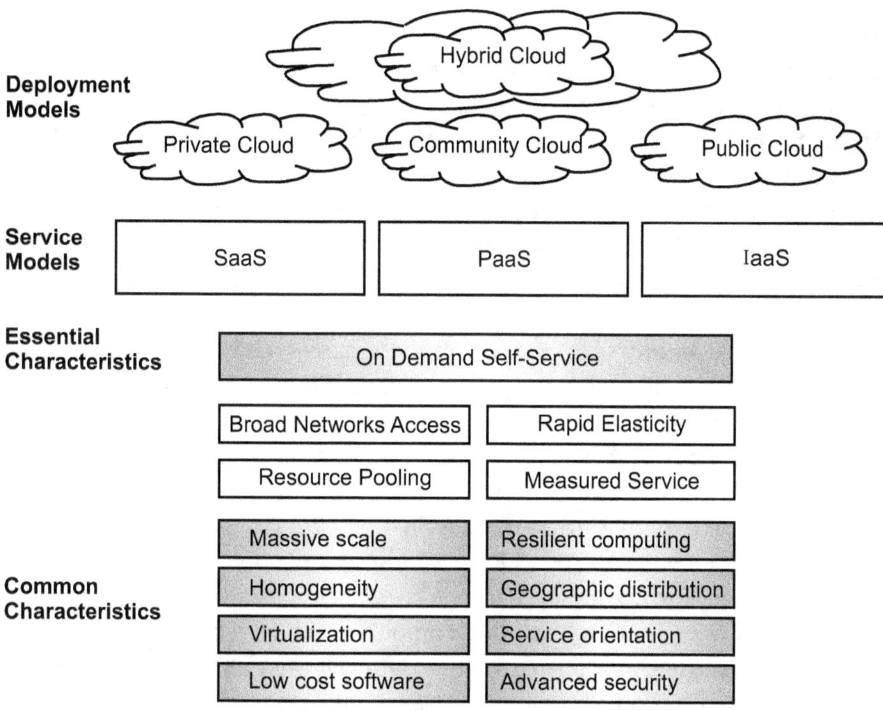

Fig. 5.7

Cloud Computing Architecture:

- Cloud Computing architecture comprises of many cloud components, which are loosely coupled. We can broadly divide the cloud architecture into two parts i.e., Front End and Back End as shown in Fig. 5.8. Each of the ends is connected through a network, usually Internet.
- The **Front End** refers to the client part of cloud computing system. It consists of interfaces and applications that are required to access the cloud computing platforms, Example - Web Browser.
- The **Back End** refers to the cloud itself. It consists of all the resources required to provide cloud computing services. It comprises of huge data storage, virtual machines, security mechanism, services, deployment models, servers, etc.

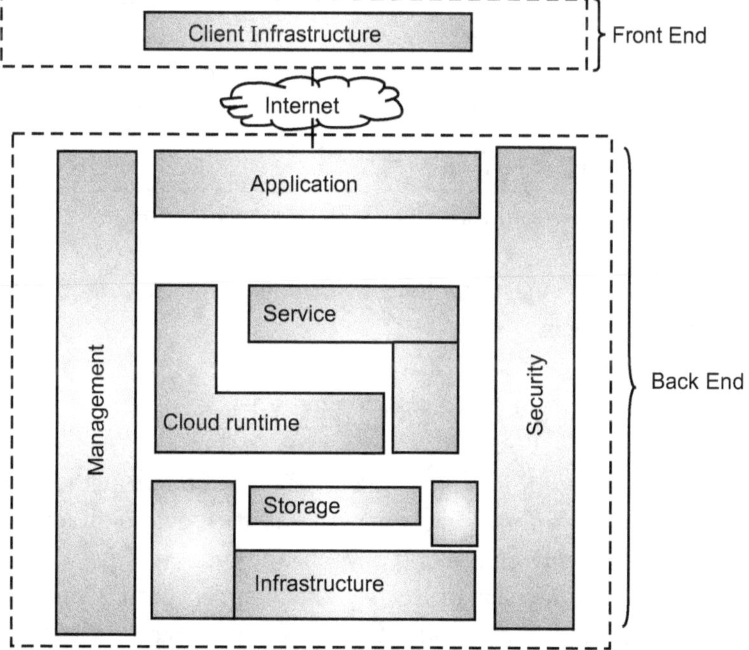

Fig. 5.8: Architecture to could computing

Advantages of Cloud Computing:

1. **Lower Costs:** Because cloud networks operate at higher efficiencies and with greater utilization, significant cost reductions are often encountered.
2. **Ease of Utilization:** Depending upon the type of service being offered, we may find that we do not require hardware or software licenses to implement the service.
3. **Quality of Service (QoS):** The QoS is something that we can obtain under contract from the vendor.
4. **Reliability:** The scale of cloud computing networks and their ability to provide load balancing and failover makes them highly reliable, often much more reliable than what we can achieve in a single organization.
5. **Outsourced IT Management:** A cloud computing deployment lets someone else manage the computing infrastructure while we manage the business. In most instances, we achieve considerable reductions in IT staffing costs.
6. **On-demand Self-Service:** A client can provision computer resources without the need for interaction with cloud service provider personnel.
7. **Broad Network Access:** Access to resources in the cloud is available over the network using standard methods in a manner that provides platform-independent access to clients of all types.

Disadvantges of Cloud Computing:

1. **Security and Privacy:** The drawback of cloud computing is how it deals with the security and privacy issues. The truth is that the expensive enterprise data will exist remotely from the corporate firewall raises severe concerns. Hacking and other problems with cloud infrastructure would dimness many clients even if not more than a single website is attacked.
2. **Service Consignment and Billing:** In cloud computing, it is hard to evaluate the cost incurred due to the demand type of the services. Financial planning and estimation of the expenditure will be very complex except that the provider has some excellent and analogous benchmark to present.
3. **Reliability and Availability Problems :** Cloud providers lack continuous service which results in regular failures. It is significant to examine the service bring offered through internal or third party tools.
4. **Stable Internet Connection:** If the system is not in the network then cloud computing is inaccessible. Cloud computing does not employ when the customer are on offline.
5. **Time Consuming Performance:** The complete information related to the program from the edge to the existing document has to be sent backward and forward from the individual computer to the computers in the cloud. If the Internet speed is low, then one cannot acquire the direct access that the customer might imagine from desktop applications.

5.6 INTRODUCTION TO IoT AND IoE

- Internet of Things (IoT) and Internet of Everything (IoE) is a new growth area for wireless connectivity and sensor sub-systems.
- IoT involves bringing wireless connectivity to consumer and industrial applications. IoE applications range across many broad categories.
- Correspondingly, the wireless eco-system associated with these applications also needs to have a standardized and inter-operable set of protocols and interfaces.

5.6.1 IoT

- The concept of 'Internet of Things' (IoT) is an inspiring vision to bring together innumerable technology advancements in computing and communications and further evolve those through innovation, to improve quality of human life by interconnecting physical and cyber worlds.
- IoT is a global infrastructure for information society enabling services by interconnecting physical and virtual things based on existing and evolving interoperable Information Communication Technologies (ICT).

- The connection of physical things to the Internet makes it possible to access remote sensor data and to control the physical world from a distance.

- IoT is a new revolution of the Internet. Objects make themselves recognizable and they get intelligence thanks to the fact that they can communicate information about themselves and they can access information that has been aggregated by other things.

- The IoT is the network of physical objects devices, vehicles, buildings and other items—embedded with electronics, software, sensors, and network connectivity that enables these objects to collect and exchange data.

- Fig. 5.9 shows conceptual view of IoT.

- The IoT is a computing concept that describes a future where every day physical objects will be connected to the Internet and will be able to identify themselves to other devices.

- The current revolution in Internet, mobile, and Machine-to-Machine (M2M) technologies can be seen as the first phase of the IoT. M2M is defined as the technologies that allow machines, typically (small) computing sensors that perform specific tasks (intelligence) to communicate or relay information as needed typically over simple protocols but more recently over Internet protocols (IP) over wireless or wire-line or even SMS.

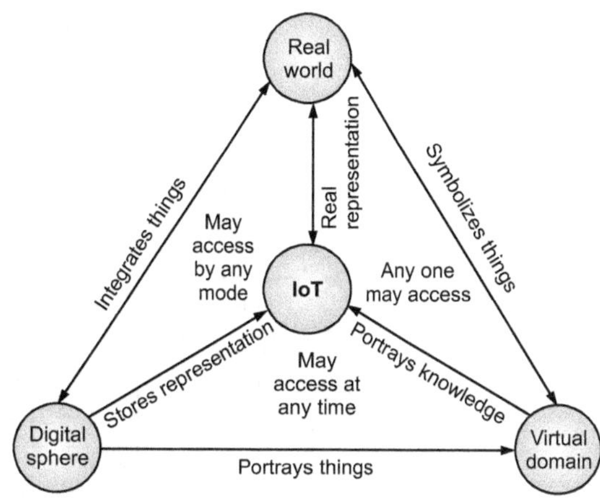

Fig. 5.9: A Conceptual View IoT

- But the Internet of Things is so much more than M2M. IoT it is about interacting with the objects around us; even static non-intelligent objects and augmenting such interactions with context as provided by geo-location, time and so on.

- In the coming years, the IoT is expected to bridge diverse technologies to enable new applications by connecting physical objects together in support of intelligent decision making.

Definition of IoT:

- Fig. 5.10 shows key interconnected elements that make up the IoT.

- Internet of Things (IoT) is defined as, "an integrated part of Future Internet and could be defined as a dynamic global network infrastructure with self-configuring capabilities based on standard and interoperable communication protocols where physical and virtual 'things' have identities, physical attributes, and virtual personalities and use intelligent interfaces, and are seamlessly integrated into the information network." **OR**

- IEEE IoT technical community defines it as, "the Internet of Things (IoT) is a self-configuring and adaptive system consisting of networks of sensors and smart objects whose purpose is to

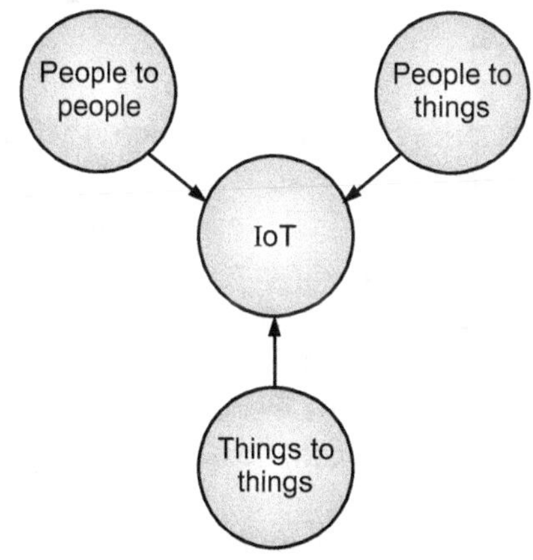

Fig. 5.10

interconnect "all" things, including everyday and industrial objects, in such a way as to make them intelligent, programmable and more capable of interacting with humans."

- Fig. 5.11 shows how the IoT works.
- Internet of Things builds on three major technology layers: Hardware (including chips and sensors), Communication (including mostly some form of wireless network), and Software (including data storage, analytics, and front end applications).

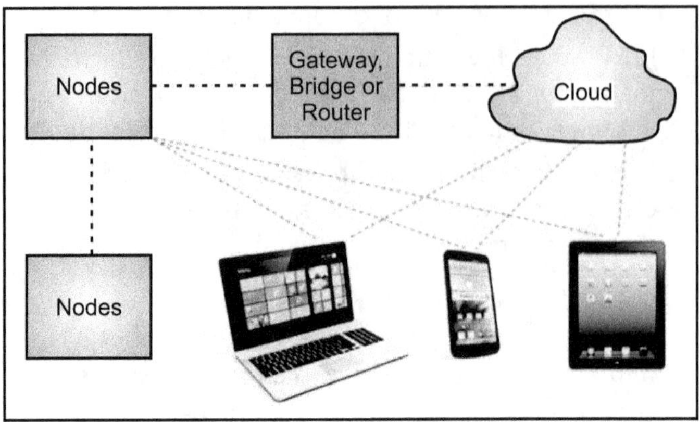

Fig. 5.11: Working of IoT

- Fig. 5.12 shows basic architecture of IoT.

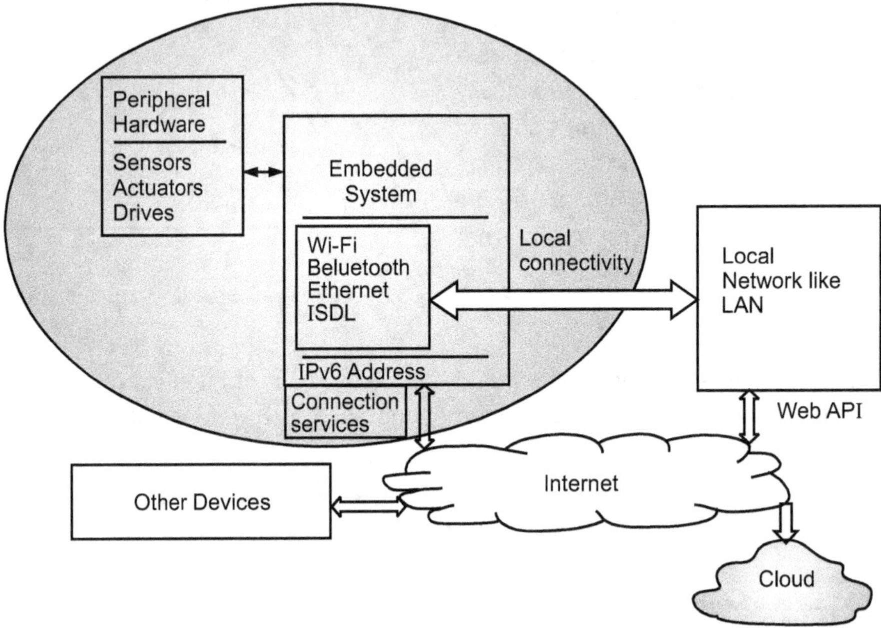

Fig. 5.12: IoT Basic Architecture

5.6.2 IoE

- From the Internet of Things (IoT), where we are today, we are just beginning to enter a new realm i.e., the Internet of Everything (IoE), where things will gain context awareness, increased processing power, and greater sensing abilities.
- The term Internet of Everything (IoE) refers to the networked interconnection of objects of diverse nature, such as electronic devices, sensors, but also physical objects and beings as well as virtual data and environments.
- The IoE consists of four grouping with are data, things, people, and process. IoE leverages data as a means to make more insightful decisions.
- IoT plays a significant role in the things of IoE as this is the network of physical devices and objects connected to the Internet for decisions making.
- The IoE connects people in more valuable and relevant ways. The process is the last part which is delivering the correct information to the right entity at the right time.
- IoE brings together people (humans), process (manages the way people, data, and things work together), data (rich information), and things (inanimate objects and devices) to make networked connections more

relevant and valuable than ever before – turning information into actions that create new capabilities, richer experiences, and unprecedented economic opportunity for organizations, businesses, individuals, and countries.

- As things add capabilities like context awareness, increased processing power etc., and as more people and new types of information are connected, IoT becomes an Internet of Everything – a network of networks where billions or even trillions of connections create unprecedented opportunities as well as new risks.

- To better understand, it is helpful to take a quick look at the evolution of the Internet as shown in Fig. 5.13. In the early 1990s, devices connected to the Internet were essentially "fixed." At its peak, this first wave reached about 200 million devices by the late 1990s. Around the year 2000, devices started to come with us. As the number of both fixed and mobile devices (including machines) ballooned, the number of things connected to the Internet increased, reaching about 10 billion this year. This second wave of Internet growth ushered in the "Age of the Device."

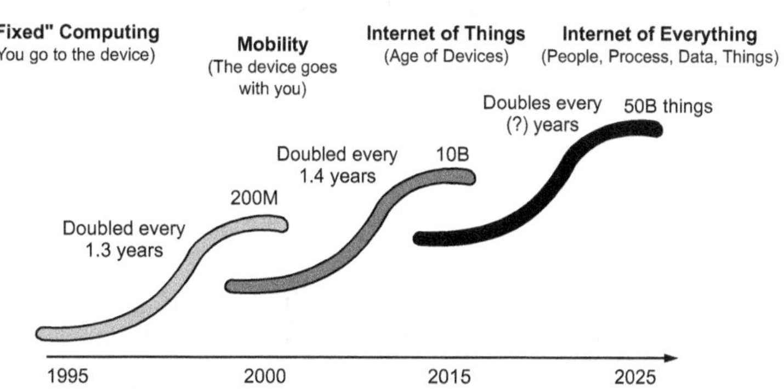

Fig. 5.13: Internet Growth is occuring in Waves

- Cisco believes the third wave of Internet growth has already begun. As the things connected to the Internet are joined by people and more intelligent data, IoE could potentially connect 50 billion people, data, and things by 2020.

- In addition, IoE further advances the power of the Internet to improve business and industry outcomes, and ultimately make people's lives better by adding to the progress of IoT.

- According to Cisco IoE is defined as, "IoE brings together people, process, data and things to make networked connections more relevant and valuable than ever before – turning information into actions that create new capabilities, richer experiences, and unprecedented economic opportunity for businesses, individuals, and countries".

- To better understand above definition, we must first break down IoEs individual components as shown in Fig. 5.14 and explained below:

1. People:

- In IoE, people will be able to connect to the Internet in innumerable ways. Today, most people connect to the Internet through their use of devices (such as PCs, tablets, TVs, and smartphones) and social networks (such as Facebook, Twitter, LinkedIn, and Pinterest).

- As the Internet evolves toward IoE, we will be connected in more relevant and valuable ways. For example, in the future, people will be able to swallow a pill that senses and reports the health of their digestive tract to a doctor over a secure Internet connection.

2. Data:

- With IoE, devices typically gather data and stream it over the Internet to a central source, where it is analyzed and processed. As the capabilities of things connected to the Internet continue to advance, they will become more intelligent by combining data into more useful information.

- Rather than just reporting raw data, connected things will soon send higher-level information back to machines, computers, and people for further evaluation and decision making.

- This transformation from data to information in IoE is important because it will allow us to make faster, more intelligent decisions, as well as control our environment more effectively.

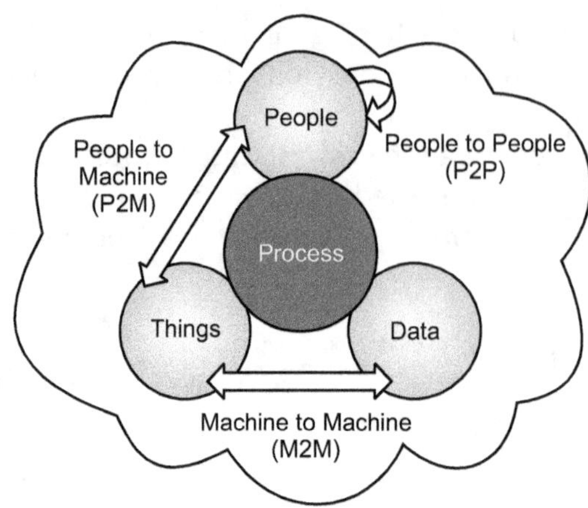

Fig. 5.14: The what, where, and how of the IoE

3. Things:

- This group is made up of physical items such as sensors, consumer devices, and enterprise assets that are connected to both the Internet and each other.
- In IoE, these things will sense more data, become context-aware, and provide more experiential information to help people and machines make more relevant and valuable decisions.
- Examples of "things" in IoE include smart sensors built into structures like bridges, and disposable sensors that will be placed on everyday items such as milk cartons.

4. Process:

- Process plays an important role in how each of these entities — people, data, and things — works with the others to deliver value in the connected world of IoE.
- With the correct process, connections become relevant and add value because the right information is delivered to the right person at the right time in the most appropriate way.

Architecture of IoE:

- Implementation of IoE environments is usually based on a standard architecture derived from IoT. This architecture consists of several layers from the data acquisition layer at the bottom to the application layer at the top.
- Fig. 5.15 presents the generic architecture for IoE.
- The architecture layered has two different divisions with an Internet layer in between to serve the purpose of a common media for communication. Two lower responsible for layers to data capturing while two layers at the top is responsible for data utilization in applications.

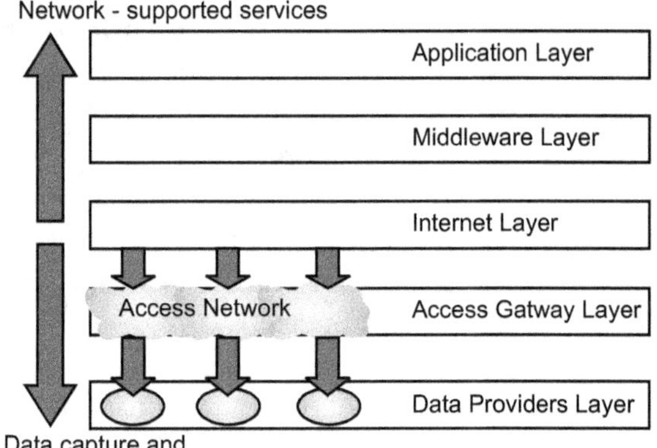

Fig. 5.15: Layered Architecture of IoE

- The functionality of IoE layers in Fig. 5.15 explained below:

    Layer 1 : Data Providers Layer consists of hardware components such as sensor networks, embedded systems, RFID (Radio Frequency Identification) tags and readers or other IoE devices in different forms. Moreover, in this layer is also present other components, like people information, that is also an IoE entity that provides data to the environment. These entities are the primary data sources deployed in the field. Many of these elements provide identification and information storage (e.g, RFID tags), information collection (e.g., sensor networks), information processing (e.g. embedded edge processors), communication, control and actuation.

Layer 2 : In **Access Gateway Layer** the first stage of data handling happens. It takes care of message routing, publishing and subscribing, and also performs cross platform communication, if required.

Layer 3 : **Middleware Layer** acts as an interface between the hardware layer at the bottom and the application layer at the top. It is responsible for critical functions such as device management and information management, and also takes care of issues like data filtering, data aggregation, semantic analysis, access control and information discovery.

Layer 4 : **Application Layer** at the top of the stack is responsible for the delivery of various services to different users/applications in IoE environments. The applications can be from different industry verticals like manufacturing, logistics, retail, environment, public safety, healthcare, food and drug, etc.

5.7 APPLICATIONS OF COMPUTERS IN VARIOUS FIELDS

- Following list demonstrates various applications of computers in today's fields:
  1. **Government:** Computers play an important role in government. Some major fields in this category are Budgets, Sales tax department, Income tax department, Computerization of voters lists, Computerization of driving licensing system, Computerization of PAN card and so on.
  2. **Banking:** Today banking is almost totally dependent on computer. Banks provide online accounting facility, which includes current balances, deposits, overdrafts, interest charges, shares, and trustee records.
  3. **Education:** The computer has provided a lot of facilities in the education system like e-learning. The computer provides a tool in the education system known as CBE (Computer Based Education).
  4. **Business:** A computer has high speed of calculation, diligence, accuracy, reliability, or versatility which made it an integrated part in all business or organizations. Computer is used in business organizations for Payroll calculations, Budgeting, Sales analysis, Financial forecasting, Managing employees database and so on.
  5. **Engineering Design:** Computers are widely used in engineering purpose. One of major areas is CAD (Computer Aided Design). Some engineering fields are:
     (i) Architectural Engineering: Computers help in planning towns, designing buildings, determining a range of buildings on a site using both 2D and 3D drawings.
     (ii) Structural Engineering: Requires stress and strain analysis for design of Ships, Buildings, Budgets, Airplanes etc.
     (iii) Industrial Engineering: Computers deal with design, implement-tation and improvement of integrated systems of people, materials and equipments.
  6. **Insurance:** Insurance companies are keeping all records up-to-date with the help of computers. Insurance companies are maintaining a database of all clients with information showing procedure to continue with policies, starting date of the policies, maturity date, interests due, survival benefits, bonus and so on.
  7. **Health Care:** Computers have become important part in hospitals, labs, and dispensaries. The computers are being used in hospitals to keep the record of doctors, patients and medicines. Some major fields of health care in which computers are used are:
     (i) Surgery: Now-a-days, computers are also used in performing surgery.
     (ii) Diagnostic System: Computers are used to collect data and identify cause of illness.
     (iii) Patient Monitoring System: These are used to check patient's signs for abnormality such as in Cardiac Arrest, ECG etc.
     (iv) Lab-diagnostic System: All tests can be done and reports are prepared by computer.
     (v) Pharma Information System: Computer checks drug-labels, expiry dates, harmful drug's side effects etc.

8. **Communication:** Communication means to convey a message, an idea, a picture or speech that is received and understood clearly and correctly by the person for whom it is meant for. Some main areas in this category are E-mail, Chatting, video-conferencing and so on.

9. **Marketing:** In marketing, uses of computer are as follows:
   (i) **Advertising:** With computers, advertising professionals create art and graphics, write and revise copy, and print and disseminate ads with the goal of selling more products or services.
   (ii) **At Home Shopping:** Home shopping has been made possible through use of computerized catalogues that provide access to product information and permit direct entry of orders to be filled by the customers.

10. **Military:** Computers are largely used in defense. Some military areas where a computer has been used are Missile control, Military communication, Military operation and planning, Smart weapons etc.

11. **Database Management:** The computer saves large/big type of information like organization data, government data, medical data, university data etc. using database. A database is an organized collection of data. It is a collection of schemas, tables, queries, reports, views, and other objects. Database designers typically organize the data to model aspects of reality in a way that supports processes requiring information, such as (for example) modelling the availability of rooms in hotels in a way that supports finding a hotel with vacancies. A Database-Management System (DBMS) is a computer software application that interacts with end-users, other applications, and the database itself to capture and analyze data. A general-purpose DBMS allows the definition, creation, querying, update, and administration of databases. There are many DBMS like MySQL, PostgreSQL, Microsoft Access, SQL Server, FileMaker, Oracle, dBASE, FoxPro and so on.

12. **Artificial Intelligence (AI):** Artificial Intelligence is a way of making a computer, a computer-controlled robot, or a software think intelligently, in the similar manner the intelligent humans think. AI is accomplished by studying how human brain thinks, and how humans learn, decide, and work while trying to solve a problem, and then using the outcomes of this study as a basis of developing intelligent software and systems. Artificial intelligence (Machine Intelligence (MI)) is intelligence exhibited by machines, rather than humans or other animals (Natural Intelligence (NI)).

13. **Data Analysis:** Data is getting bigger and more diversified every day. Therefore, analyzing and processing data to advance human knowledge. The computer technology like big data which collets large amounts of unstructured raw data, retrieved from different sources to a data product useful for organizations.

Practice Questions

1. What is IT? Define it.
2. Explain current IT tools in detail.
3. What is meant by social networking? Name some common social networking sites used today.
4. What is cloud computing?
5. Enlist advantages and disadvantages of social networking.
6. With the help of diagram describe architecture of cloud computing.
7. What is IoT?
8. What is IoE. Explain its layered architecture.
9. Define the following terms:
   (i) Mobile computing (ii) Cloud (iii) Cloud computing (iv) IoT (v) IoE (vi) Social networking.
10. Describe cloud computing model diagrammatically.
11. Explain mobile computing functions in detail.
12. What are the advantages and disadvantages of mobile computing?
13. With help of diagram describe IoT architecture.
14. What is cloud? Enlist various cloud models in detail.
15. What are the applications of computers in following fields:
    (i) Bank (ii) Data Analysis (iii) University (iv) Database management (v) AI (vi) Medical.

■■■